AF389458

Kinetoplastid Phylogenomics and Evolution

Kinetoplastid Phylogenomics and Evolution

Editors

Vyacheslav Yurchenko
Dmitri Maslov

MDPI • Basel • Beijing • Wuhan • Barcelona • Belgrade • Manchester • Tokyo • Cluj • Tianjin

Editors
Vyacheslav Yurchenko
Life Science Research Centre
University of Ostrava
Ostrava
Czech Republic

Dmitri Maslov
Department of Molecular, Cell
and Systems Biology
University of California
Riverside
Riverside
United States

Editorial Office
MDPI
St. Alban-Anlage 66
4052 Basel, Switzerland

This is a reprint of articles from the Special Issue published online in the open access journal *Pathogens* (ISSN 2076-0817) (available at: www.mdpi.com/journal/pathogens/special_issues/Kinetoplastid_phylogenomics_evolution).

For citation purposes, cite each article independently as indicated on the article page online and as indicated below:

LastName, A.A.; LastName, B.B.; LastName, C.C. Article Title. *Journal Name* **Year**, *Volume Number*, Page Range.

ISBN 978-3-0365-3053-6 (Hbk)
ISBN 978-3-0365-3052-9 (PDF)

Contents

Vyacheslav Yurchenko, Anzhelika Butenko and Alexei Y. Kostygov
Genomics of Trypanosomatidae: Where We Stand and What Needs to Be Done?
Reprinted from: *Pathogens* **2021**, *10*, 1124, doi:10.3390/pathogens10091124 . . . . **1**

Alexa Kaufer, Damien Stark and John Ellis
Evolutionary Insight into the Trypanosomatidae Using Alignment-Free Phylogenomics of the Kinetoplast
Reprinted from: *Pathogens* **2019**, *8*, 157, doi:10.3390/pathogens8030157 . . . . **17**

Fred R. Opperdoes, Anzhelika Butenko, Alexandra Zakharova, Evgeny S. Gerasimov, Sara L. Zimmer and Julius Lukeš et al.
The Remarkable Metabolism of *Vickermania ingenoplastis*: Genomic Predictions
Reprinted from: *Pathogens* **2021**, *10*, 68, doi:10.3390/pathogens10010068 . . . . **35**

Michel Tibayrenc and Francisco J. Ayala
Genomics and High-Resolution Typing Confirm Predominant Clonal Evolution Down to a Microevolutionary Scale in *Trypanosoma cruzi*
Reprinted from: *Pathogens* **2020**, *9*, 356, doi:10.3390/pathogens9050356 . . . . **47**

Tomáš Skalický, João M. P. Alves, Anderson C. Morais, Jana Režnarová, Anzhelika Butenko and Julius Lukeš et al.
Endosymbiont Capture, a Repeated Process of Endosymbiont Transfer with Replacement in Trypanosomatids *Angomonas* spp.
Reprinted from: *Pathogens* **2021**, *10*, 702, doi:10.3390/pathogens10060702 . . . . **61**

Hina Durrani, Marshall Hampton, Jon N. Rumbley and Sara L. Zimmer
A Global Analysis of Enzyme Compartmentalization to Glycosomes
Reprinted from: *Pathogens* **2020**, *9*, 281, doi:10.3390/pathogens9040281 . . . . **75**

Ingrid Škodová-Sveráková, Kristína Záhonová, Barbora Bučková, Zoltán Füssy, Vyacheslav Yurchenko and Julius Lukeš
Catalase and Ascorbate Peroxidase in Euglenozoan Protists
Reprinted from: *Pathogens* **2020**, *9*, 317, doi:10.3390/pathogens9040317 . . . . **95**

Ryo Harada, Yoshihisa Hirakawa, Akinori Yabuki, Yuichiro Kashiyama, Moe Maruyama and Ryo Onuma et al.
Inventory and Evolution of Mitochondrion-localized Family A DNA Polymerases in Euglenozoa
Reprinted from: *Pathogens* **2020**, *9*, 257, doi:10.3390/pathogens9040257 . . . . **111**

Evgeny S. Gerasimov, Ksenia A. Zamyatnina, Nadezda S. Matveeva, Yulia A. Rudenskaya, Natalya Kraeva and Alexander A. Kolesnikov et al.
Common Structural Patterns in the Maxicircle Divergent Region of Trypanosomatidae
Reprinted from: *Pathogens* **2020**, *9*, 100, doi:10.3390/pathogens9020100 . . . . **127**

Igor B. Rogozin, Arzuv Charyyeva, Ivan A. Sidorenko, Vladimir N. Babenko and Vyacheslav Yurchenko
Frequent Recombination Events in *Leishmania donovani*: Mining Population Data
Reprinted from: *Pathogens* **2020**, *9*, 572, doi:10.3390/pathogens9070572 . . . . **145**

Review

Genomics of Trypanosomatidae: Where We Stand and What Needs to Be Done?

Vyacheslav Yurchenko [1,2,*], **Anzhelika Butenko** [1,3] and **Alexei Y. Kostygov** [1,4,*]

1   Life Science Research Centre, Faculty of Science, University of Ostrava, 710 00 Ostrava, Czech Republic; anzhelika.butenko@paru.cas.cz
2   Martsinovsky Institute of Medical Parasitology, Tropical and Vector Borne Diseases, Sechenov University, 119435 Moscow, Russia
3   Institute of Parasitology, Biology Centre, Czech Academy of Sciences, 370 05 České Budějovice, Czech Republic
4   Zoological Institute of the Russian Academy of Sciences, 190121 St. Petersburg, Russia
*   Correspondence: vyacheslav.yurchenko@osu.cz (V.Y.); kostygov@gmail.com (A.Y.K.)

Abstract: Trypanosomatids are easy to cultivate and they are (in many cases) amenable to genetic manipulation. Genome sequencing has become a standard tool routinely used in the study of these flagellates. In this review, we summarize the current state of the field and our vision of what needs to be done in order to achieve a more comprehensive picture of trypanosomatid evolution. This will also help to illuminate the lineage-specific proteins and pathways, which can be used as potential targets in treating diseases caused by these parasites.

Keywords: trypanosomatids; next-generation sequencing; genomics

Citation: Yurchenko, V.; Butenko, A.; Kostygov, A.Y. Genomics of Trypanosomatidae: Where We Stand and What Needs to Be Done? *Pathogens* **2021**, *10*, 1124. https://doi.org/10.3390/pathogens 10091124

Academic Editor: Kimberly Paul

Received: 22 July 2021
Accepted: 31 August 2021
Published: 2 September 2021

Publisher's Note: MDPI stays neutral with regard to jurisdictional claims in published maps and institutional affiliations.

1. Introduction

The flagellates of the family Trypanosomatidae represent one of the most evolutionarily successful groups of parasitic protists, adapted to an extremely wide range of hosts—from various animals (mainly insects and vertebrates) to flowering plants and even ciliates. Depending on whether their life cycle includes a single host or there is an obligate alternation between two different hosts, trypanosomatids are subdivided into monoxenous (predominantly insect parasites) and dixenous (typically insect-transmitted parasites of vertebrates or plants) [1]. Most research efforts have been focused on studying dixenous trypanosomatids of the genera *Trypanosoma* and *Leishmania*, which cause severe (often fatal) diseases in humans and domestic animals. Therefore, sequencing of trypanosomatid genomes started from the three important human pathogens: *Trypanosoma brucei*, *T. cruzi*, and *Leishmania major* [2–4]. A comparative study has shown that despite differences in genome size and gene content, these species share a relatively high level of gene order conservation (synteny) and overall genomic organization: most protein-coding genes are intron-less and form conserved polycistronic gene clusters, whereas species-specific genes predominate sub-telomeric or internal non-syntenic chromosomal regions [5]. The subsequent genomic studies expectedly focused on other species of these two genera with the clear preference for *Leishmania*, since it contains more species infective to humans. At the time of writing this review, the assembled genome sequences for multiple isolates of 24 species of *Leishmania* and about a dozen species and subspecies of the genus *Trypanosoma* are available in public databases (Table S1).

However, the diversity of trypanosomatids is predominantly represented by monoxenous parasites, from which their dixenous kin have originated at least three times independently [1]. These cases are *Leishmania* (along with *Endotrypanum* and *Porcisia*) spp. within subfamily Leishmaniinae, *Phytomonas* spp. in the subfamily Herpetomonadinae, and *Trypanosoma* spp. constituting a separate early-diverging lineage (Figure 1). The research interest in the monoxenous trypanosomatids has significantly increased in the last decade; of note, 12 out of the 19 currently recognized genera of these flagellates have been described within this short period [1].

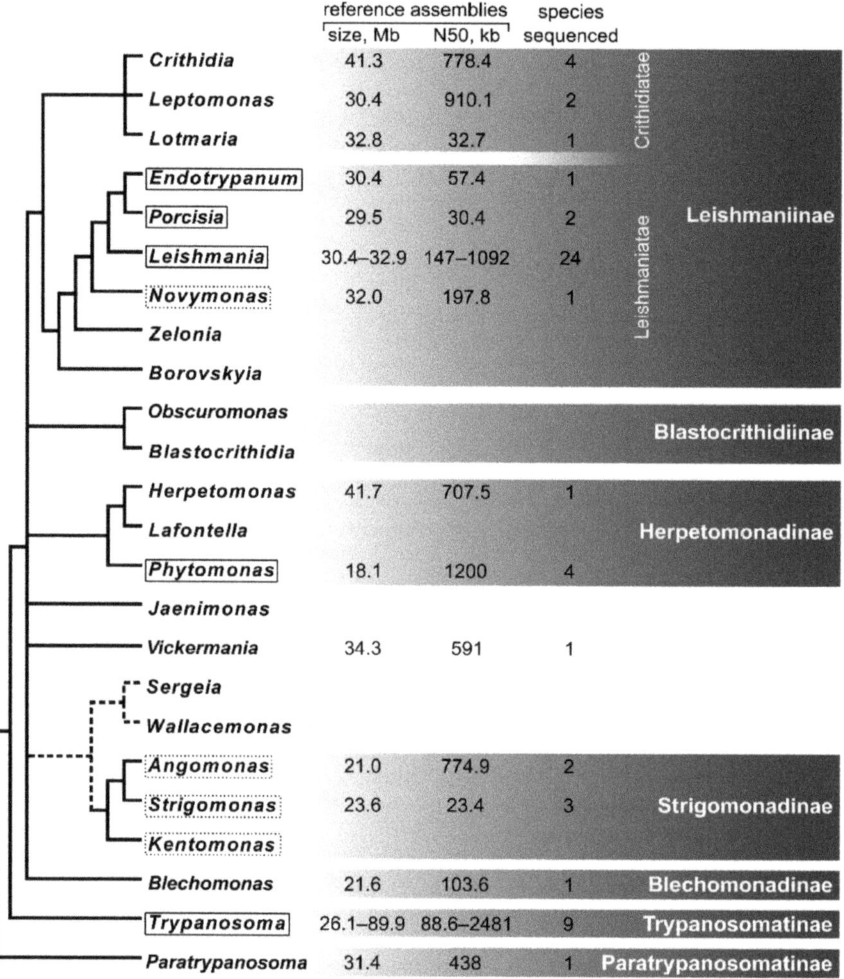

Figure 1. Schematic phylogenetic tree of Trypanosomatidae with a summary on sequenced genomes. Solid and dashed boxes mark dixenous and endosymbiont-bearing genera, respectively.

The studies of insect-dwelling flagellates are important for better understanding not only the biology of their dixenous relatives, but also eukaryotic evolution in general [6]. For example, the members of the genus *Blastocrithidia* evolved an idiosyncratic genetic code with all three stop codons used for coding amino acids [7]. Some trypanosomatids, namely *Novymonas* and the three genera of the subfamily Strigomonadinae (*Angomonas*, *Strigomonas*, and *Kentomonas*) harbor intracellular bacterial symbionts [8–10]. These endosymbionts complement the metabolic requirements of their flagellate hosts with pathways responsible for the synthesis of amino acids, vitamins, and heme [11–14]. The unusual genus *Vickermania* became biflagellate by disrupting the processes of cell division and flagellum duplication to resist the fly midgut peristaltic flow in the absence of an opportunity to attach to the intestinal wall [15]. Various monoxenous trypanosomatids independently acquired thermotolerance, a prerequisite of the transition to dixeny, and some of them have even been documented in vertebrates [16–18]. Below, we review the current state of genomic research in trypanosomatids with a focus on monoxenous species. The taxonomy is presented in accordance with [1].

2. *Trypanosoma* spp.

The first trypanosome, whose genome had been sequenced and analyzed, was the agent of African animal trypanosomiasis—*T. brucei brucei* [3,19] (Table S1). The studies of human-infective *T. b. gambiense* and *T. b. rhodesiense* demonstrated extremely high similarity of the genomes in all three subspecies, conservation of the variant surface glycoprotein (VSG) repertoire, and only rare segmental duplications [20,21]. In *T. b. evansi*, mechanically transmitted by insects and lacking kinetoplast, the procyclin-associated genes needed for the development in the vector have been lost or disrupted, and the γ-subunit of ATP synthase, which is involved in generation of the mitochondrial membrane potential in the absence of kDNA, has mutated [22,23]. The comparison of the genomes of all the above subspecies did not allow identification of factors leading to pathogenicity in humans. Two draft genome assemblies of *T. b. equiperdum*, which is dyskinetoplastic (lacks part of its kDNA) due to the loss of the vector part of its life cycle, have been published with no accompanied analysis [24,25]. Several studies of the genome of the tsetse-transmitted *T. congolense* focused on the analysis of its VSG repertoire and its comparison to that of *T. brucei* [26–29]. They revealed several important differences in the organization and functioning of the VSG expression sites, including the absence of conserved repeats flanking the VSG loci and the scarcity of expression site associated genes in *T. congolense*, and the scale of recombination. *Trypanosoma vivax* genome encodes the most diverse VSG repertoire among all investigated trypanosomes [26,30].

The studies of the *T. cruzi* genome involved numerous strains of this species, allowing to improve the quality of the existing assemblies and providing a deeper insight into its population structure [31–41]. A recent genome analysis of two *T. cruzi* strains revealed that the rapid evolution of gene families involved in immune evasion is one of the major contributors to the intraspecific genome variation in this species [42]. Interestingly, despite the shorter overall length, multiple genes were acquired by lateral gene transfer and some gene families underwent expansions in the genome of a bat-infecting species *T. marinkellei*, which is closely related to *T. cruzi* [43]. Genomes of human non-pathogenic *T. rangeli* and the bat parasite *T. conorhini*, representing a clade related to that of *T. cruzi*, have less retrotransposons and multigene family copies, but more genes involved in the biosynthesis of carbohydrates [44,45]. The crocodile-infecting species *T. grayi* was shown to lack surface proteins (mucins and VSGs), which are characteristic for other trypanosomes investigated thus far [46]. The genome analysis of ruminant-parasitizing *T. theileri* revealed several new families of surface proteins, as well as a general conservation of core cellular metabolic pathways [47].

<u>What needs to be done</u>: The genus *Trypanosoma* corresponds rather to a subfamily than to a single genus—it is very speciose (over 500 described species) and diverse. According to the latest taxonomical revision, it includes sixteen subgenera and several undescribed lineages of the same level [1]. Only a few of these have been analyzed to date, and this significantly limits our understanding of the evolution of parasitism in this group (Table S1). Surprisingly, the genome of one of the most common trypanosome species, flea-transmitted *T. lewisi*, which typically inhabits rats [48], but occasionally infects humans [49], has not been analyzed yet. Of special interest would be the genomic analyses of anuran trypanosomes (subgenus *Trypanosoma*), which gave rise to the parasites of fish and may represent the ancestral group for all terrestrial subgenera [50]. The representatives of this subgenus are expected to keep archaic traits of genomic organization, inherent to the common ancestor of trypanosomes, and their study using NGS might shed light on the origin and evolution of some important gene families, such as VSGs, procyclins, mucins, etc.

3. Dixenous Leishmaniinae

Out of the four *Leishmania* subgenera, i.e., *Leishmania, Mundinia, Sauroleishmania*, and *Viannia*, early genomic studies have focused on the first one (in particular, *L. major, L. donovani, L. infantum, L. mexicana*), and only *L. (V.) braziliensis* was used for comparison. Those studies revealed extremely high synteny levels, interspecific differences in the gene content, and associations of some genes with drug resistance phenotype [2,51,52] (Table S1). More *L. (Leishmania)* species and strains were analyzed later [53–63].

Later on, the subgenus *Viannia* started to receive more attention. Comparative genomic analysis of *L. braziliensis* and *L. peruviana* demonstrated substantial differences in gene content, chromosome copy number, as well as numerous SNPs and indels [64–66]. Sequencing of *L. panamensis* genome uncovered several mobile elements absent from the genomes of *L. (Leishmania)*, along with a higher number of pseudogenes compared to the latter [67]. The study of *L. naiffi* and *L. guyanensis* genomes identified common features of the subgenus *Viannia*, such as aneuploidy, the presence of about 20 subgenus-specific gene families, and a high content of TATE transposons [68,69].

The early genomic study of a lizard parasite *L. (Sauroleishmania) tarentolae* demonstrated the loss of genes involved in oxidative stress protection and vesicular-mediated protein transport, as well as those expressed in *L. (Leishmania)* amastigotes. Meanwhile, the surface glycoprotein GP63 and promastigote surface antigen PSA31C gene families are expanded in this species [70,71]. Other studies of a species from this subgenus—*L. adleri* infecting rodents and lizards—has identified gene amplification, changes in chromosome copy number, and chromosome fission events [72,73].

The genome assemblies of *L. (Mundinia)* spp. were found to be similar in size to those of *Sauroleishmania*, but smaller than those of *Leishmania* and *Viannia*, due to multiple gene losses and gene family contractions [74]. The absence or reduction in the number of lipophosphoglycan-modifying side chain galactosyltransferases and arabinosyltransferases, as well as β-amastins has confirmed previous reports on the differences in cell surface architecture in *L. (Mundinia)* and other *Leishmania* spp. [75–77].

Endotrypanum monterogeii and *Porcisia* spp., being dixenous parasites of sloths and porcupines, respectively, represent the closest known relatives of *Leishmania*. The recently published analysis of their genomic sequences shed light on the evolution of pathogenicity in dixenous Leishmaniinae, which appears to be shaped mainly by changes in the amastin repertoire [78].

L. donovani and *L. braziliensis*, are the only trypanosomatids, to which single-cell genome sequencing approach has been applied thus far [79]. While the respective methods are widely used in human and cancer research, their application is restricted to just a handful of pathogenic species, including some apicomplexans and *Leishmania* [80]. Single-cell genome sequencing is instrumental in investigation of the haplotype diversity and de novo mutations in populations of pathogens. It allowed to characterize the karyotypes of *L. braziliensis* cells demonstrating mosaic aneuploidy [79]. A combination of multiple types of omics data originating from single trypanosomatid cells will provide a holistic view on the interactions of these pathogens with their hosts.

<u>What needs to be done</u>: The genus *Leishmania* is not as speciose as *Trypanosoma*, and the genomes for most of its representatives have been already sequenced with the exception of the poorly studied subgenus *Sauroleishmania*, for which 19 species have been described [81]. The peculiarities of the life cycles of these lizard-dwelling flagellates, such as their presence in the host gut and ability to infect a wide range of the mononuclear cells, erythrocytes, and thrombocytes [82,83], warrant further studies. Meanwhile, only one species has been analyzed for the genus *Endotrypanum*—*E. monterogeii*, and adding at least *E. colombiensis* (previously classified into *Leishmania* [84]), which can infect humans, would be important for understanding the pathogenesis of these flagellates. In addition, several genome assemblies of *Leishmania* spp. are available in public databases waiting to be analyzed and put into the context of comparative studies [85–88].

4. Monoxenous Leishmaniinae

Genomes for several monoxenous representatives of the subfamily Leishmaniinae have been sequenced and analyzed. The study of *Lotmaria passim*, *Crithidia bombi*, and *C. expoeki*, parasitizing agriculturally important Hymenoptera (honeybees and bumblebees), demonstrated numerous examples of horizontal gene transfer [89,90]. Genomic analysis of the latter two species at the population level has revealed that different strains vary considerably in terms of single nucleotide polymorphisms and gene copy number with a pattern fitting a scenario of rapid host-parasite coevolution, where the selective advantage of a given parasite strain is only temporary [91]. The genome and transcriptome sequencing of *Leptomonas seymouri*, the species repeatedly found in clinical samples along with *Leishmania donovani* [92], has allowed identifying its pre-adaptations to dixeny [17]. The genomic data of *Leptomonas pyrrhocoris*, an omnipresent parasite of firebugs, which has been proposed as a new model trypanosomatid species, were used to find new virulence factors of *Leishmania* [93]. The transcriptomic study of *Crithidia thermophila* showed a clear distinction in the mechanisms of thermotolerance in this species and *L. seymouri* [16]. The *C. fasciculata* RNA-seq data were used to elucidate potential mechanisms for insect-specific adhesion in trypanosomatids [94]. The available genomic data of *C. acanthocephali* made possible the comparative analysis of the endosymbiont-bearing and aposymbiotic species [14]. Two species closely related to *C. fasciculata* have been recently reported from human infections and their genomes have been sequenced [18,95]. The genome of the endosymbiont-bearing *Novymonas esmeraldas*, the closest known relative of dixenous Leishmaniinae, revealed a very similar gene content to the latter with the large number of GP63 proteases and pteridin/biopterin transporters, recognized virulence factors of *Leishmania* spp. Owing to the presence of the endosymbiont, this flagellate became prototrophic for all amino acids, heme, and most vitamins, i.e., even more independent of the presence of essential nutrients in the host than Strigomonadinae [12,96].

<u>What needs to be done</u>: Sequencing of additional species belonging to the non-monophyletic genera *Crithidia* and *Leptomonas* will help to delineate the entangled taxonomy of the infrafamily Crithidiatae (Figure 1). In addition, this lineage presents good examples of species with narrow and broad host specificity, which would be interesting to compare from the genomic point of view (e.g., *L. pyrrhocoris* is restricted to firebugs [97], while various species of true bugs and flies are documented for *C. brevicula* [98,99]). Although *Novymonas* is the closest relative of dixenous Leishmaniinae, the acquisition of endosymbionts resulted in very specific adaptations. Therefore, sequencing the genomes of other monoxenous trypanosomatids of the infrafamily Leishmaniatae (genera *Zelonia* [84] and *Borovskyia* [100]) is needed to illuminate the evolutionary origin and molecular signatures of dixenous Leishmaniinae.

5. Herpetomonadinae

The less studied lineage Herpetomonadinae is another subfamily containing dixenous parasites (plant-dwelling *Phytomonas* spp.) along with their monoxenous relatives (Figure 1). Some of the latter appear to be on the way to dixeny, as judged by their detection in plants [101,102] or vertebrates [103]. The analysis of four available genomes of *Phytomonas* spp. (those are *Phytomonas* spp. (isolates EM1 and Hart1) [104], *P. serpens* (isolate 9T) [105], and *P. françai* [106]) revealed additional peculiarities of these plant-inhabiting flagellates, such as significant genome streamlining at the expense of intergenic regions, mobile elements and narrowed gene repertoires, as well as the absence of some electron transport chain proteins. The only *Herpetomonas* species whose genome has been sequenced to date is *H. muscarum* [14,107]. It was used as a reference for the comparative analyses either with endosymbiont-bearing or dixenous trypanosomatids, therefore it is not clear what are its own peculiarities.

<u>What needs to be done</u>: Of special interest would be genomic studies of the speciose genus *Herpetomonas*, which actively explores various ecological niches. Ancestrally, these flagellates are parasites of (brachyceran) flies, but some of them switched to parasitism in

true bugs, cockroaches, mosquitoes, or biting midges, while one species, *H. samuelpessoai*, demonstrates an astonishing ecological plasticity and has been isolated also from plants and even a human patient [1]. These should shed light on the adaptation of trypanosomatids to different hosts and environments. Although the genomes of *Phytomonas* spp. have been already investigated, the analysis was restricted to only four species inhabiting the phloem, latex or fruit and representing the "crown" of this lineage. Thus, the genomic features observed in these flagellates represent a derived state and it is still not clear what has allowed these flagellates to become dixenous. Therefore, the genomes of some early-branching species, such as *P. lipae* and *P. oxycareni* infecting seeds [108,109] need to be analyzed and compared with those of the two closely related monoxenous genera *Herpetomonas* and *Lafontella* [110], which would serve as outgroups. Of special interest would be to study the genomes of the secondarily monoxenous *P. nordicus* [111] (to identify genomic features associated with dixeny in this genus) and its closely related species— *P. borealis*, possessing a bacterial endosymbiont, the relationship with which is likely distinct from those in *Novymonas* and Strigomonadinae [112].

6. Strigomonadinae

The genomes of endosymbiont-bearing Strigomonadinae and their intracellular bacteria (*Ca.* Kinetoplastibacterium spp.) have been studied quite intensively. A series of papers characterized the genomes of *Angomonas deanei*, *A. desouzai*, *Strigomonas oncopelti*, *S. galati*, and *S. culicis*, as well as the metabolic interactions with their symbiotic partners [13,14,113]. It was demonstrated that the amino acid biosynthetic pathways are interlaced between the endosymbionts and their flagellate hosts and that many genes had been acquired by Strigomonadinae from various groups of bacteria. The importance of Strigomonadinae led to the establishment of the first genetically-trackable system in the model species, *A. deanei* [114]. A recent study using genomic data of two *A. ambiguus* strains, *A. deanei*, and their endosymbionts demonstrated that bacteria from the latter species repeatedly replaced bacteria in the former [115].

<u>What needs to be done</u>: The genus *Kentomonas* represent the earliest branch within the subfamily [8] and, therefore, may keep in its genome some archaic traits inherent to the common ancestor of the subfamily. In addition, it has been shown to differ from its cousins in the dependence of external source of heme (or its precursors) [11] and may also diverge in other aspects of its metabolism. Hence, a genomic analysis of this trypanosomatid is warranted.

7. Other Monoxenous Lineages

There are three more monoxenous species, whose genomes have been sequenced and analyzed. Genome sequencing of the early-diverging *Paratrypanosoma confusum* and a representative of the flea-parasitizing genus *Blechomonas ayalai* has allowed to draw preliminary conclusions concerning the evolution of metabolic pathways in the family Trypanosomatidae [116,117]. The most recent addition to the collection of trypanosomatid genomes was that of *Vickermania ingenoplastis*, a species lacking mitochondrial respiratory complexes III and IV and, thus, mainly relying on glycolysis, similarly to *Phytomonas* spp. However, in contrast to the plant trypanosomatids, the genome of this flagellate did not shrink, but experienced a substantial expansion of some protein families, in particular, the glycolytic enzymes [118].

<u>What needs to be done</u>: Representatives of numerous trypanosomatid genera have not been sequenced and some of them have not even been studied since their original description. (1) *Blastocrithidia* and *Obscuromonas* of the subfamily Blastocrithidiinae (Figure 1) share a unique resistant developmental stage—the cyst-like amastigote [119]. Moreover, some of them demonstrate quite a complex development in insects, comparable to that in dixenous parasites [120,121]. It would be interesting to find the genomic basis of these peculiarities. (2) *Jaenimonas drosophilae* inhabits fruit flies and have been proposed as a model to study the insect immune response to trypanosomatid parasites [122]. Sequencing

the genome of this parasite would ease using it as such and understanding its intimate relationships with the host. (3) The genus *Sergeia* parasitizes biting midges and sandflies [123] and thus represents a good model to study the challenges faced and solutions used by trypanosomatids in blood-sucking nematoceran Diptera. Importantly, the same host groups are used by medically relevant *Leishmania* spp. and, therefore, finding parallels in the genome evolution between them and *Sergeia* might provide additional information on the biology of the former [124]. (4) The symbiont-free genus *Wallacemonas* is closely related to Strigomonadinae (Figure 1) and is similar to them in morphology and lifestyle [119]. Thus, it represents a promising reference to reconstruct the metabolism of the ancestors of these endosymbiont-bearing flagellates and answers the question of why some trypanosomatids need endosymbionts, while others successfully live without them in the same hosts.

8. Other Applications of the Trypanosomatid Genomic Data

The availability of multiple representative genome sequences from various Trypanosomatidae enabled a robust analysis of the evolution of different gene families in this group. Some examples are provided below. The analysis of amastins, a large family of surface glycoproteins expressed primarily in amastigotes, revealed that δ-amastin subfamily is restricted to the dixenous Leishmaniinae and its expansion has likely happened in the ancestor of the genus *Leishmania* [78,125]. The repertoire of adenylate cyclases has expanded in dixenous trypanosomatids and many genes encoding these proteins pseudogenized in those subspecies of *T. brucei*, which lost the ability to develop in insects [126]. The analysis of myosin gene family suggested that these proteins were already diversified in the kinetoplastid common ancestor and secondarily, lost multiple times afterwards [127]. Genomic studies revealed that at least three trypanosomatid lineages—Leishmaniinae, Blastocrithidiinae, and *Vickermania*—independently acquired catalase from different groups of bacteria, whereas dixenous Leishmaniinae secondarily lost it [118,128,129]. The study of tubulin gene arrays demonstrated that while in the majority of trypanosomatid lineages and in the free-living bodonids that the α- and β-tubulin genes are alternated, in Leishmaniinae, these multicopy genes are organized in homogeneous (α-only and β-only) stretches [130]. The analysis of the evolution of trypanosomatid UDP-glycosyltransferases, the superfamily of enzymes participating in the modification of various surface macromolecules, showed their independent diversification in distinct groups of these parasites. Interestingly, one of the ancient lineages of these enzymes present in the free-living *Bodo saltans* has been lost from all trypanosomatids except stercorarian trypanosomes [131]. Side chain galactosyl and arabinosyltransferases of that large superfamily ensure lipophosphoglycan modifications needed for *Leishmania* attachment and detachment inside insects [132–134]. The analysis demonstrated differences in the repertoires of these enzymes between the subgenera *Leishmania* and *Viannia* correlating with the affinity of the flagellates to different intestinal sections of their different insect hosts [135,136]. In *Leptomonas pyrrhocoris*, which does not attach to the intestinal wall of its firebug host, the orthologs of these genes showed early divergence and expansion, suggesting distinct functions [137].

The analysis of gene families and comparative genomic studies discussed above can be hampered by the absence of contiguous assemblies with well-resolved repetitive regions. Although trypanosomatid genomes are relatively small (typically around 20–30 Mb), they contain many repeats and, therefore, it is challenging to obtain a chromosome-level assembly based on short sequencing reads [138]. For several trypanosomatid genera, more contiguous hybrid assemblies based on the combination of short and long sequencing reads have become available (Table S1). Application of a combination of long read sequencing and genome-wide chromosome conformation capture (Hi-C) enabled haplotype-specific assembly of *T. brucei* 427 Lister genome and revealed that antigen-encoding sub-telomeric regions are folded into distinct compact structures [139]. For *T. cruzi*, the trypanosomatid having the largest genome sequenced so far, a newer assembly obtained using Nanopore data, led to a significant increase of the number of identified single-copy orthologs and repetitive transposable elements as well as overall estimated genome size [37]. By far, the most contiguous genome assembly, which we suggest to use as a new reference for this species, was obtained recently using a combination of PacBio Single-Molecule Real-Time sequencing and proximity ligation methods [42]. One more example of a substantial quality improvement is the recently published Nanopore-based genome assembly for *Angomonas deanei*, which identified new chromosome-level features such as a supernumerary chromosome, a long inversion and a translocation [140]. After careful annotation of such genome assemblies based on multiple types of evidence (including transcriptomic and proteomic data), the trypanosomatid research community should consider using these new assemblies instead of the old references based solely on short reads and sometimes erroneous annotations.

Although no DNA viruses have been reported in trypanosomatids so far, the available genomic data for several *Leptomonas pyrrhocoris* strains has allowed identification of an endogenous viral element related to *Leppyr*TLV1 (a tombus-like single-stranded positive sense RNA virus), which was apparently captured via reverse transcription and integrated into the trypanosomatid genome [141].

Finally, next-generation sequencing data can be used for analyzing the composition and (to some extent) function of the kinetoplast. In this respect, the kinetoplast genomes of the two model species, dixenous *T. brucei* and monoxenous *Leptomonas pyrrhocoris*, have been scrutinized. Their analyses revealed novel non-canonical mechanisms, as well as species-specific differences in RNA editing [142,143]. Such studies can delineate not only the structure of maxicircles and minicircles [144–147], but also predict the guide RNA repertoire in a given species [143,148,149]. As judged from pre-genomic studies carried out on single genes, different lineages of trypanosomatids possess distinct kDNA editing patterns [150]. This, along with the abovementioned degradation of kDNA in two *T. brucei* subspecies, demonstrates the underestimated importance of the kinetoplast genome in the trypanosomatid development. Performing comparative studies on the editing using whole kinetoplast genomes with a wide range of trypanosomatid phylogroups should shed light on their particular life strategies and allow better understanding of the evolution of this fascinating group of parasites.

9. Conclusions

A fair number of trypanosomatid genomes have been sequenced and there is a significant progress in understanding their evolution, structure, and function. Nevertheless, many questions still remain unanswered and more of them arise, as new representatives of this group of flagellates are discovered and/or analyzed in broadscale biodiversity assays.

The relatively small size of trypanosomatid genomes makes these parasites an attractive model to study how the evolution of traits and genomes are correlated. This is further facilitated by the possibility to cultivate and genetically modify many trypanosomatids, combined with a knowledge of their diversity. However, as judged from the environmental screens (for example, refs. [151,152] and many others), there are still taxa of the generic level and above to be described. Meanwhile, the lack of data on the biology of many trypanosomatid groups still represents an important obstacle in interpreting the observed genomic differences, therefore, more data on trypanosomatid development, strategies of transmission, host-parasite interactions, etc., are needed.

Supplementary Materials: The following are available online at https://www.mdpi.com/article/10.3390/pathogens10091124/s1, Table S1: Available genomic data for Trypanosomatidae.

Author Contributions: Conceptualization, V.Y. and A.Y.K.; formal analysis, A.B. and A.Y.K.; data curation, all authors; writing—original draft preparation, all authors; writing—review and editing, all authors; funding acquisition, V.Y. All authors have read and agreed to the published version of the manuscript.

Funding: This research was funded by the European Regional Funds (CZ.02.1.01/16_019/0000759 to V.Y. and A.Y.K.), the Grant Agency of Czech Republic (20-07186S to V.Y. and A.Y.K. and 21-09283S to V.Y.), and the State Assignment AAAA-A19-119031390116-9 for ZIN RAS to A.Y.K.

Institutional Review Board Statement: Not applicable.

Informed Consent Statement: Not applicable.

Acknowledgments: We thank Evgeny Gerasimov (Moscow State University) for the discussion on kDNA analyses.

Conflicts of Interest: The authors declare no conflict of interest. The funders had no role in the design of the study; in the collection, analyses, or interpretation of data; in the writing of the manuscript, or in the decision to publish the results.

References

1. Kostygov, A.Y.; Karnkowska, A.; Votýpka, J.; Tashyreva, D.; Maciszewski, K.; Yurchenko, V.; Lukeš, J. Euglenozoa: Taxonomy, diversity and ecology, symbioses and viruses. *Open Biol.* **2021**, *11*, 200407. [CrossRef]
2. Ivens, A.C.; Peacock, C.S.; Worthey, E.A.; Murphy, L.; Aggarwal, G.; Berriman, M.; Sisk, E.; Rajandream, M.A.; Adlem, E.; Aert, R.; et al. The genome of the kinetoplastid parasite, *Leishmania major*. *Science* **2005**, *309*, 436–442. [CrossRef] [PubMed]
3. Berriman, M.; Ghedin, E.; Hertz-Fowler, C.; Blandin, G.; Renauld, H.; Bartholomeu, D.C.; Lennard, N.J.; Caler, E.; Hamlin, N.E.; Haas, B.; et al. The genome of the African trypanosome *Trypanosoma brucei*. *Science* **2005**, *309*, 416–422. [CrossRef]
4. El-Sayed, N.M.; Myler, P.J.; Bartholomeu, D.C.; Nilsson, D.; Aggarwal, G.; Tran, A.N.; Ghedin, E.; Worthey, E.A.; Delcher, A.L.; Blandin, G.; et al. The genome sequence of *Trypanosoma cruzi*, etiologic agent of Chagas disease. *Science* **2005**, *309*, 409–415. [CrossRef]
5. El-Sayed, N.M.; Myler, P.J.; Blandin, G.; Berriman, M.; Crabtree, J.; Aggarwal, G.; Caler, E.; Renauld, H.; Worthey, E.A.; Hertz-Fowler, C.; et al. Comparative genomics of trypanosomatid parasitic protozoa. *Science* **2005**, *309*, 404–409. [CrossRef] [PubMed]
6. Lukeš, J.; Butenko, A.; Hashimi, H.; Maslov, D.A.; Votýpka, J.; Yurchenko, V. Trypanosomatids are much more than just trypanosomes: Clues from the expanded family tree. *Trends Parasitol.* **2018**, *34*, 466–480. [CrossRef] [PubMed]
7. Záhonová, K.; Kostygov, A.; Ševčíková, T.; Yurchenko, V.; Eliáš, M. An unprecedented non-canonical nuclear genetic code with all three termination codons reassigned as sense codons. *Curr. Biol.* **2016**, *26*, 2364–2369. [CrossRef] [PubMed]
8. Votýpka, J.; Kostygov, A.Y.; Kraeva, N.; Grybchuk-Ieremenko, A.; Tesařová, M.; Grybchuk, D.; Lukeš, J.; Yurchenko, V. *Kentomonas* gen. n.; a new genus of endosymbiont-containing trypanosomatids of Strigomonadinae subfam. n. *Protist* **2014**, *165*, 825–838. [CrossRef]
9. Kostygov, A.; Dobáková, E.; Grybchuk-Ieremenko, A.; Váhala, D.; Maslov, D.A.; Votýpka, J.; Lukeš, J.; Yurchenko, V. Novel trypanosomatid—Bacterium association: Evolution of endosymbiosis in action. *mBio* **2016**, *7*, e01985-15. [CrossRef] [PubMed]
10. Teixeira, M.M.; Borghesan, T.C.; Ferreira, R.C.; Santos, M.A.; Takata, C.S.; Campaner, M.; Nunes, V.L.; Milder, R.V.; de Souza, W.; Camargo, E.P. Phylogenetic validation of the genera *Angomonas* and *Strigomonas* of trypanosomatids harboring bacterial endosymbionts with the description of new species of trypanosomatids and of proteobacterial symbionts. *Protist* **2011**, *162*, 503–524. [CrossRef]
11. Silva, F.M.; Kostygov, A.Y.; Spodareva, V.V.; Butenko, A.; Tossou, R.; Lukes, J.; Yurchenko, V.; Alves, J.M.P. The reduced genome of *Candidatus* Kinetoplastibacterium sorsogonicusi, the endosymbiont of *Kentomonas sorsogonicus* (Trypanosomatidae): Loss of the haem-synthesis pathway. *Parasitology* **2018**, *145*, 1287–1293. [CrossRef] [PubMed]
12. Kostygov, A.Y.; Butenko, A.; Nenarokova, A.; Tashyreva, D.; Flegontov, P.; Lukeš, J.; Yurchenko, V. Genome of *Ca.* Pandoraea novymonadis, an endosymbiotic bacterium of the trypanosomatid *Novymonas esmeraldas*. *Front. Microbiol.* **2017**, *8*, 1940. [CrossRef]
13. Alves, J.M.; Serrano, M.G.; Maia da Silva, F.; Voegtly, L.J.; Matveyev, A.V.; Teixeira, M.M.; Camargo, E.P.; Buck, G.A. Genome evolution and phylogenomic analysis of *Candidatus* Kinetoplastibacterium, the beta-proteobacterial endosymbionts of *Strigomonas* and *Angomonas*. *Genome Biol. Evol.* **2013**, *5*, 338–350. [CrossRef]
14. Alves, J.M.; Klein, C.C.; da Silva, F.M.; Costa-Martins, A.G.; Serrano, M.G.; Buck, G.A.; Vasconcelos, A.T.; Sagot, M.F.; Teixeira, M.M.; Motta, M.C.; et al. Endosymbiosis in trypanosomatids: The genomic cooperation between bacterium and host in the synthesis of essential amino acids is heavily influenced by multiple horizontal gene transfers. *BMC Evol. Biol.* **2013**, *13*, 190. [CrossRef] [PubMed]
15. Kostygov, A.; Frolov, A.O.; Malysheva, M.N.; Ganyukova, A.I.; Chistyakova, L.V.; Tashyreva, D.; Tesařová, M.; Spodareva, V.V.; Režnarová, J.; Macedo, D.H.; et al. *Vickermania* gen. nov.; trypanosomatids that use two joined flagella to resist midgut peristaltic flow within the fly host. *BMC Biol.* **2020**, *18*, 187. [CrossRef] [PubMed]
16. Ishemgulova, A.; Butenko, A.; Kortišová, L.; Boucinha, C.; Grybchuk-Ieremenko, A.; Morelli, K.A.; Tesařová, M.; Kraeva, N.; Grybchuk, D.; Pánek, T.; et al. Molecular mechanisms of thermal resistance of the insect trypanosomatid *Crithidia thermophila*. *PLoS ONE* **2017**, *12*, e0174165. [CrossRef]
17. Kraeva, N.; Butenko, A.; Hlaváčová, J.; Kostygov, A.; Myškova, J.; Grybchuk, D.; Leštinová, T.; Votýpka, J.; Volf, P.; Opperdoes, F.; et al. *Leptomonas seymouri*: Adaptations to the dixenous life cycle analyzed by genome sequencing, transcriptome profiling and co-infection with *Leishmania donovani*. *PLoS Pathog.* **2015**, *11*, e1005127. [CrossRef]

18. Maruyama, S.R.; de Santana, A.K.M.; Takamiya, N.T.; Takahashi, T.Y.; Rogerio, L.A.; Oliveira, C.A.B.; Milanezi, C.M.; Trombela, V.A.; Cruz, A.K.; Jesus, A.R.; et al. Non-*Leishmania* parasite in fatal visceral leishmaniasis–like disease, Brazil. *Emerg. Infect. Dis.* **2019**, *25*, 2088–2092. [CrossRef]

19. Brems, S.; Guilbride, D.L.; Gundlesdodjir-Planck, D.; Busold, C.; Luu, V.D.; Schanne, M.; Hoheisel, J.; Clayton, C. The transcriptomes of *Trypanosoma brucei* Lister 427 and TREU927 bloodstream and procyclic trypomastigotes. *Mol. Biochem. Parasitol.* **2005**, *139*, 163–172. [CrossRef]

20. Jackson, A.P.; Sanders, M.; Berry, A.; McQuillan, J.; Aslett, M.A.; Quail, M.A.; Chukualim, B.; Capewell, P.; MacLeod, A.; Melville, S.E.; et al. The genome sequence of *Trypanosoma brucei gambiense*, causative agent of chronic human african trypanosomiasis. *PLoS Negl. Trop. Dis.* **2010**, *4*, e658. [CrossRef]

21. Sistrom, M.; Evans, B.; Benoit, J.; Balmer, O.; Aksoy, S.; Caccone, A. *De novo* genome assembly shows genome wide similarity between *Trypanosoma brucei brucei* and *Trypanosoma brucei rhodesiense*. *PLoS ONE* **2016**, *11*, e0147660.

22. Carnes, J.; Anupama, A.; Balmer, O.; Jackson, A.; Lewis, M.; Brown, R.; Cestari, I.; Desquesnes, M.; Gendrin, C.; Hertz-Fowler, C.; et al. Genome and phylogenetic analyses of *Trypanosoma evansi* reveal extensive similarity to *T. brucei* and multiple independent origins for dyskinetoplasty. *PLoS Negl. Trop. Dis.* **2015**, *9*, e3404. [CrossRef] [PubMed]

23. Zheng, L.; Jiang, N.; Sang, X.; Zhang, N.; Zhang, K.; Chen, H.; Yang, N.; Feng, Y.; Chen, R.; Suo, X.; et al. In-depth analysis of the genome of Trypanosoma evansi, an etiologic agent of surra. *Sci. China Life Sci.* **2019**, *62*, 406–419. [CrossRef]

24. Davaasuren, B.; Yamagishi, J.; Mizushima, D.; Narantsatsral, S.; Otgonsuren, D.; Myagmarsuren, P.; Battsetseg, B.; Battur, B.; Inoue, N.; Suganuma, K. Draft genome sequence of Trypanosoma equiperdum strain IVM-t1. *Microbiol. Resour. Announc.* **2019**, *8*, e01119-18. [CrossRef] [PubMed]

25. Hébert, L.; Moumen, B.; Madeline, A.; Steinbiss, S.; Lakhdar, L.; Van Reet, N.; Buscher, P.; Laugier, C.; Cauchard, J.; Petry, S. First draft genome sequence of the dourine causative agent: *Trypanosoma Equiperdum* strain OVI. *J. Genom.* **2017**, *5*, 1–3. [CrossRef] [PubMed]

26. Jackson, A.P.; Berry, A.; Aslett, M.; Allison, H.C.; Burton, P.; Vavrova-Anderson, J.; Brown, R.; Browne, H.; Corton, N.; Hauser, H.; et al. Antigenic diversity is generated by distinct evolutionary mechanisms in African trypanosome species. *Proc. Natl. Acad. Sci. USA* **2012**, *109*, 3416–3421. [CrossRef]

27. Abbas, A.H.; Silva Pereira, S.; D'Archivio, S.; Wickstead, B.; Morrison, L.J.; Hall, N.; Hertz-Fowler, C.; Darby, A.C.; Jackson, A.P. The structure of a conserved telomeric region associated with variant antigen loci in the blood parasite *Trypanosoma congolense*. *Genome Biol. Evol.* **2018**, *10*, 2458–2473. [CrossRef]

28. Silvester, E.; Ivens, A.; Matthews, K.R. A gene expression comparison of *Trypanosoma brucei* and *Trypanosoma congolense* in the bloodstream of the mammalian host reveals species-specific adaptations to density-dependent development. *PLoS Negl. Trop. Dis.* **2018**, *12*, e0006863. [CrossRef]

29. Awuoche, E.O.; Weiss, B.L.; Mireji, P.O.; Vigneron, A.; Nyambega, B.; Murilla, G.; Aksoy, S. Expression profiling of *Trypanosoma congolense* genes during development in the tsetse fly vector *Glossina morsitans* morsitans. *Parasit. Vectors* **2018**, *11*, 380. [CrossRef] [PubMed]

30. Greif, G.; Ponce de Leon, M.; Lamolle, G.; Rodriguez, M.; Pineyro, D.; Tavares-Marques, L.M.; Reyna-Bello, A.; Robello, C.; Alvarez-Valin, F. Transcriptome analysis of the bloodstream stage from the parasite *Trypanosoma vivax*. *BMC Genom.* **2013**, *14*, 149. [CrossRef]

31. Callejas-Hernández, F.; Rastrojo, A.; Poveda, C.; Girones, N.; Fresno, M. Genomic assemblies of newly sequenced *Trypanosoma cruzi* strains reveal new genomic expansion and greater complexity. *Sci. Rep.* **2018**, *8*, 14631. [CrossRef] [PubMed]

32. Callejas-Hernández, F.; Gironès, N.; Fresno, M. Genome sequence of *Trypanosoma cruzi* strain Bug2148. *Genome Announc.* **2018**, *6*, e01497-17. [CrossRef] [PubMed]

33. Gómez, I.; Rastrojo, A.; Sanchez-Luque, F.J.; Lorenzo-Diaz, F.; Macias, F.; Valladares, B.; Aguado, B.; Requena, J.M.; Lopez, M.C.; Thomas, M.C. Draft genome sequence of the *Trypanosoma cruzi* B. M. Lopez strain (TcIa), isolated from a Colombian patient. *Microbiol. Resour. Announc.* **2020**, *9*, e00031-20. [PubMed]

34. Gómez, I.; Rastrojo, A.; Lorenzo-Diaz, F.; Sanchez-Luque, F.J.; Macias, F.; Aguado, B.; Valladares, B.; Requena, J.M.; Lopez, M.C.; Thomas, M.C. *Trypanosoma cruzi* Ikiakarora (TcIII) draft genome sequence. *Microbiol. Resour. Announc.* **2020**, *9*, e00453-20. [CrossRef] [PubMed]

35. Berná, L.; Rodriguez, M.; Chiribao, M.L.; Parodi-Talice, A.; Pita, S.; Rijo, G.; Alvarez-Valin, F.; Robello, C. Expanding an expanded genome: Long-read sequencing of *Trypanosoma cruzi*. *Microb. Genom.* **2018**, *4*, e000177. [CrossRef]

36. DeCuir, J.; Tu, W.; Dumonteil, E.; Herrera, C. Sequence of *Trypanosoma cruzi* reference strain SC43 nuclear genome and kinetoplast maxicircle confirms a strong genetic structure among closely related parasite discrete typing units. *Genome* **2021**, *64*, 525–531. [CrossRef] [PubMed]

37. Díaz-Viraqué, F.; Pita, S.; Greif, G.; de Souza, R.C.M.; Iraola, G.; Robello, C. Nanopore sequencing significantly improves genome assembly of the protozoan parasite *Trypanosoma cruzi*. *Genome Biol. Evol.* **2019**, *11*, 1952–1957. [CrossRef] [PubMed]

38. Franzén, O.; Ochaya, S.; Sherwood, E.; Lewis, M.D.; Llewellyn, M.S.; Miles, M.A.; Andersson, B. Shotgun sequencing analysis of *Trypanosoma cruzi* I Sylvio X10/1 and comparison with *T. cruzi* VI CL Brener. *PLoS Negl. Trop. Dis.* **2011**, *5*, e984. [CrossRef]

39. Grisard, E.C.; Teixeira, S.M.; de Almeida, L.G.; Stoco, P.H.; Gerber, A.L.; Talavera-Lopez, C.; Lima, O.C.; Andersson, B.; de Vasconcelos, A.T. *Trypanosoma cruzi* clone Dm28c draft genome sequence. *Genome Announc.* **2014**, *2*, e01114-13. [CrossRef]

40. Reis-Cunha, J.L.; Rodrigues-Luiz, G.F.; Valdivia, H.O.; Baptista, R.P.; Mendes, T.A.; de Morais, G.L.; Guedes, R.; Macedo, A.M.; Bern, C.; Gilman, R.H.; et al. Chromosomal copy number variation reveals differential levels of genomic plasticity in distinct *Trypanosoma cruzi* strains. *BMC Genom.* **2015**, *16*, 499. [CrossRef]

41. Baptista, R.P.; Reis-Cunha, J.L.; DeBarry, J.D.; Chiari, E.; Kissinger, J.C.; Bartholomeu, D.C.; Macedo, A.M. Assembly of highly repetitive genomes using short reads: The genome of discrete typing unit III *Trypanosoma cruzi* strain 231. *Microb. Genom.* **2018**, *4*, e000156. [CrossRef] [PubMed]

42. Wang, W.; Peng, D.; Baptista, R.P.; Li, Y.; Kissinger, J.C.; Tarleton, R.L. Strain-specific genome evolution in *Trypanosoma cruzi*, the agent of Chagas disease. *PLoS Pathog.* **2021**, *17*, e1009254. [CrossRef] [PubMed]

43. Franzén, O.; Talavera-Lopez, C.; Ochaya, S.; Butler, C.E.; Messenger, L.A.; Lewis, M.D.; Llewellyn, M.S.; Marinkelle, C.J.; Tyler, K.M.; Miles, M.A.; et al. Comparative genomic analysis of human infective *Trypanosoma cruzi* lineages with the bat-restricted subspecies *T. cruzi marinkellei*. *BMC Genom.* **2012**, *13*, 531.

44. Bradwell, K.R.; Koparde, V.N.; Matveyev, A.V.; Serrano, M.G.; Alves, J.M.P.; Parikh, H.; Huang, B.; Lee, V.; Espinosa-Alvarez, O.; Ortiz, P.A.; et al. Genomic comparison of *Trypanosoma conorhini* and *Trypanosoma rangeli* to *Trypanosoma cruzi* strains of high and low virulence. *BMC Genom.* **2018**, *19*, 770. [CrossRef] [PubMed]

45. Stoco, P.H.; Wagner, G.; Talavera-Lopez, C.; Gerber, A.; Zaha, A.; Thompson, C.E.; Bartholomeu, D.C.; Luckemeyer, D.D.; Bahia, D.; Loreto, E.; et al. Genome of the avirulent human-infective trypanosome *Trypanosoma rangeli*. *PLoS Negl. Trop. Dis.* **2014**, *8*, e3176. [CrossRef]

46. Kelly, S.; Ivens, A.; Manna, P.T.; Gibson, W.; Field, M.C. A draft genome for the African crocodilian trypanosome *Trypanosoma grayi*. *Sci. Data* **2014**, *1*, 140024. [CrossRef]

47. Kelly, S.; Ivens, A.; Mott, G.A.; O'Neill, E.; Emms, D.; Macleod, O.; Voorheis, P.; Tyler, K.; Clark, M.; Matthews, J.; et al. An alternative strategy for trypanosome survival in the mammalian bloodstream revealed through genome and transcriptome analysis of the ubiquitous bovine parasite *Trypanosoma (Megatrypanum) theileri*. *Genome Biol. Evol.* **2017**, *9*, 2093–2109. [CrossRef]

48. Hoare, C.A. *The Trypanosomes of Mammals*; Blackwell Scientific Publications: Oxford, UK, 1972; p. 768.

49. Truc, P.; Buscher, P.; Cuny, G.; Gonzatti, M.I.; Jannin, J.; Joshi, P.; Juyal, P.; Lun, Z.R.; Mattioli, R.; Pays, E.; et al. Atypical human infections by animal trypanosomes. *PLoS Negl. Trop. Dis.* **2013**, *7*, e2256. [CrossRef]

50. Spodareva, V.V.; Grybchuk-Ieremenko, A.; Losev, A.; Votýpka, J.; Lukeš, J.; Yurchenko, V.; Kostygov, A.Y. Diversity and evolution of anuran trypanosomes: Insights from the study of European species. *Parasit. Vectors* **2018**, *11*, 447. [CrossRef] [PubMed]

51. Rogers, M.B.; Hilley, J.D.; Dickens, N.J.; Wilkes, J.; Bates, P.A.; Depledge, D.P.; Harris, D.; Her, Y.; Herzyk, P.; Imamura, H.; et al. Chromosome and gene copy number variation allow major structural change between species and strains of *Leishmania*. *Genome Res.* **2011**, *21*, 2129–2142. [CrossRef] [PubMed]

52. Peacock, C.S.; Seeger, K.; Harris, D.; Murphy, L.; Ruiz, J.C.; Quail, M.A.; Peters, N.; Adlem, E.; Tivey, A.; Aslett, M.; et al. Comparative genomic analysis of three *Leishmania* species that cause diverse human disease. *Nat. Genet.* **2007**, *39*, 839–847. [CrossRef] [PubMed]

53. Real, F.; Vidal, R.O.; Carazzolle, M.F.; Mondego, J.M.; Costa, G.G.; Herai, R.H.; Wurtele, M.; de Carvalho, L.M.; Carmona e Ferreira, R.; Mortara, R.A.; et al. The genome sequence of *Leishmania (Leishmania) amazonensis*: Functional annotation and extended analysis of gene models. *DNA Res.* **2013**, *20*, 567–581. [CrossRef]

54. Tschoeke, D.A.; Nunes, G.L.; Jardim, R.; Lima, J.; Dumaresq, A.S.; Gomes, M.R.; de Mattos Pereira, L.; Loureiro, D.R.; Stoco, P.H.; de Matos Guedes, H.L.; et al. The comparative genomics and phylogenomics of *Leishmania amazonensis* parasite. *Evol. Bioinform. Online* **2014**, *10*, 131–153. [CrossRef] [PubMed]

55. Valdivia, H.O.; Almeida, L.V.; Roatt, B.M.; Reis-Cunha, J.L.; Pereira, A.A.; Gontijo, C.; Fujiwara, R.T.; Reis, A.B.; Sanders, M.J.; Cotton, J.A.; et al. Comparative genomics of canine-isolated *Leishmania (Leishmania) amazonensis* from an endemic focus of visceral leishmaniasis in Governador Valadares, southeastern Brazil. *Sci. Rep.* **2017**, *7*, 40804. [CrossRef]

56. Aoki, J.I.; Muxel, S.M.; Zampieri, R.A.; Laranjeira-Silva, M.F.; Muller, K.E.; Nerland, A.H.; Floeter-Winter, L.M. RNA-seq transcriptional profiling of *Leishmania amazonensis* reveals an arginase-dependent gene expression regulation. *PLoS Negl. Trop. Dis.* **2017**, *11*, e0006026. [CrossRef]

57. Patino, L.H.; Muskus, C.; Ramirez, J.D. Transcriptional responses of *Leishmania (Leishmania) amazonensis* in the presence of trivalent sodium stibogluconate. *Parasit. Vectors* **2019**, *12*, 348. [CrossRef]

58. Downing, T.; Imamura, H.; Decuypere, S.; Clark, T.G.; Coombs, G.H.; Cotton, J.A.; Hilley, J.D.; de Doncker, S.; Maes, I.; Mottram, J.C.; et al. Whole genome sequencing of multiple *Leishmania donovani* clinical isolates provides insights into population structure and mechanisms of drug resistance. *Genome Res.* **2011**, *21*, 2143–2156. [CrossRef] [PubMed]

59. Singh, N.; Chikara, S.; Sundar, S. SOLiD sequencing of genomes of clinical isolates of *Leishmania donovani* from India confirm *Leptomonas* co-infection and raise some key questions. *PLoS ONE* **2013**, *8*, e55738. [CrossRef]

60. Franssen, S.U.; Durrant, C.; Stark, O.; Moser, B.; Downing, T.; Imamura, H.; Dujardin, J.C.; Sanders, M.J.; Mauricio, I.; Miles, M.A.; et al. Global genome diversity of the *Leishmania donovani* complex. *eLife* **2020**, *9*, e51243. [CrossRef]

61. González-de la Fuente, S.; Peiro-Pastor, R.; Rastrojo, A.; Moreno, J.; Carrasco-Ramiro, F.; Requena, J.M.; Aguado, B. Resequencing of the *Leishmania infantum* (strain JPCM5) genome and *de novo* assembly into 36 contigs. *Sci. Rep.* **2017**, *7*, 18050. [CrossRef] [PubMed]

62. Ishemgulova, A.; Hlaváčová, J.; Majerová, K.; Butenko, A.; Lukeš, J.; Votýpka, J.; Volf, P.; Yurchenko, V. CRISPR/Cas9 in *Leishmania mexicana*: A case study of LmxBTN1. *PLoS ONE* **2018**, *13*, e0192723. [CrossRef] [PubMed]

63. Iantorno, S.A.; Durrant, C.; Khan, A.; Sanders, M.J.; Beverley, S.M.; Warren, W.C.; Berriman, M.; Sacks, D.L.; Cotton, J.A.; Grigg, M.E. Gene expression in *Leishmania* is regulated predominantly by gene dosage. *mBio* **2017**, *8*, e01393-17. [CrossRef]
64. Valdivia, H.O.; Reis-Cunha, J.L.; Rodrigues-Luiz, G.F.; Baptista, R.P.; Baldeviano, G.C.; Gerbasi, R.V.; Dobson, D.E.; Pratlong, F.; Bastien, P.; Lescano, A.G.; et al. Comparative genomic analysis of *Leishmania* (*Viannia*) *peruviana* and *Leishmania* (*Viannia*) *braziliensis*. *BMC Genom.* **2015**, *16*, 715. [CrossRef] [PubMed]
65. González-de la Fuente, S.; Camacho, E.; Peiro-Pastor, R.; Rastrojo, A.; Carrasco-Ramiro, F.; Aguado, B.; Requena, J.M. Complete and *de novo* assembly of the *Leishmania braziliensis* (M2904) genome. *Mem. Inst. Oswaldo Cruz.* **2018**, *114*, e180438. [CrossRef]
66. Ruy, P.C.; Monteiro-Teles, N.M.; Miserani Magalhaes, R.D.; Freitas-Castro, F.; Dias, L.; Aquino Defina, T.P.; Rosas De Vasconcelos, E.J.; Myler, P.J.; Kaysel Cruz, A. Comparative transcriptomics in *Leishmania braziliensis*: Disclosing differential gene expression of coding and putative noncoding RNAs across developmental stages. *RNA Biol.* **2019**, *16*, 639–660. [CrossRef] [PubMed]
67. Llanes, A.; Restrepo, C.M.; Del Vecchio, G.; Anguizola, F.J.; Lleonart, R. The genome of *Leishmania panamensis*: Insights into genomics of the *L.* (*Viannia*) subgenus. *Sci. Rep.* **2015**, *5*, 8550. [CrossRef] [PubMed]
68. Coughlan, S.; Taylor, A.S.; Feane, E.; Sanders, M.; Schonian, G.; Cotton, J.A.; Downing, T. *Leishmania naiffi* and *Leishmania guyanensis* reference genomes highlight genome structure and gene evolution in the *Viannia* subgenus. *R. Soc. Open Sci.* **2018**, *5*, 172212. [CrossRef] [PubMed]
69. Batra, D.; Lin, W.; Rowe, L.A.; Sheth, M.; Zheng, Y.; Loparev, V.; de Almeida, M. Draft genome sequence of French Guiana *Leishmania* (*Viannia*) *guyanensis* strain 204–365, assembled using long reads. *Microbiol. Resour. Announc.* **2018**, *7*, e01421-18. [CrossRef]
70. Raymond, F.; Boisvert, S.; Roy, G.; Ritt, J.F.; Legare, D.; Isnard, A.; Stanke, M.; Olivier, M.; Tremblay, M.J.; Papadopoulou, B.; et al. Genome sequencing of the lizard parasite *Leishmania tarentolae* reveals loss of genes associated to the intracellular stage of human pathogenic species. *Nucleic Acids Res.* **2012**, *40*, 1131–1147. [CrossRef]
71. Goto, Y.; Kuroki, A.; Suzuki, K.; Yamagishi, J. Draft genome sequence of *Leishmania tarentolae* Parrot Tar II, obtained by single-molecule real-time sequencing. *Microbiol. Resour. Announc.* **2020**, *9*, e00050-20. [CrossRef]
72. Coughlan, S.; Mulhair, P.; Sanders, M.; Schonian, G.; Cotton, J.A.; Downing, T. The genome of *Leishmania adleri* from a mammalian host highlights chromosome fission in *Sauroleishmania*. *Sci. Rep.* **2017**, *7*, 43747. [CrossRef] [PubMed]
73. Harkins, K.M.; Schwartz, R.S.; Cartwright, R.A.; Stone, A.C. Phylogenomic reconstruction supports supercontinent origins for *Leishmania*. *Infect. Genet. Evol.* **2016**, *38*, 101–109. [CrossRef]
74. Butenko, A.; Kostygov, A.Y.; Sádlová, J.; Kleschenko, Y.; Bečvář, T.; Podešvová, L.; Macedo, D.H.; Žihala, D.; Lukeš, J.; Bates, P.A.; et al. Comparative genomics of *Leishmania* (*Mundinia*). *BMC Genom.* **2019**, *20*, 726. [CrossRef] [PubMed]
75. Paranaiba, L.F.; Pinheiro, L.J.; Torrecilhas, A.C.; Macedo, D.H.; Menezes-Neto, A.; Tafuri, W.L.; Soares, R.P. *Leishmania enriettii* (Muniz & Medina, 1948): A highly diverse parasite is here to stay. *PLoS Pathog.* **2017**, *13*, e1006303.
76. Paranaiba, L.F.; de Assis, R.R.; Nogueira, P.M.; Torrecilhas, A.C.; Campos, J.H.; Silveira, A.C.; Martins-Filho, O.A.; Pessoa, N.L.; Campos, M.A.; Parreiras, P.M.; et al. *Leishmania enriettii*: Biochemical characterisation of lipophosphoglycans (LPGs) and glycoinositolphospholipids (GIPLs) and infectivity to *Cavia porcellus*. *Parasit. Vectors* **2015**, *8*, 31. [CrossRef] [PubMed]
77. De Assis, R.R.; Ibraim, I.C.; Nogueira, P.M.; Soares, R.P.; Turco, S.J. Glycoconjugates in New World species of *Leishmania*: Polymorphisms in lipophosphoglycan and glycoinositolphospholipids and interaction with hosts. *Biochim. Biophys. Acta* **2012**, *1820*, 1354–1365. [CrossRef] [PubMed]
78. Albanaz, A.T.S.; Gerasimov, E.S.; Shaw, J.J.; Sádlová, J.; Lukeš, J.; Volf, P.; Opperdoes, F.R.; Kostygov, A.Y.; Butenko, A.; Yurchenko, V. Genome analysis of *Endotrypanum* and *Porcisia* spp.; closest phylogenetic relatives of *Leishmania*, highlights the role of amastins in shaping pathogenicity. *Genes* **2021**, *12*, 444. [CrossRef]
79. Imamura, H.; Monsieurs, P.; Jara, M.; Sanders, M.; Maes, I.; Vanaerschot, M.; Berriman, M.; Cotton, J.A.; Dujardin, J.C.; Domagalska, M.A. Evaluation of whole genome amplification and bioinformatic methods for the characterization of *Leishmania* genomes at a single cell level. *Sci. Rep.* **2020**, *10*, 15043. [CrossRef]
80. Dia, A.; Cheeseman, I.H. Single-cell genome sequencing of protozoan parasites. *Trends Parasitol.* **2021**, in press. [CrossRef]
81. Akhoundi, M.; Kuhls, K.; Cannet, A.; Votýpka, J.; Marty, P.; Delaunay, P.; Sereno, D. A historical overview of the classification, evolution, and dispersion of *Leishmania* parasites and sandflies. *PLoS Negl. Trop. Dis.* **2016**, *10*, e0004349. [CrossRef]
82. Klatt, S.; Simpson, L.; Maslov, D.A.; Konthur, Z. *Leishmania tarentolae*: Taxonomic classification and its application as a promising biotechnological expression host. *PLoS Negl. Trop. Dis.* **2019**, *13*, e0007424. [CrossRef]
83. Telford, S.R. *Hemoparasites of the Reptilia: Color Atlas and Text*; CRC Press: Boca Raton, FL, USA, 2009; Volume xv, 376p.
84. Espinosa, O.A.; Serrano, M.G.; Camargo, E.P.; Teixeira, M.M.; Shaw, J.J. An appraisal of the taxonomy and nomenclature of trypanosomatids presently classified as *Leishmania* and *Endotrypanum*. *Parasitology* **2018**, *145*, 430–442. [CrossRef]
85. Warren, W.C.; Akopyants, N.S.; Dobson, D.E.; Hertz-Fowler, C.; Lye, L.F.; Myler, P.J.; Ramasamy, G.; Shanmugasundram, A.; Silva-Franco, F.; Steinbiss, S.; et al. Genome assemblies across the diverse evolutionary spectrum of Leishmania protozoan parasites. *bioRxiv* **2021**. [CrossRef]
86. Batra, D.; Lin, W.; Narayanan, V.; Rowe, L.A.; Sheth, M.; Zheng, Y.; Loparev, V.; de Almeida, M. Draft genome sequences of *Leishmania* (*Leishmania*) *amazonensis*, *Leishmania* (*Leishmania*) *mexicana*, and *Leishmania* (*Leishmania*) *aethiopica*, potential etiological agents of diffuse cutaneous leishmaniasis. *Microbiol. Resour. Announc.* **2019**, *8*, e00269-19. [CrossRef]

87. Almutairi, H.; Urbaniak, M.D.; Bates, M.D.; Jariyapan, N.; Al-Salem, W.S.; Dillon, R.J.; Bates, P.A.; Gatherer, D. Chromosome-scale assembly of the complete genome sequence of *Leishmania* (*Mundinia*) *martiniquensis*, isolate LSCM1, strain LV760. *Microbiol. Resour. Announc.* **2021**, *10*, e0005821. [CrossRef]
88. Lin, W.; Batra, D.; Narayanan, V.; Rowe, L.A.; Sheth, M.; Zheng, Y.; Juieng, P.; Loparev, V.; de Almeida, M. First draft genome sequence of *Leishmania* (*Viannia*) *lainsoni* strain 216–34, isolated from a Peruvian clinical case. *Microbiol. Resour. Announc.* **2019**, *8*, e01524-18. [CrossRef]
89. Runckel, C.; DeRisi, J.; Flenniken, M.L. A draft genome of the honey bee trypanosomatid parasite *Crithidia mellificae*. *PLoS ONE* **2014**, *9*, e95057. [CrossRef] [PubMed]
90. Schmid-Hempel, P.; Aebi, M.; Barribeau, S.; Kitajima, T.; du Plessis, L.; Schmid-Hempel, R.; Zoller, S. The genomes of *Crithidia bombi* and *C. expoeki*, common parasites of bumblebees. *PLoS ONE* **2018**, *13*, e0189738. [CrossRef] [PubMed]
91. Gerasimov, E.; Zemp, N.; Schmid-Hempel, R.; Schmid-Hempel, P.; Yurchenko, V. Genomic variation among strains of *Crithidia bombi* and *C. expoeki*. *mSphere* **2019**, *4*, e00482-19. [CrossRef] [PubMed]
92. Ghosh, S.; Banerjee, P.; Sarkar, A.; Datta, S.; Chatterjee, M. Coinfection of *Leptomonas seymouri* and *Leishmania donovani* in Indian leishmaniasis. *J. Clin. Microbiol.* **2012**, *50*, 2774–2778. [CrossRef] [PubMed]
93. Flegontov, P.; Butenko, A.; Firsov, S.; Kraeva, N.; Eliáš, M.; Field, M.C.; Filatov, D.; Flegontova, O.; Gerasimov, E.S.; Hlaváčová, J.; et al. Genome of *Leptomonas pyrrhocoris*: A high-quality reference for monoxenous trypanosomatids and new insights into evolution of *Leishmania*. *Sci. Rep.* **2016**, *6*, 23704. [CrossRef] [PubMed]
94. Filosa, J.N.; Berry, C.T.; Ruthel, G.; Beverley, S.M.; Warren, W.C.; Tomlinson, C.; Myler, P.J.; Dudkin, E.A.; Povelones, M.L.; Povelones, M. Dramatic changes in gene expression in different forms of *Crithidia fasciculata* reveal potential mechanisms for insect-specific adhesion in kinetoplastid parasites. *PLoS Negl. Trop. Dis.* **2019**, *13*, e0007570. [CrossRef]
95. Ghobakhloo, N.; Motazedian, M.H.; Naderi, S.; Sepideh, E. Isolation of *Crithidia* spp. from lesions of immunocompetent patients with suspected cutaneous leishmaniasis in Iran. *Trop. Med. Int. Health* **2018**, *24*, 116–126. [CrossRef]
96. Zakharova, A.; Saura, A.; Butenko, A.; Podešvová, L.; Warmusová, S.; Kostygov, A.Y.; Nenarokova, A.; Lukeš, J.; Opperdoes, F.R.; Yurchenko, V. A new model trypanosomatid *Novymonas esmeraldas*: Genomic perception of its "*Candidatus* Pandoraea novymonadis" endosymbiont. *mBio* **2021**, *12*, e01606-21. [CrossRef]
97. Votýpka, J.; Klepetková, H.; Yurchenko, V.Y.; Horák, A.; Lukeš, J.; Maslov, D.A. Cosmopolitan distribution of a trypanosomatid *Leptomonas pyrrhocoris*. *Protist* **2012**, *163*, 616–631. [CrossRef] [PubMed]
98. Kostygov, A.Y.; Grybchuk-Ieremenko, A.; Malysheva, M.N.; Frolov, A.O.; Yurchenko, V. Molecular revision of the genus *Wallaceina*. *Protist* **2014**, *165*, 594–604. [CrossRef] [PubMed]
99. Ganyukova, A.I.; Zolotarev, A.V.; Frolov, A.O. Geographical distribution and host range of monoxenous trypanosomatid *Crithidia brevicula* (Frolov et Malysheva, 1989) in the northern regions of Eurasia. *Protistology* **2020**, *14*, 70–78. [CrossRef]
100. Kostygov, A.Y.; Yurchenko, V. Revised classification of the subfamily Leishmaniinae (Trypanosomatidae). *Folia Parasitol.* **2017**, *64*, 020. [CrossRef]
101. Marin, C.; Fabre, S.; Sanchez-Moreno, M.; Dollet, M. *Herpetomonas* spp. isolated from tomato fruits (*Lycopersicon esculentum*) in southern Spain. *Exp. Parasitol.* **2007**, *116*, 88–90. [CrossRef]
102. Fiorini, J.E.; Takata, C.S.; Teofilo, V.M.; Nascimento, L.C.; Faria-e-Silva, P.M.; Soares, M.J.; Teixeira, M.M.; De Souza, W. Morphological, biochemical and molecular characterization of *Herpetomonas samuelpessoai camargoi* n. subsp.; a trypanosomatid isolated from the flower of the squash *Cucurbita moschata*. *J. Eukaryot. Microbiol.* **2001**, *48*, 62–69. [CrossRef]
103. Morio, F.; Reynes, J.; Dollet, M.; Pratlong, F.; Dedet, J.P.; Ravel, C. Isolation of a protozoan parasite genetically related to the insect trypanosomatid *Herpetomonas samuelpessoai* from a human immunodeficiency virus-positive patient. *J. Clin. Microbiol.* **2008**, *46*, 3845–3847. [CrossRef]
104. Porcel, B.M.; Denoeud, F.; Opperdoes, F.R.; Noel, B.; Madoui, M.-A.; Hammarton, T.C.; Field, M.C.; Da Silva, C.; Couloux, A.; Poulain, J.; et al. The streamlined genome of *Phytomonas* spp. relative to human pathogenic kinetoplastids reveals a parasite tailored for plants. *PLoS Genet.* **2014**, *10*, e1004007. [CrossRef]
105. Kořený, L.; Sobotka, R.; Kovářová, J.; Gnipová, A.; Flegontov, P.; Horváth, A.; Oborník, M.; Ayala, F.J.; Lukeš, J. Aerobic kinetoplastid flagellate *Phytomonas* does not require heme for viability. *Proc. Natl. Acad. Sci. USA* **2012**, *109*, 3808–3813. [CrossRef] [PubMed]
106. Butler, C.E.; Jaskowska, E.; Kelly, S. Genome sequence of *Phytomonas françai*, a cassava (*Manihot esculenta*) latex parasite. *Genome Announc.* **2017**, *5*, e01266-16. [CrossRef]
107. Sloan, M.A.; Brooks, K.; Otto, T.D.; Sanders, M.J.; Cotton, J.A.; Ligoxygakis, P. Transcriptional and genomic parallels between the monoxenous parasite *Herpetomonas muscarum* and *Leishmania*. *PLoS Genet.* **2019**, *15*, e1008452. [CrossRef] [PubMed]
108. Frolov, A.O.; Malysheva, M.N.; Ganyukova, A.I.; Spodareva, V.V.; Yurchenko, V.; Kostygov, A.Y. Development of *Phytomonas lipae* sp. n. (Kinetoplastea: Trypanosomatidae) in the true bug *Coreus marginatus* (Heteroptera: Coreidae) and insights into the evolution of life cycles in the genus *Phytomonas*. *PLoS ONE* **2019**, *14*, e0214484. [CrossRef]
109. Seward, E.A.; Votýpka, J.; Kment, P.; Lukeš, J.; Kelly, S. Description of *Phytomonas oxycareni* n. sp. from the salivary glands of *Oxycarenus lavaterae*. *Protist* **2017**, *168*, 71–79. [CrossRef]
110. Yurchenko, V.; Kostygov, A.; Havlová, J.; Grybchuk-Ieremenko, A.; Ševčíková, T.; Lukeš, J.; Ševčík, J.; Votýpka, J. Diversity of trypanosomatids in cockroaches and the description of *Herpetomonas tarakana* sp. n. *J. Eukaryot. Microbiol.* **2016**, *63*, 198–209. [CrossRef]

111. Frolov, A.O.; Malysheva, M.N.; Yurchenko, V.; Kostygov, A.Y. Back to monoxeny: *Phytomonas nordicus* descended from dixenous plant parasites. *Eur. J. Protistol.* **2016**, *52*, 1–10. [CrossRef]
112. Ganyukova, A.I.; Frolov, A.O.; Malysheva, M.N.; Spodareva, V.V.; Yurchenko, V.; Kostygov, A.Y. A novel endosymbiont-containing trypanosomatid *Phytomonas borealis* sp. n. from the predatory bug *Picromerus bidens* (Heteroptera: Pentatomidae). *Folia Parasitol.* **2020**, *67*, 4. [CrossRef] [PubMed]
113. Motta, M.C.; Martins, A.C.; de Souza, S.S.; Catta-Preta, C.M.; Silva, R.; Klein, C.C.; de Almeida, L.G.; de Lima Cunha, O.; Ciapina, L.P.; Brocchi, M.; et al. Predicting the proteins of *Angomonas deanei, Strigomonas culicis* and their respective endosymbionts reveals new aspects of the trypanosomatidae family. *PLoS ONE* **2013**, *8*, e60209. [CrossRef]
114. Morales, J.; Kokkori, S.; Weidauer, D.; Chapman, J.; Goltsman, E.; Rokhsar, D.; Grossman, A.R.; Nowack, E.C. Development of a toolbox to dissect host-endosymbiont interactions and protein trafficking in the trypanosomatid *Angomonas deanei*. *BMC Evol. Biol.* **2016**, *16*, 247. [CrossRef]
115. Skalický, T.; Alves, J.M.P.; Morais, A.C.; Režnarová, J.; Butenko, A.; Lukeš, J.; Serrano, M.G.; Buck, G.A.; Teixeira, M.M.G.; Camargo, E.P.; et al. Endosymbiont capture, a repeated process of endosymbiont transfer with replacement in trypanosomatids *Angomonas* spp. *Pathogens* **2021**, *10*, 702. [CrossRef] [PubMed]
116. Skalický, T.; Dobáková, E.; Wheeler, R.J.; Tesařová, M.; Flegontov, P.; Jirsová, D.; Votýpka, J.; Yurchenko, V.; Ayala, F.J.; Lukeš, J. Extensive flagellar remodeling during the complex life cycle of *Paratrypanosoma*, an early-branching trypanosomatid. *Proc. Natl. Acad. Sci. USA* **2017**, *114*, 11757–11762. [CrossRef] [PubMed]
117. Opperdoes, F.R.; Butenko, A.; Flegontov, P.; Yurchenko, V.; Lukeš, J. Comparative metabolism of free-living *Bodo saltans* and parasitic trypanosomatids. *J. Eukaryot. Microbiol.* **2016**, *63*, 657–678. [CrossRef] [PubMed]
118. Opperdoes, F.R.; Butenko, A.; Zakharova, A.; Gerasimov, E.S.; Zimmer, S.L.; Lukeš, J.; Yurchenko, V. The remarkable metabolism of *Vickermania ingenoplastis*: Genomic predictions. *Pathogens* **2021**, *10*, 68. [CrossRef] [PubMed]
119. Frolov, A.O.; Kostygov, A.Y.; Yurchenko, V. Development of monoxenous trypanosomatids and phytomonads in insects. *Trends Parasitol.* **2021**, *37*, 538–551. [CrossRef] [PubMed]
120. Frolov, A.O.; Malysheva, M.N.; Ganyukova, A.I.; Spodareva, V.V.; Kralova, J.; Yurchenko, V.; Kostygov, A.Y. If host is refractory, insistent parasite goes berserk: Trypanosomatid *Blastocrithidia raabei* in the dock bug *Coreus marginatus*. *PLoS ONE* **2020**, *15*, e0227832. [CrossRef] [PubMed]
121. Frolov, A.O.; Malysheva, M.N.; Ganyukova, A.I.; Yurchenko, V.; Kostygov, A.Y. Obligate development of *Blastocrithidia papi* (Trypanosomatidae) in the Malpighian tubules of *Pyrrhocoris apterus* (Hemiptera) and coordination of host-parasite life cycles. *PLoS ONE* **2018**, *13*, e0204467.
122. Hamilton, P.T.; Votýpka, J.; Dostalova, A.; Yurchenko, V.; Bird, N.H.; Lukeš, J.; Lemaitre, B.; Perlman, S.J. Infection dynamics and immune response in a newly described *Drosophila*-trypanosomatid association. *mBio* **2015**, *6*, e01356-15. [CrossRef]
123. Svobodová, M.; Zídková, L.; Čepička, I.; Oborník, M.; Lukeš, J.; Votýpka, J. *Sergeia podlipaevi* gen. nov.; sp. nov. (Trypanosomatidae, Kinetoplastida), a parasite of biting midges (Ceratopogonidae, Diptera). *Int. J. Syst. Evol. Microbiol.* **2007**, *57 Pt 2*, 423–432. [CrossRef]
124. Dvorák, V.; Shaw, J.J.; Volf, P. Parasite Biology: The Vectors. In *The leishmaniases: Old Neglected Tropical Diseases*; Bruschi, F., Gradoni, L., Eds.; Springer: Cham, Switzerland, 2018; pp. 31–77.
125. Jackson, A.P. The evolution of amastin surface glycoproteins in trypanosomatid parasites. *Mol. Biol. Evol.* **2010**, *27*, 33–45. [CrossRef]
126. Durante, I.M.; Butenko, A.; Rašková, V.; Charyyeva, A.; Svobodová, M.; Yurchenko, V.; Hashimi, H.; Lukeš, J. Large-scale phylogenetic analysis of trypanosomatid adenylate cyclases reveals associations with extracellular lifestyle and host-pathogen interplay. *Genome Biol. Evol.* **2020**, *12*, 2403–2416. [CrossRef]
127. De Souza, D.A.S.; Pavoni, D.P.; Krieger, M.A.; Ludwig, A. Evolutionary analyses of myosin genes in trypanosomatids show a history of expansion, secondary losses and neofunctionalization. *Sci. Rep.* **2018**, *8*, 1376. [CrossRef]
128. Bianchi, C.; Kostygov, A.Y.; Kraeva, N.; Záhonová, K.; Horáková, E.; Sobotka, R.; Lukeš, J.; Yurchenko, V. An enigmatic catalase of *Blastocrithidia*. *Mol. Biochem. Parasitol.* **2019**, *232*, 111199. [CrossRef] [PubMed]
129. Kraeva, N.; Horáková, E.; Kostygov, A.; Kořený, L.; Butenko, A.; Yurchenko, V.; Lukeš, J. Catalase in Leishmaniinae: With me or against me? *Infect. Genet. Evol.* **2017**, *50*, 121–127. [CrossRef] [PubMed]
130. Jackson, A.P.; Vaughan, S.; Gull, K. Evolution of tubulin gene arrays in trypanosomatid parasites: Genomic restructuring in *Leishmania*. *BMC Genom.* **2006**, *7*, 261. [CrossRef]
131. Silva Pereira, S.; Jackson, A.P. UDP-glycosyltransferase genes in trypanosomatid genomes have diversified independently to meet the distinct developmental needs of parasite adaptations. *BMC Evol. Biol.* **2018**, *18*, 31. [CrossRef]
132. Dobson, D.E.; Scholtes, L.D.; Valdez, K.E.; Sullivan, D.R.; Mengeling, B.J.; Cilmi, S.; Turco, S.J.; Beverley, S.M. Functional identification of galactosyltransferases (SCGs) required for species-specific modifications of the lipophosphoglycan adhesin controlling *Leishmania major*-sand fly interactions. *J. Biol. Chem.* **2003**, *278*, 15523–15531. [CrossRef] [PubMed]
133. Beverley, S.M.; Turco, S.J. Lipophosphoglycan (LPG) and the identification of virulence genes in the protozoan parasite *Leishmania*. *Trends Microbiol.* **1998**, *6*, 35–40. [CrossRef]
134. Sacks, D.L. *Leishmania*-sand fly interactions controlling species-specific vector competence. *Cell Microbiol.* **2001**, *3*, 189–196. [CrossRef]

135. Soares, R.P.; Margonari, C.; Secundino, N.C.; Macedo, M.E.; da Costa, S.M.; Rangel, E.F.; Pimenta, P.F.; Turco, S.J. Differential midgut attachment of *Leishmania* (*Viannia*) *braziliensis* in the sand flies *Lutzomyia* (*Nyssomyia*) *whitmani* and *Lutzomyia* (*Nyssomyia*) *intermedia*. *J. Biomed. Biotechnol.* **2010**, *2010*, 439174. [CrossRef]

136. Azevedo, L.G.; de Queiroz, A.T.L.; Barral, A.; Santos, L.A.; Ramos, P.I.P. Proteins involved in the biosynthesis of lipophosphoglycan in *Leishmania*: A comparative genomic and evolutionary analysis. *Parasit. Vectors* **2020**, *13*, 44. [CrossRef] [PubMed]

137. Butenko, A.; Vieira, T.D.S.; Frolov, A.O.; Opperdoes, F.R.; Soares, R.P.; Kostygov, A.Y.; Lukeš, J.; Yurchenko, V. *Leptomonas pyrrhocoris*: Genomic insight into parasite's physiology. *Curr. Genom.* **2018**, *19*, 150–156. [CrossRef] [PubMed]

138. Díaz-Viraqué, F.; Greif, G.; Berna, L.; Robello, C. Nanopore long read DNA sequencing of protozoan parasites: Hybrid genome assembly of Trypanosoma cruzi. In *Parasite Genom.*; de Pablos, L.M., Sotillo, J., Eds.; Humana: New York, NY, USA, 2021; pp. 3–13.

139. Müller, L.S.M.; Cosentino, R.O.; Förstner, K.U.; Guizetti, J.; Wedel, C.; Kaplan, N.; Janzen, C.J.; Arampatzi, P.; Vogel, J.; Steinbiss, S.; et al. Genome organization and DNA accessibility control antigenic variation in trypanosomes. *Nature* **2018**, *563*, 121–125. [CrossRef] [PubMed]

140. Davey, J.W.; Catta-Preta, C.M.C.; James, S.; Forrester, S.; Motta, M.C.M.; Ashton, P.D.; Mottram, J.C. Chromosomal assembly of the nuclear genome of the endosymbiont-bearing trypanosomatid *Angomonas deanei*. G3 *(Bethesda)* **2021**, *11*, jkaa018. [CrossRef] [PubMed]

141. Grybchuk, D.; Akopyants, N.S.; Kostygov, A.Y.; Konovalovas, A.; Lye, L.F.; Dobson, D.E.; Zangger, H.; Fasel, N.; Butenko, A.; Frolov, A.O.; et al. Viral discovery and diversity in trypanosomatid protozoa with a focus on relatives of the human parasite *Leishmania*. *Proc. Natl. Acad. Sci. USA* **2018**, *115*, E506–E515. [CrossRef] [PubMed]

142. Gerasimov, E.S.; Gasparyan, A.A.; Kaurov, I.; Tichý, B.; Logacheva, M.D.; Kolesnikov, A.A.; Lukeš, J.; Yurchenko, V.; Zimmer, S.L.; Flegontov, P. Trypanosomatid mitochondrial RNA editing: Dramatically complex transcript repertoires revealed with a dedicated mapping tool. *Nucleic Acids Res.* **2018**, *46*, 765–781. [CrossRef] [PubMed]

143. Cooper, S.; Wadsworth, E.S.; Ochsenreiter, T.; Ivens, A.; Savill, N.J.; Schnaufer, A. Assembly and annotation of the mitochondrial minicircle genome of a differentiation-competent strain of *Trypanosoma brucei*. *Nucleic Acids Res.* **2019**, *47*, 11304–11325. [CrossRef]

144. Li, S.J.; Zhang, X.; Lukeš, J.; Li, B.Q.; Wang, J.F.; Qu, L.H.; Hide, G.; Lai, D.H.; Lun, Z.R. Novel organization of mitochondrial minicircles and guide RNAs in the zoonotic pathogen *Trypanosoma lewisi*. *Nucleic Acids Res.* **2020**, *48*, 9747–9761. [CrossRef]

145. Greif, G.; Rodriguez, M.; Reyna-Bello, A.; Robello, C.; Alvarez-Valin, F. Kinetoplast adaptations in American strains from *Trypanosoma vivax*. *Mutat. Res.* **2015**, *773*, 69–82. [CrossRef] [PubMed]

146. Callejas-Hernández, F.; Herreros-Cabello, A.; Del Moral-Salmoral, J.; Fresno, M.; Gironès, N. The complete mitochondrial DNA of *Trypanosoma cruzi*: Maxicircles and minicircles. *Front. Cell Infect. Microbiol.* **2021**, *11*, 672448. [CrossRef]

147. Camacho, E.; Rastrojo, A.; Sanchiz, A.; Gonzalez-de la Fuente, S.; Aguado, B.; Requena, J.M. *Leishmania* mitochondrial genomes: Maxicircle structure and heterogeneity of minicircles. *Genes* **2019**, *10*, 758. [CrossRef] [PubMed]

148. Gerasimov, E.S.; Gasparyan, A.A.; Afonin, D.A.; Zimmer, S.L.; Kraeva, N.; Lukeš, J.; Yurchenko, V.; Kolesnikov, A. Complete minicircle genome of *Leptomonas pyrrhocoris* reveals sources of its non-canonical mitochondrial RNA editing events. *Nucleic Acids Res.* **2021**, *49*, 3354–3370. [CrossRef] [PubMed]

149. Rusman, F.; Floridia-Yapur, N.; Tomasini, N.; Diosque, P. Guide RNA repertoires in the main lineages of *Trypanosoma cruzi*: High diversity and variable redundancy among strains. *Front. Cell Infect. Microbiol.* **2021**, *11*, 663416. [CrossRef] [PubMed]

150. Gerasimov, E.S.; Kostygov, A.Y.; Yan, S.; Kolesnikov, A.A. From cryptogene to gene? ND8 editing domain reduction in insect trypanosomatids. *Eur. J. Protistol.* **2012**, *48*, 185–193. [CrossRef]

151. Králová, J.; Grybchuk-Ieremenko, A.; Votýpka, J.; Novotný, V.; Kment, P.; Lukeš, J.; Yurchenko, V.; Kostygov, A.Y. Insect trypanosomatids in Papua New Guinea: High endemism and diversity. *Int. J. Parasitol.* **2019**, *49*, 1075–1086. [CrossRef]

152. Týč, J.; Votýpka, J.; Klepetková, H.; Šuláková, H.; Jirků, M.; Lukeš, J. Growing diversity of trypanosomatid parasites of flies (Diptera: Brachycera): Frequent cosmopolitism and moderate host specificity. *Mol. Phylogenet. Evol.* **2013**, *69*, 255–264. [CrossRef]

 *pathogens*

Article

Evolutionary Insight into the Trypanosomatidae Using Alignment-Free Phylogenomics of the Kinetoplast

Alexa Kaufer [1,*]**, Damien Stark** [2] **and John Ellis** [1] 

[1] School of Life Sciences, University of Technology Sydney, Ultimo, NSW 2007, Australia; john.ellis@uts.edu.au

[2] Department of Microbiology, St Vincent's Hospital Sydney, Darlinghurst, NSW 2010, Australia; damien.stark@svha.org.au

* Correspondence: Alexa.Kaufer@student.uts.edu.au

Received: 23 August 2019; Accepted: 13 September 2019; Published: 18 September 2019

Abstract: Advancements in next-generation sequencing techniques have led to a substantial increase in the genomic information available for analyses in evolutionary biology. As such, this data requires the exponential growth in bioinformatic methods and expertise required to understand such vast quantities of genomic data. Alignment-free phylogenomics offer an alternative approach for large-scale analyses that may have the potential to address these challenges. The evolutionary relationships between various species within the trypanosomatid family, specifically members belonging to the genera *Leishmania* and *Trypanosoma* have been extensively studies over the last 30 years. However, there is a need for a more exhaustive analysis of the Trypanosomatidae, summarising the evolutionary patterns amongst the entire family of these important protists. The mitochondrial DNA of the trypanosomatids, better known as the kinetoplast, represents a valuable taxonomic marker given its unique presence across all kinetoplastid protozoans. The aim of this study was to validate the reliability and robustness of alignment-free approaches for phylogenomic analyses and its applicability to reconstruct the evolutionary relationships between the trypanosomatid family. In the present study, alignment-free analyses demonstrated the strength of these methods, particularly when dealing with large datasets compared to the traditional phylogenetic approaches. We present a maxicircle genome phylogeny of 46 species spanning the trypanosomatid family, demonstrating the superiority of the maxicircle for the analysis and taxonomic resolution of the Trypanosomatidae.

Keywords: Trypanosomatidae; kinetoplast; second-generation sequencing; third-generation sequencing; alignment-free phylogenetics

1. Introduction

Protozoan flagellates of the trypanosomatid family (syn. Trypanosomatidae) are obligate, unicellular parasites that infect a wide array of vertebrates, invertebrates and plants [1,2]. Protozoan parasites of this diverse family are predominately monoxenous (i.e., those restricted to a single, mainly invertebrate lifecycle), however the better known dixenous members (i.e., those with an invertebrate and vertebrate host) such as *Leishmania* and *Trypanosoma* are the causative agents of some of the most important neglected tropical diseases (NTD) including leishmaniasis, human African sleeping sickness (HAT) and Chagas disease [3–5]. The trypanosomatid family belongs to a distinct evolutionary lineage of eukaryotes within the class Kinetoplastida [6]. The current stance on kinetoplastid phylogeny is that the dixenous organisms evolved from the monoxenous members of the Trypanosomatidae several times throughout history, leading to the independent emergence of the genera *Trypanosoma*, *Phytomonas* and a group that unites *Leishmania*, *Porcisia* and *Endotrypanum* [5,7,8].

Traditionally, phylogenetic relationships within the kinetoplastids, including the trypanosomatid family have been predominately based on the analysis of the SSU rRNA genes [7,9]. The evolutionary rate of substitutions in genes is considered one of the most important factors that influence the informativeness and robustness of phylogenetic analyses [10]. Advancements in molecular biology have demonstrated that slow-evolving genes (like the SSU rRNA) are not reliable markers for the deep level resolution of species in order to determine the exact branching within the trypanosomatid family [7]. Nowadays, a multi-marker approach using concatenated sequences of multiple genes are becoming the preferred choice and are routinely employed for the phylogenetic inference of related organisms [5].

Exclusive to kinetoplastid protozoans, the mitochondrial DNA of trypanosomatids is an extensive network of DNA circles which are condensed into a periflagellar structure known as the kinetoplast (kDNA) [11,12]. The kDNA consists of approximately 10,000 minicircles ranging from 0.5 kb to 10 kb and 20–50 larger maxicircles ranging from 20 to 40 kb [13,14]. Recent analyses of *Trypanosoma brucei* demonstrated that the kDNA constitutes only 4.18 megabase pairs (Mbp) of the trypanosomatids' total genome size of 77.7 Mbp [12]. Containing the mitochondrial homologues common to other eukaryotes, the maxicircle kDNA consists of two regions; a coding region containing the protein-coding genes and a highly repetitive non-coding region termed the divergent region (DR) [15–17]. In recent years, the maxicircle genome has become well-established as a superior taxonomic marker for the evolutionary analyses between related organisms of the Leishmaniinae and *Trypanosoma* [11,18,19]. Following suit with this rationale, the maxicircle kDNA should provide a more resolute model to investigate the genetic relationships between the entire trypanosomatid family, providing an all-encompassing analysis on the origins and biology of the Trypanosomatidae.

Over the past decade, second-generation sequencing (SGS) and third-generation sequencing (TGS) have become the leading technologies in the typing and evolutionary analyses of related organisms [20]. While short-read SGS techniques such as Illumina platforms have revolutionised biomedical research, their limitations, specifically their short-read (SR) lengths, make them inadequately suited for the assembly of complex and highly repetitive genomic regions [21]. Long-read TGS techniques such as PacBio offer longer read lengths (average >10 kb) than those of SGS, making it an ideal candidate for complex genomes [22]. However, a long-read (LR) length is hindered by a higher error rate of approximately 11–15% compared to that of SGS [22,23]. Despite this, the advantages of SGS and TGS are complementary, offering an alternative 'hybrid' strategy that makes use of both technologies to overcome the drawbacks of each method alone. These hybrid methodologies using long (and inaccurate) and short (and accurate) have proven to be extremely useful in producing high-quality, accurate assemblies [23,24].

Next-generation sequence data has paved the way for the use of phylogenomic approaches for the analysis of evolutionary relationships. Phylogenomics is the junction between evolution and genomics, using the comparative analysis of genome scale data for the reconstruction of evolutionary histories between organisms [25]. However, analyses with larger datasets across a wider breadth of taxa are becoming increasingly computationally infeasible [26]. The vast increase in genetic information necessitates the exponential growth in bioinformatic methods and expertise required to understand such immense quantities of genome-scale data. Thus, for large-scale analyses of genomes (i.e., phylogenomics), alignment-free (AF) methods of phylogenetic inference have been increasingly employed over the last few years [27]. Alignment-free software was first introduced nearly a decade ago but received little attention due to the traditional belief of its inferior resolution to multiple sequence alignment (MSA) based methods. However, due to the immense quantities of genome-scale data being produced, recent years have seen a surge in publications using AF applications for the phylogenetic analysis of organisms [20,26–32].

Traditionally, MSA-phylogenetics is based on the correspondence of individual nucleotides or amino acids that are in the same order between the species analysed [33]. Alignment-based methods for the most part yield excellent results when the dataset of sequences can be reliably aligned, however in certain instances, alignment-based sequence analyses can become problematic [34]. When sequences are divergent (i.e., analysis of the entire trypanosomatid family), the accuracy of sequence alignments decreases rapidly when the sequence identity falls below a certain critical point [33]. Second, alignment-based protocols are computationally intensive on a genome-scale [33], which is problematic with the increasing trend of whole-genome sequencing (WGS) replacing the use of single or few genes for the phylogenetic inference of related organisms.

The increasing trend of alignment-free sequence analyses offers an alternative approach for large-scale comparisons that may have the potential to address these limitations. One such alignment-free method introduced by Sims et al. [35] uses a measure based on the divergence between feature frequency profiles (FFPs), where the features (syn. word length) called k-mers are short nucleotide or amino-acid sequences of length k. For example, two sequences x = TTAAGG and y = AAGGCC and a feature length or k-mer size of three nucleotides produces K_3^x = (TTA, TAA, AAG, AGG) and K_3^y = (AAG, AGG, GGC, GCC) [33]. The frequency of these sub-sequences of a defined length or k-mers (k) are counted via a sliding frame implementation and used to calculate distance scores, which are subsequently used to generate a phylogenetic tree [35].

The aim of this paper is two-fold. First, the validation of the FFP protocol using the trypanosomatid family, in order to determine the limitations and variables in which the method provides reliable and robust results. Second, the application of this method to establish and summarise the current evolutionary relationships of the Trypanosomatidae. Rather than using data from the entire kinetoplast (minicircles and maxicircles), we focus on assembled genomes of the maxicircle only. In our previous study, we demonstrated that the maxicircle represents a superior, phylogenetically informative marker for studying the evolutionary relationships of the Leishmaniinae [19]. In this study, we expand on this concept and use the maxicircle kDNA to analyse not only the members of the Leishmaniinae, but representatives of the entire trypanosomatid family to provide an updated scheme for the taxonomic classification and resolution of the Trypanosomatidae.

In this study we applied a traditional MSA and AF method to the coding region of the maxicircle from a dataset of 46 trypanosomatid species. In addition to sequences already available from online databases and our previous study [19], twelve maxicircle genomes were assembled using both SR and SR/LR hybrid assemblies from raw Illumina and PacBio sequence data freely available from online databases (Sequence Read Archive). As a contribution to these efforts, this study contributes to our understanding of the phylogenomic relationships of the trypanosomatid family, demonstrating the power and robustness of the alignment-free analysis method based on the divergence between feature frequency profiles.

2. Materials and Methods

2.1. Samples

The trypanosomatid species used in this study are listed in Supplementary Materials S1 (S1 file).

2.2. Genome Assembly and Sequence Analysis

To obtain the complete maxicircle genome, processed reads were assembled from WGS data freely available through the Sequence Read Archive (SRA) on NCBI. Four paired-end Illumina and eight hybrid assemblies used a combination of long-read (PacBio) and short-read (Illumina) data to generate a hybrid contig using SPAdes version 3.12.0 [36]. For the hybrid assemblies, the datasets analysed were selected on the basis that both Illumina and PacBio sequence reads were available from the same author/provider and available for identical isolates/species. The maxicircle kDNA assembled from the Illumina assemblies (*Leishmania macropodum*, *Leishmania martiniquensis*, *Trypanosoma grayi* and *Phytomonas françai*) and hybrid assemblies (*Leishmania aethiopica*, *Leishmania amazonensis*, *Leishmania braziliensis*, *Leishmania guyanensis*, *Leishmania infantum*, *Leishmania mexicana*, *Leishmania tropica* and *Trypanosoma brucei rhodesiense*) were identified through BLAST analysis using NCBI BLAST software [37].

Annotation, gene identification and sequence analysis of the maxicircle from whole-genome sequencing was completed using Geneious version 11.0.2 [38]. The GC% content and GC skew was visualised with the software DNAPlotter [39].

In a given DNA sequence, we measure skewed mononucleotide frequencies by:

$$\text{ATS} = \frac{f_A - f_T}{f_A + f_T} \text{ and GCS} = \frac{f_G - f_C}{f_G + f_C}$$

where f_N denotes the observed frequency of nucleotides A, T, G and C [40].

2.3. Comparative Analysis of the Non-Coding Divergent Region of the Trypanosomatid Family

Self-dot plot analyses were generated in Geneious using the EMBOSS 6.5.7 software Dottup suite add-on [38]. The repetitive portion of the non-coding divergent region of 42 trypanosomatid species were visualised as a dense block of identity in a dot matrix homology search comparing each species against its own DNA sequence. The divergent region of *Endotrypanum herreri*, *Trypanosoma vivax* (Liem strain), *Trypanosoma copemani* and *Trypanosoma cruzi* (Silvio strain) were not available from the sequence reads for subsequent analyses. Additionally, REPuter was also used to identify repeat sequences including direct and palindromic repeats within the divergent region of these 42 trypanosomatid species [41]. A minimum repeat size of 60 bp and 15 bp was chosen respectively for each repeat identification. Tandem repeats in all five species were identified using Tandem Repeats Finder version 4.09 with default settings [42].

2.4. MSA Phylogenetic Analysis

A phylogenetic analysis was performed using the entire coding region from the maxicircle of 46 trypanosomatid species to investigate the evolutionary relationships between members of the Trypansomatidae. All sequences were aligned using the MUSCLE algorithm implemented in the Seaview software package [43]. Phylogenetic relationships were inferred using the maximum likelihood optimality criterion using PhyML version 3.0. For ML trees, the best-fit model of evolution, GTR+I+G was selected using jModelTest 2.1 under the Bayesian information criterion [44]. Bootstrap support for clade topologies was estimated following the analysis of 1000 pseudo-replicate datasets using a heuristic tree search.

2.5. Alignment-Free FFP Analysis

The alignment-free FFP analysis was performed on the maxicircle coding region genome of 46 trypanosomatid species using the command line software FFP version 3.1.9 [35]. To determine the optimal length of k-mer, the FFPvprof utility was used to determine the average lower word length limit and the FFPvreprof utlity was used to determine the average upper word limit of all 46 trypanosomatid species [20]. The optimum k-mer length can be identified from the overlapping region of length ranges (i.e., average of the upper and lower length limit). To determine the lower k- mer length limit, FFPvprof calculates the number of k-mers (and length) that occur at least twice between the dataset. To determine the upper k-mer length limit, FFPreprof calculates the relative entropy frequency (REF) between the expected and observed frequencies of specified k-mer lengths. Following optimisation, the UNIX-style command line pipeline was used to generate a distance matrix, which is used for the subsequent phylogenomic inference using the program Phylip version 3.697 [45]. Bootstrap support for clade topologies was estimated using the FFPboot utility following the analysis of 1000 pseudo-replicate datasets.

2.6. Estimating Divergence Time

Divergence dates of the trypanosomatid family were estimated using the Realtime method [1] and the General Time Reversible model [2] of the MEGA7 package. The maximum likelihood of this timetree was computed using one calibration constraint; the divergence of *Leishmania enriettii* and *Leishmania macropodum* approximately 40 MYA [46,47]. The maxicircle sequence of the monoxenous trypanosomatid *Paratrypanosoma confusum* served as an outgroup.

3. Results

3.1. Assembly of Data from Illumina and Hybrid Illumina/Pacbio Reads

The initial assemblies of Illumina paired-end reads using the SPAdes assembler resulted in the generation of twelve maxicircle genomes from the trypanosomatid family. Of these, we used the same Illumina dataset of eight species; *L. aethiopica*, *L. amazonensis*, *L. braziliensis*, *L. guyanensis*, *L. infantum*, *L. mexicana*, *L. tropica* and *T. brucei rhodesiense* in addition to their corresponding PacBio data for a hybrid assembly. The assemblies from both short-read and combination short- and long-read data resulted in a high-quality assembly of the maxicircle kDNA. A comparison of the final contig generated from paired-end reads to the hybrid revealed a 100% sequence identity between the coding region, reflecting the excellent accuracy of the maxicircle genome assembled from whole gene sequence generated by both SGS and TGS (Table 1). Comparison of the non-coding divergent region revealed a sequence identity >99% between the contigs generated from paired-end reads to the hybrid. The long-read lengths of the PacBio data were extremely valuable for the de novo assembly of the maxicircle, allowing us to overcome the problems caused by the repetitive nature of the divergent region. The difference in sequence identity of the divergent region is due to the increased resolution of the hybrid approach, yielding a more accurate estimate of the length of the divergent region than that of Illumina reads alone. Compared to that of the hybrid approach, SGS reads of *L. aethiopica*, *L. amazonensis*, *L. braziliensis*, *L. guyanensis*, *L. infantum*, *L. mexicana*, *L. tropica* and *T. brucei rhodesiense* assembled only 20%, 11%, 20%, 43%, 49%, 46%, 96% and 88% of the divergent region produced from TGS respectively.

Table 1. Comparison of the short-read Illumina and hybrid Illumina/PacBio assembly of various trypanosomatid species showing output length and sequence identity (%).

	Coding Region		Non-Coding Region		Overall	
	Size (bp)	% Identity	Size (bp)	% Identity	Size (bp)	% Identity
L. aethiopica (Illumina)	16 210	100	304	100	17 264	100
L. aethiopica (Hybrid)	16 210		1552		18 571	
L. amazonensis (Illumina)	16 381	100	527	99.7	17 941	100
L. amazonensis (Hybrid)	16 381		4621		23 616	
L. braziliensis (Illumina)	16 232	100	1121	98.5	18 118	99.9
L. braziliensis (Hybrid)	16 232		5745		23 012	
L. guyanensis (Illumina)	16 235	100	1607	100	18 753	100
L. guyanensis (Hybrid)	16 235		3694		20 986	
L. infantum (Illumina)	16 277	100	628	99.5	18 287	100
L. infantum (Hybrid)	16 277		1280		18 637	
L. mexicana (Illumina)	16 472	100	736	99.7	17 946	100
L. mexicana (Hybrid)	16 472		1606		18 696	
L. tropica (Illumina)	16 229	100	1582	97	18 800	99.5
L. tropica (Hybrid)	16 229		1644		19 020	
T. brucei rhodesiense (Illumina)	14 905	100	2785	97	18 200	99.5
T. brucei rhodesiense (Hybrid)	14 905		3131		18 583	

3.2. Genomic Organisation and Patterns of the Maxicircle

Sequence analysis of the maxicircle from representative species of the genera *Leishmania*, *Porcisia*, *Endotrypanum*, *Zelonia*, *Leptomonas*, *Crithidia*, *Herpetomonas*, *Angomonas*, *Blechomonas*, *Trypanosoma* and *Paratrypanosoma* revealed a conserved region typical for the maxicircle. The conserved region includes; 12S rRNA (large subunit), 9S rRNA (small subunit), seven subunits of NADH dehydrogenase,

(ND8, ND9, ND7, ND1, ND3, ND5 and ND5), three subunits of cytochrome *c* oxidase (COI, COII and COIII), one subunit of the cytochrome bc_1 complex (CYb) a single ribosomal protein (RPS12) and four open-reading frames whose role is unknown (MURF1, MURF2, MURF4 and MURF5). There was also two G-rich (G3 and G4) or C-rich (CR3 and CR4) pan-edited cryptogenes in *Leishmania* and *Trypanosoma* species respectively. Analysis of the *Phytomonas françai* maxicircle revealed that the genes for subunits I, II and III (COI, COII and COIII) of cytochrome *c* oxidase and cytochrome bc_1 complex (CYb) were missing (Figure 1).

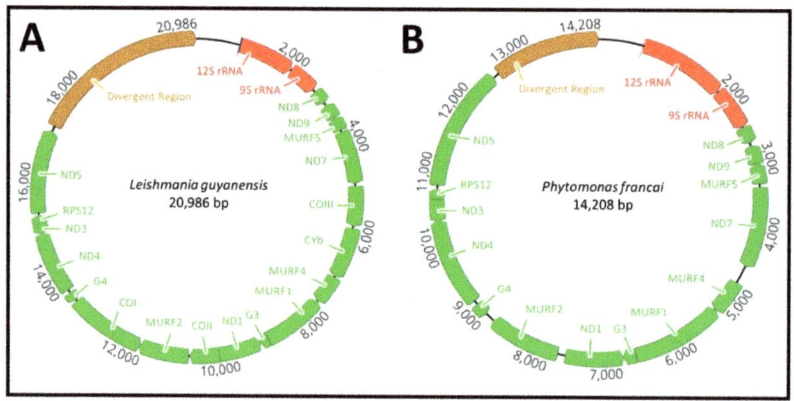

Figure 1. Graphical map of *Leishmania guyanensis* and *Phytomonas françai* maxicircle genomes assembled from Illumina and PacBio sequencing. Comparison of the maxicircle genome of (**A**) *Leishmania guyanensis* and (**B**) *Phytomonas françai*. *Phytomonas* spp. is characterised by the loss of the three subunits of the cytochrome *c* oxidase (COI, COII and COIII) and single subunit of cytochrome *b* (CYb).

The overall nucleotide frequency and skew throughout the maxicircle genome of all 46 trypanosomatid species is shown in Supplementary Materials S2. The overall AT-richness of the maxicircle ranged from 58–82%, reflecting the AT-overabundance of the mitochondrial genome as a whole. Of the clinically important *L. braziliensis*, *T. brucei rhodesiense* and *T. cruzi* (CL strain), the GC skew and GC% content is represented in the top panel of Figure 2. From the GC content, it was observed that the AT-richness of *T. brucei rhodesiense* and *T. cruzi* is primarily due to the repeat region, where the GC% of the divergent region is predominately below average (dark blue lines). The AT-richness of *L. braziliensis* is still high (80%), although the GC plot demonstrates that the AT-richness in the divergent region is to a lesser extent less than that of *T. brucei rhodesiense* and *T. cruzi*, with the AT-abundance seen throughout the entire genome. The AT skew is shown in the bottom panel of Figure 2. The AT skew of all three species demonstrates the bias towards a T-rich sequence that is more pronounced in the coding region compared to that of the DR, which demonstrates a bias towards an A-rich nucleotide frequency.

A dot matrix search of the entire maxicircle sequence (left panel) and divergent region (right panel) of *L. braziliensis*, *T. brucei rhodesiense* and *T. cruzi* (CL strain), versus itself is shown in Figure 3. The comparative analysis for the remaining trypanosomatid species can be found in Supplementary Materials S3. Based on the left panel of each graph, we can see a clear distinction between the coding and non-coding region of the maxicircle sequence, which is highlighted in a black outline. The box structures located on and residing symmetrically around the line of identity are indicative of a clustered organisation of repetitive sequences. Multiple diagonal lines seen in all three panels are indicative of direct repeats. The length of the divergent region of the various trypanosomatid species is significantly variable and thus responsible for the size difference of the complete maxicircle genome of the kinetoplast.

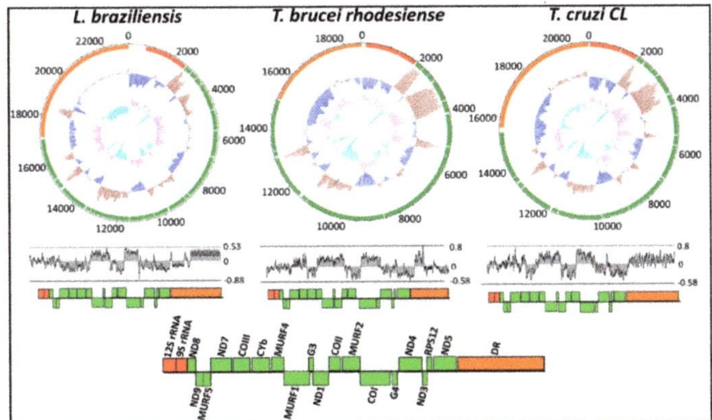

Figure 2. Circular confirmation of GC plot and GC skew of the maxicircle kinetoplast for *Leishmania braziliensis, Trypanosoma brucei rhodesiense and Trypanosoma cruzi.* The outer ring indicates gene arrangement and gene distribution. The middle circle represents the GC plot showing GC% content (dark blue for below-average and red for above-average) and the inner circle represents the GC skew (light blue for positive and orange for negative). The AT skew and corresponding gene arrangement and distribution of each species is shown below the circular plot.

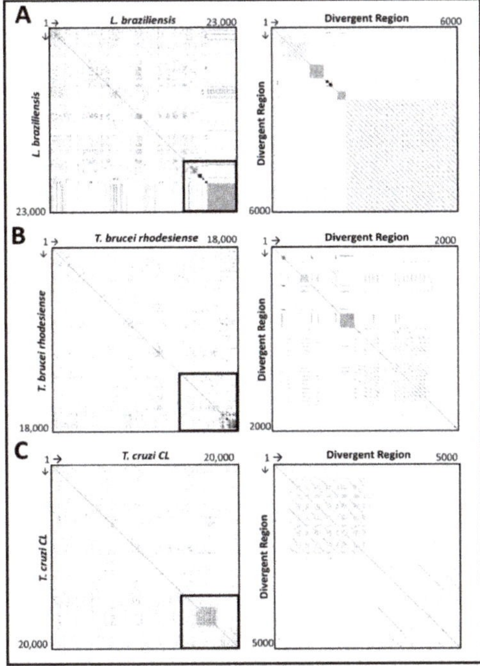

Figure 3. Self Dottup plot comparative analysis of the entire maxicircle genome (right panel) and divergent region (left panel) of *Leishmania braziliensis* (**A**), *Trypanosoma brucei rhodesiense* (**B**) and *Trypanosoma cruzi* (**C**) against their own sequences. The sequence of each species is placed on the axis and full identity over a 10 bp-long window is represented by a dot. The main diagonal line represents the sequence's alignment with itself and the lines about the main diagonal represent repetitive patterns within the maxicircle sequence.

The repeat analysis demonstrated that a different number of the various repeats were observed in the species compared (Figure 4). In these divergent region sequences, the repeat analysis detected tandem, palindromic, forward repeats. The various repeats found within the non-coding region differed significantly between species (Supplementary Materials S4). Following suit with the overall composition of the maxicircle, the forward, tandem and palindromic repeats all demonstrated a strong bias towards an AT-rich repeat in all trypanosomatid species compared. Tandem repeats identified were highly abundant and densely interspersed throughout the divergent region. The majority of tandem repeats are arranged in clusters composed of predominately short A_nT_n repeats.

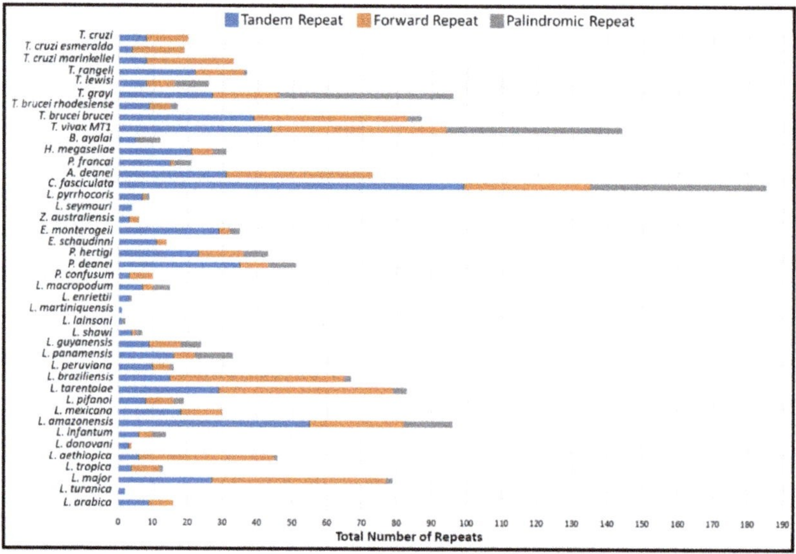

Figure 4. Analysis of total repeated sequences in the maxicircle divergent region of various trypanosomatid species. Totals of three repeat types; (Blue) Number of tandem repeats; (Orange) Number of forward repeats; (Grey) Number of palindromic repeats.

3.3. Phylogenetic Analyses

To assess how FFP-based phylogenies compares to the traditionally aligned-phylogenies, the phylogenetic relationships showing the genetic distance between members of the Trypanosomatidae based on the MSA- and AF-approach is shown in Figure 5A,B. The sequences of 46 trypanosomatid species were used to determine the optimal word length or k-mer for subsequent alignment-free analyses. From the FFPvprof and FFPreprof analyses, the average lower world length (k-mer) limit of the 45 species was 14 ($k = 14$) and the average upper world length limit was 30 ($k = 30$) leading to the selection of k-mer lengths of 22 (Supplementary Materials S5).

In the inferred phylogenies, all *Leishmania* and *Trypanosoma* species formed two separate, strongly supported monophyletic clades. The FFP analysis showed the same subgrouping of the Trypanosomatidae, with all six subfamilies and eleven genera corresponding to the PhyML-generated phylogenetic tree. The subfamily Leishmaniinae unites the monoxenous protozoans of insects (genera *Zelonia*, *Leptomonas* and *Crithidia*) and dixenous protozoans of insects and vertebrates (genera *Leishmania*, *Endotrypanum* and *Porcisia*). The monoxenous subfamilies Phytomonadinae, Strigononadinae and Blechomonadinae include parasites of the genera *Herpetomonas*, *Angomonas* and *Blechomonas* respectively. In agreement with a recent study [11], despite *T. rangeli* and *T. lewisi* belonging to the same subgenus *Herpetosoma*, *T. rangeli* is more closely related to *T. cruzi*, clustering with the Schizotrypanum clade of trypanosomes.

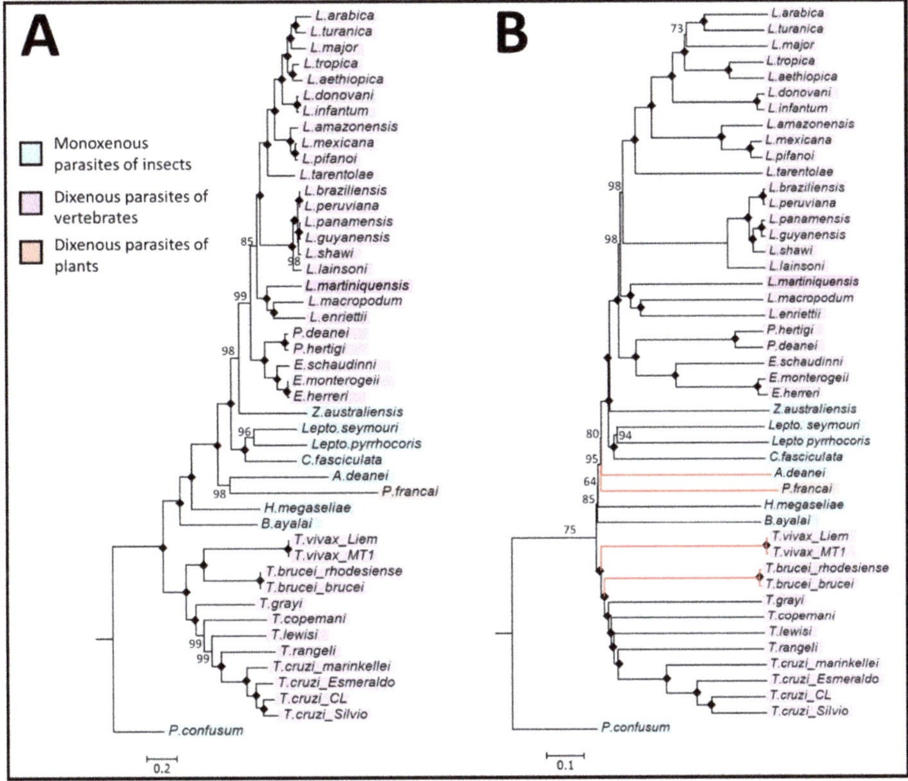

Figure 5. Inferred evolutionary relationships between species of the trypanosomatid family using aligned and alignment-free analysis of the maxicircle coding region. (**A**) PhyML-derived phylogeny showing the relationships between members of the trypanosomatid family using a multiple sequence alignment analysis. The maximum likelihood optimality criterion was used for the phylogenetic inference of the dataset with 1000 bootstrap samples. (**B**) Featue frequency profiles (FPP)-estimated phylogeny between species of the trypanosomatid family using an alignment-free analysis of the maxicircle coding region. Branches in red represent species whose clustering differs from those in (A). In both (A) and (B) a black diamond highlights a node that obtained a bootstrap value of 100% confidence and the scale bars depict the number of nucleotide substitutions per position.

The discrepancies observed were the branching of the Salivarian trypanosomes, where *T. brucei brucei* and *T. brucei rhodesiense* formed a clade basal to the *T. vivax* clade and between the monoxenous *A. deanei and P. françai*. The distance matrix generated from the alignment-free analysis is listed in Supplementary Materials S6.

3.4. Divergence Time Estimates of the Trypanosomatidae

The node depicting the separation of *L. macropodum* and *L. enriettii* was chosen as a calibration marker. The divergence date was established at an average of 40 MYA, which is the time period when Australia and South America became separated, representing a minimum date of divergence [46,47]. From this calibration point, a common ancestor to the Trypanosomatinae subfamily and specifically the *T. cruzi* clade (*T. cruzi* and *T. rangeli*) was predicted to have appeared 150 MYA and 84 MYA respectively (Figure 6). The dixenous members of the Leishmaniinae was predicted to have first appeared 95 MYA, arising from a monoxenous ancestor.

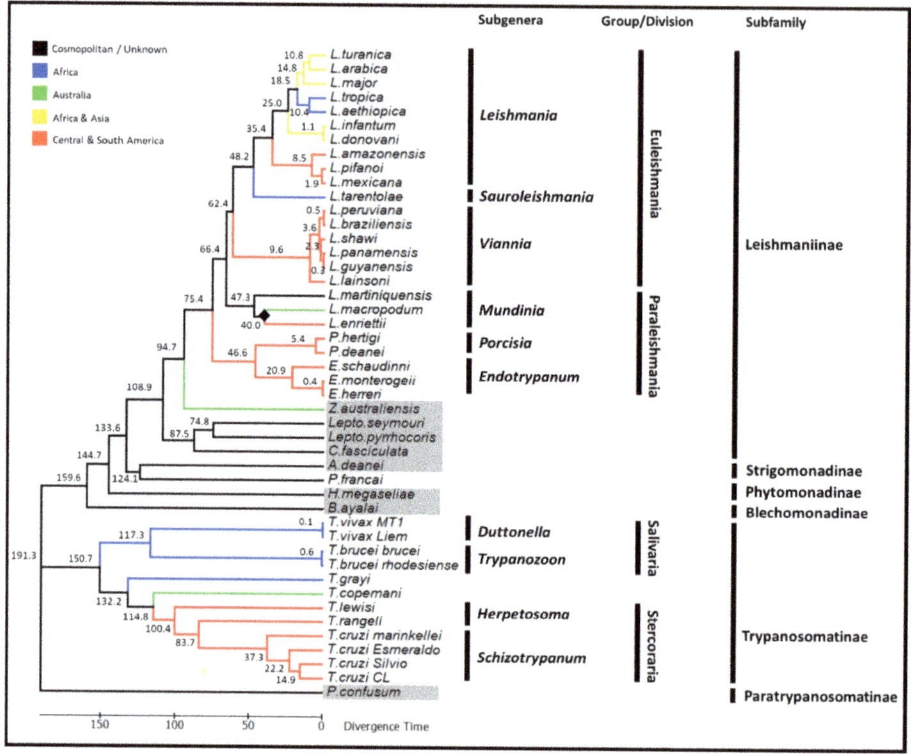

Figure 6. Phylogenetic time tree demonstrating the complex evolutionary relationships of the trypanosomatid family using the coding region of the maxicircle kDNA. The maximum likelihood of this tree was inferred using the GTR model. The time tree was calculated using a single calibration date, illustrated by a solid black diamond (the separation of *Leishmania macropodum* and *Leishmania enriettii* approximately 40 million years ago). Predicted divergence times of the various species are displayed on the nodes of the tree. Monoxenous trypanosomatid species are highlighted in grey.

4. Discussion

In this study we applied the alignment-free analysis of frequency feature profiles (FFP) for the phylogenomic analysis of the trypanosomatid family based on the entire maxicircle coding region genome. For closely related organisms i.e., within a single genus, alignment-based methods are preferred because in the most part, their genetic similarities allow a relatively straight-forward alignment of high-quality [33]. However, in the context of looking at an entire family of organisms (such as the Trypanosomatidae), the various members exhibit greater variation than species within a single genus. Knowledge of these evolutionary processes demonstrates that there is going to be genetic discrepancies (i.e., the loss of maxicircle genes in *Phytomonas* spp.) which is going to negatively impact on the quality of any sequence alignment. Due to these less conserved sequences and scale of the dataset (greater than 15 kb across 46 species), the alignment-free method is preferred for the analysis of distant relatives, allowing the analysis of organisms which exhibit greater variation in the number and order of genetic elements.

In accordance with published standards in the assessment of robustness (i.e., bootstrapping), the percentile method validates the accuracy of a node, with a confidence value of >60% in support of the observed clade [48]. The MSA and AF trees showed extremely similar and robust topologies (Figure 5), demonstrating the suitability of the alignment-free FFP method for phylogenetic analyses of the Trypanosomatidae. All major clades (*Leishmania, Endotrypanum, Porcisia, Zelonia, Leptomonas, Crithidia, Herpetomonas, Phytomonas, Angomonas, Blechomonas, Trypanosoma* and *Paratrypanosoma*, cluster with their respective subgenera within the trypanosomatid family with high bootstrap confidence. Analysing a large dataset from an entire family of parasites, a greater divergence is expected than a typical phylogenetic analysis of a few, closely related genera, which can impact on the accuracy of the sequence alignment. In addition, the overall running times of alignment-free phylogenetic inference were shorter (run time less than one day compared to >3 days for alignment and PhyML analysis), making it an attractive approach for large datasets >15 kb. However, important issues need to be taken into consideration for AF-approaches, namely that optimal k-mer lengths must be determined to establish accurate phylogenomic relationships between related organisms. Nevertheless, having the advantage of not requiring pre-alignment of a large collection of maxicircle genome sequences, FFP analysis allowed the rapid, accurate and robust phylogenetic analysis of a large group of diverse species within the Trypanosomatidae.

Kinetoplastid parasites of the trypanosomatid family possess a unique, single, densely-packed periflagellar network of DNA circles known as the kinetoplast [12,49]. In our analysis, both paired- end short reads and a hybrid protocol using short- and long-read data resulted in the high- quality assembly of the maxicircle genome (Table 1). The maxicircle sequences generated from SGS and hybrid assemblies resulted in 100% identity between the coding region and >99% identity between the divergent region (due to the difference in sequence length of the divergent region between Illumina and hybrid assemblies) of each respective species. The hybrid protocol offered an alternative approach, capable of overcoming the difficulties present in assembling complex genomic regions from short sequencing reads. Our hybrid assemblies resulted in the resolution of the divergent region (DR), generating highly contiguous, accurate assemblies of the DR, even when these regions contain a large number of near-identical repeats.

Analysis of the repetitive sequences demonstrated that the various trypanosomatid species have a variable number of repeat arrays throughout their divergent region (Figure 4 and Supplementary Materials S4). The non-coding divergent region of the maxicircle remains the most poorly studied region of kinetoplast DNA [16,50,51]. From our analysis, we can see that the overall DR structure is composed almost entirely of various repeat arrays, with a clear distinction between the coding-region of the maxicircle kDNA (Figures 3 and 4). The high repeat content of the divergent region posed a substantial obstacle to the assembly and output length of the maxicircle kDNA with Illumina short- reads alone. Compared to that of the hybrid approach, SGS reads of *L. aethiopica, L. amazonensis, L. braziliensis, L. guyanensis, L. infantum, L. mexicana, L. tropica* and *T. brucei rhodesiense* assembled only 20%, 11%, 20%, 43%, 49%, 46%, 96% and 88% of the divergent region produced from TGS respectively.

Protozoan mitochondrial genomes often display higher AT% than mitochondrial genomes from metazoans [12,52]. Thus, it is expected that in the mitochondrial homologue in trypanosomatids; the kinetoplast will have a low GC%. The extreme overabundance of AT% in the maxicircle genome of trypanosomatids is primarily due to the repetitive portion of the non-coding divergent region (Figure 2). The AT skew demonstrates the bias towards an AT-rich maxicircle genome of trypanosomatid species.

The vast majority of maxicircles in the trypanosomatid family contain the same set of genes and genomic organisation pattern. The notable exception to this is the plant parasite *Phytomonas françai*, which lacks the respiratory chain complexes III and IV, including the three subunits of the cytochrome *c* oxidase (COI, COII and COIII) and single cytochrome *b* (CYb), respectively (Figure 1). The respiratory chain complexes III and IV are required to maintain a complete electron transport chain in trypanosomatid parasites, playing a critical role in the biochemical production and synthesis of adenosine triphosphate (ATP) [53,54]. It is speculated that the absence of these subunits is related to the adaptation of *Phytomonas* spp. to the carbohydrate-rich medium of the host plant environment [55]. In the presence of this carbohydrate-rich medium, the cytochrome-mediated respiration complexes of *Phytomonas* were lost after their function became nonessential and the presence of the glycolysis metabolic pathway in glycosomes alone was sufficient for ATP production [55,56]. The feeding behaviour of the insect host is also speculated to have played a role in the loss of cytochromes in *Phytomonas* [57]. Phytophagous hemipteran insects feed exclusively on the sap and juices of plants that are rich in carbohydrates. This suggests a passive, mechanical transmission from plant to insect host that does not require a metabolic shift from a carbohydrate to an amino acid metabolism [58]. Previous studies have also demonstrated the absence of these respective subunits in *Phytomonas serpens* and *Phytomonas* sp. HART1, demonstrating this respiratory deficiency is characteristic of these dixenous plant trypanosomatids [55,57,58].

A second notable exception is the mammalian infecting *Trypanosoma brucei evansi* and *Trypanosoma brucei equiperdum* [59]. Originally considered separate species due to their differences in host-range, transmission and pathogenicity, recent data demonstrates that *T. b. evansi* and *T. b. equiperdum* represent dyskinetoplastic subspecies of *T. brucei*, characterized by the complete or partial loss of their maxicircle kDNA, although their taxonomic status is often debated [60–62]. Recent studies have shown that mutations located in the nuclear-bound ATPase subunit γ of some *T. b. evansi* and *T. b. equiperdum* were found to compensate for the total or partial absence of maxicircle kDNA, demonstrating an important nuclear/kinetoplast interaction for the viability of these species [61].

The relationships between the monoxenous trypanosomatids and their dixenous relatives have been extensively debated over the last few decades [63,64]. It was speculated that dixenous parasitism of the Trypanosomatidae has independently evolved several times over the course of history from the monoxenous ancestors, giving rise to the *Trypanosoma*, *Leishmania*, *Endotrypanum*, *Porcisia* and *Phytomonas* lineages [63]. As such, it is impossible to answer questions surrounding the origins of the trypanosomatid family without studying the non-pathogenic, monoxenous relatives, which is reflected by the rising number of papers published in this field [46,65–69]. From the molecular data, several major clades can be identified including; *Leishmania*, *Endotrypanum*, *Porcisia*, *Zelonia*, *Leptomonas*, *Crithidia*, *Herpetomonas*, *Angomonas*, *Blechomonas*, *Trypanosoma* and *Paratrypanosoma*. All clades generated support a recent proposal on the classification of the trypanosomatids, specifically the taxonomic validity of the *Endotrypanum/Porcisia* genera and the establishment of the subfamily Trypanosomatinae to encompass species of the trypanosomes [5,70]. Molecular clock analyses suggest that at least three lineages independently acquired the ability to infect two hosts (i.e., dixenous parasitism) including plants (*Phytomonas* spp.) and vertebrates (*Trypanosoma* and *Leishmania/Porcisia/Endotrypanum*) approximately 133, 190 and 95 MYA respectively. Thus, our analyses provide strong support for the multiple and independent origins of the dixenous life-style of the trypanosomatids.

The supercontinents origin of the dixenous Leishmaniinae, where the dixenous members first emerged from monoxenous ancestors during the continental separation of Gondwana is now widely accepted [46,47,63]. As suggested by the supercontinents hypothesis of dixenous parasitism, the earliest dixenous members of the Leishmaniinae first emerged in the late Cretaceous period between 77–140 MYA, during the predicted breakup of Gondwana [19]. Based on our phylogenetic analysis, the dixenous genera *Leishmania*, *Endotrypanum* and *Porcisia* emerged as distinct monophyletic lineages from a common monoxenous ancestor approximately 95 million years ago (Figure 6). Detailed extensively in our previous work [19], in summary the first emergence of dixenous parasitism within the Leishmaniinae subfamily coincides with when the radiation of mammals first began during the Cretaceous period [71].

The phylogenetic analysis of the trypanosomatid family supported the monophyletic lineage of the trypanosomes, having evolved from a monoxenous ancestor with high statistical support (Figure 5). Trypanosomes found in mammals (and humans) are grouped into two sections: Stercoraria (including members of the subgenera *Herpetosoma* and *Schizotrypanum*), which develops in the posterior portion of the insect host digestive tract and Salivaria (including members of the subgenera *Duttonella* and *Trypanozoon*), which develops in the anterior region of the insect digestive tract [58]. Previous analyses proposed an early divergence of salivarian trypanosomes, which involved the ancient split of the trypanosomes into one clade containing all salivarian species and the other branch containing all non-salivarian lineages [72–74]. Based on the SSU rRNA, it was proposed that the emergence of the salivarian lineage first appeared approximately 300 MYA, although the authors were cautious about these date estimates due to the use of the SSU rRNA as the taxonomic marker [72]. Consistent with an ancient salivarian divergence, molecular clock analyses based on the maxicircle phylogenies suggest that the salivarian trypanosomes separated from other trypanosomes approximately 150 MYA (Figure 6). It is speculated that early African trypanosomes were most likely gut parasites or commensals of early insects, however the appearance of tsetse flies approximately 35 MYA facilitated the transmission to mammals by these blood-feeding insects [75]. In addition, the use of the maxicircle coding region as a taxonomic marker also provided strong support for the placement of *Trypanosoma vivax* with other African salivarian trypanosomes including *Trypanosoma brucei* [64].

The analysis of the coding region of the maxicircle also sheds light on the evolution of the *T. cruzi* clade, which encompasses most African, American and European trypanosomes from bats and terrestrial mammals [74]. In recent years, the southern super-continent trypanosome hypothesis has come under scrutiny, with some suggesting it can no longer be considered correct that *T. cruzi* first emerged in the New World and *T. brucei* in the Old World following the continental split of Africa and South America 100 MYA [47,76–78]. The new theory gaining the greatest support proposes that the *T. cruzi* clade (*T. cruzi* and *T. rangeli*) evolved more recently than originally thought within a broader clade of bat trypanosomes [73]. Our analysis suggests that the common ancestor of the *T. cruzi* clade first appeared approximately 84 million years ago. This timeframe provides strong additional support for the 'bat-seeding' hypothesis, suggesting that the common ancestor of the *T. cruzi* clade is likely to have evolved following the diversification of bats approximately 70–58 million years ago [79,80]. Using a geological time point i.e., the divergence of *L. enriettii* and *L. macropodum* as the calibration date facilitates suggestions on the emergence of the *T. cruzi* clade independently of the diversification of bats. This is an important advancement for the 'bat-seeding' hypothesis, previously not feasible in the absence of WGS data for *L. macropodum*, as it eliminates any bias or skew that may be introduced through using the estimated time of divergence of the host species (i.e., a secondary calibration) [80].

5. Conclusions

In conclusion, the use of the entire coding region of the maxicircle kinetoplast DNA in alignment-free analyses provided an exceptionally robust evolutionary insight into the relationships and patterns within the trypanosomatid family. We believe the present study demonstrates the applicability of AF-based approaches such as FFP to produce reliable and robust phylogenomic analyses to establish and summarise the current evolutionary relationships of the Trypanosomatidae. We suggest that future researchers aiming to analyse this diverse and widespread relationship should consider using this described approach to deal with the ever-increasing amount of sequence data that are becoming available. Ultimately, knowledge of these deep-rooted lineages is exceptionally useful for future analysis involving reconstruction of any evolutionary scenario involving these flagellated protozoans.

Supplementary Materials: The following are available online at http://www.mdpi.com/2076-0817/8/3/157/s1, Table S1: List of trypanosomatid species used in this study. Table S2: The overall nucleotide frequency and skew throughout the maxicircle genome of various trypanosomatid species. Figure S3: Self Dottup plot comparative analysis of the entire maxicircle genome (right panel) and divergent region (left panel) of various trypanosomatid species. Figure S4: Analysis of repeated sequences in the maxicircle divergent region. Figure S5: Determination of optimal feature/k-mer length (k) for 46 trypanosomatid species. Table S6: Comparison of distance matrices showing genetic differences between maxicircle kDNA of various trypanosomatid species. Below diagonal: genetic distance calculated from multiple sequence-alignment method, above diagonal: genetic distance calculated from alignment-free FFP method.

Author Contributions: A.K. conceived, wrote and edited all drafts. J.E. provided major suggestions on content and edits of the manuscript. D.S. provided additional feedback on edits and further suggestions on contents. All authors read and approved the final version of the manuscript.

Funding: This research received no external funding.

Conflicts of Interest: The authors declare no conflict of interest.

References

1. Fraga, J.; Fernández-Calienes, A.; Montalvo, A.M.; Maes, I.; Deborggraeve, S.; Büscher, P.; Dujardin, J.C.; Van der Auwera, G. Phylogenetic analysis of the *Trypanosoma* genus based on the heat-shock protein 70 gene. *Infect. Genet. Evol.* **2016**, *43*, 165–172. [CrossRef] [PubMed]
2. Kostygov, A.Y.; Yurchenko, V. Revised classification of the subfamily Leishmaniinae (Trypanosomatidae). *Folia Parasitol.* **2017**, *64*. [CrossRef] [PubMed]
3. Jackson, A.P. Genome evolution in trypanosomatid parasites. *Parasitology* **2015**, *142*, S40–S56. [CrossRef] [PubMed]
4. Kaufer, A.; Ellis, J.; Stark, D.; Barratt, J. The evolution of trypanosomatid taxonomy. *Parasites Vectors* **2017**, *10*, 287. [CrossRef] [PubMed]
5. Maslov, D.A.; Opperdoes, F.R.; Kostygov, A.Y.; Hashimi, H.; Lukes, J.; Yurchenko, V. Recent advances in trypanosomatid research: Genome organization, expression, metabolism, taxonomy and evolution. *Parasitology* **2018**, *146*, 1–27. [CrossRef] [PubMed]
6. Ghedin, E.; Bringaud, F.; Peterson, J.; Myler, P.; Berriman, M.; Ivens, A.; Andersson, B.; Bontempi, E.; Eisen, J.; Angiuoli, S.; et al. Gene synteny and evolution of genome architecture in trypanosomatids. *Mol. Biochem. Parasitol.* **2004**, *134*, 183–191. [CrossRef] [PubMed]
7. Hamilton, P.B.; Stevens, J.R.; Gaunt, M.W.; Gidley, J.; Gibson, W.C. Trypanosomes are monophyletic: Evidence from genes for glyceraldehyde phosphate dehydrogenase and small subunit ribosomal RNA. *Int. J. Parasitol.* **2004**, *34*, 1393–1404. [CrossRef]
8. Lukeš, J.; Skalický, T.; Týč, J.; Votýpka, J.; Yurchenko, V. Evolution of parasitism in kinetoplastid flagellates. *Mol. Biochem. Parasitol.* **2014**, *195*, 115–122. [CrossRef]
9. Deschamps, P.; Lara, E.; Marande, W.; López-García, P.; Ekelund, F.; Moreira, D. Phylogenomic Analysis of Kinetoplastids Supports That Trypanosomatids Arose from within Bodonids. *Mol. Biol. Evol.* **2011**, *28*, 53–58. [CrossRef]
10. Hilu, K.W.; Black, C.M.; Oza, D. Impact of Gene Molecular Evolution on Phylogenetic Reconstruction: A Case Study in the Rosids (Superorder Rosanae, Angiosperms). *PLoS ONE* **2014**, *9*, e99725. [CrossRef]
11. Lin, R.H.; Lai, D.H.; Zheng, L.L.; Wu, J.; Lukes, J.; Hide, G.; Lun, Z.R. Analysis of the mitochondrial maxicircle of *Trypanosoma lewisi*, a neglected human pathogen. *Parasites Vectors* **2015**, *8*, 665. [CrossRef] [PubMed]
12. Lukeš, J.; Wheeler, R.; Jirsova, D.; David, V.; Archibald, J.M. Massive mitochondrial DNA content in diplonemid and kinetoplastid protists. *IUBMB Life* **2018**, *70*, 1267–1274. [CrossRef] [PubMed]
13. Gerasimov, E.S.; Gasparyan, A.A.; Litus, I.A.; Logacheva, M.D.; Kolesnikov, A.A. Minicircle Kinetoplast Genome of Insect Trypanosomatid *Leptomonas pyrrhocoris*. *Biochemistry* **2017**, *82*, 572–578. [CrossRef] [PubMed]
14. Telleria, J.; Lafay, B.; Virreira, M.; Barnabe, C.; Tibayrenc, M.; Svoboda, M. Trypanosoma cruzi: Sequence analysis of the variable region of kinetoplast minicircles. *Exp. Parasitol.* **2006**, *114*, 279–288. [CrossRef] [PubMed]
15. De Vries, B.F.; Mulder, E.; Brakenhoff, J.P.; Sloof, P.; Benne, R. The variable region of the *Trypanosoma brucei* kinetoplast maxicircle: Sequence and transcript analysis of a repetitive and a non-repetitive fragment. *Mol. Biochem. Parasitol.* **1988**, *27*, 71–82. [CrossRef]
16. Flegontov, P.N.; Strelkova, M.V.; Kolesnikov, A.A. The *Leishmania major* maxicircle divergent region is variable in different isolates and cell types. *Mol. Biochem. Parasitol.* **2006**, *146*, 173–179. [CrossRef] [PubMed]
17. Muhich, M.L.; Simpson, L.; Simpson, A.M. Comparison of maxicircle DNAs of *Leishmania tarentolae* and *Trypanosoma brucei*. *Proc. Natl. Acad. Sci. USA* **1983**, *80*, 4060–4064. [CrossRef]
18. Botero, A.; Kapeller, I.; Cooper, C.; Clode, P.L.; Shlomai, J.; Thompson, R.C.A. The kinetoplast DNA of the Australian trypanosome, *Trypanosoma copemani*, shares features with *Trypanosoma cruzi* and *Trypanosoma lewisi*. *Int. J. Parasitol.* **2018**, *48*, 691–700. [CrossRef]
19. Kaufer, A.; Barratt, J.; Stark, D.; Ellis, J. The complete coding region of the maxicircle as a superior phylogenetic marker for exploring evolutionary relationships between members of the *Leishmaniinae*. *Infect. Genet. Evol.* **2019**, *70*, 90–100. [CrossRef]
20. Van Vliet, A.H.M.; Kusters, J.G. Use of Alignment-Free Phylogenetics for Rapid Genome Sequence-Based Typing of Helicobacter pylori Virulence Markers and Antibiotic Susceptibility. *J. Clin. Microbiol.* **2015**, *53*, 2877–2888. [CrossRef]

21. Fu, S.; Wang, A.; Au, K.F. A comparative evaluation of hybrid error correction methods for error-prone long reads. *Genome Biol.* **2019**, *20*, 26. [CrossRef] [PubMed]

22. Rhoads, A.; Au, K.F. PacBio Sequencing and Its Applications. *Genom. Proteom. Bioinform.* **2015**, *13*, 278–289. [CrossRef] [PubMed]

23. Antipov, D.; Korobeynikov, A.; McLean, J.S.; Pevzner, P.A. HybridSPAdes: An algorithm for hybrid assembly of short and long reads. *Bioinformatics* **2016**, *32*, 1009–1015. [CrossRef] [PubMed]

24. Koren, S.; Schatz, M.C.; Walenz, B.P.; Martin, J.; Howard, J.T.; Ganapathy, G.; Wang, Z.; Rasko, D.A.; McCombie, W.R.; Jarvis, E.D.; et al. Hybrid error correction and de novo assembly of single-molecule sequencing reads. *Nat. Biotechnol.* **2012**, *30*, 693–700. [CrossRef] [PubMed]

25. Patane, J.S.L.; Martins, J., Jr.; Setubal, J.C. Phylogenomics. *Methods Mol. Biol.* **2018**, *1704*, 103–187. [PubMed]

26. Amado Cattaneo, R.M.; Diambra, L.; McCarthy, A.N. Phylogenomics of tomato chloroplasts using assembly and alignment-free method. *Mitochondrial DNA Part A* **2018**, *29*, 1128–1138. [CrossRef] [PubMed]

27. Zhang, Q.; Jun, S.R.; Leuze, M.; Ussery, D.; Nookaew, I. Viral Phylogenomics Using an Alignment-Free Method: A Three-Step Approach to Determine Optimal Length of k-mer. *Sci. Rep.* **2017**, *7*, 40712. [CrossRef] [PubMed]

28. Bernard, G.; Chan, C.X.; Chan, Y.B.; Chua, X.Y.; Cong, Y.; Hogan, J.M.; Maetschke, S.R.; Ragan, M.A. Alignment-free inference of hierarchical and reticulate phylogenomic relationships. *Brief. Bioinform.* **2019**, *20*, 426–435. [CrossRef]

29. Bernard, G.; Chan, C.X.; Ragan, M.A. Alignment-free microbial phylogenomics under scenarios of sequence divergence, genome rearrangement and lateral genetic transfer. *Sci. Rep.* **2016**, *6*, 28970. [CrossRef]

30. Murray, K.D.; Webers, C.; Ong, C.S.; Borevitz, J.; Warthmann, N. kWIP: The k-mer weighted inner product, a de novo estimator of genetic similarity. *PLoS Comput. Biol.* **2017**, *13*, e1005727. [CrossRef]

31. Fan, H.; Ives, A.R.; Surget-Groba, Y. Reconstructing phylogeny from reduced-representation genome sequencing data without assembly or alignment. *Mol. Ecol. Resour.* **2018**, *18*, 1482–1491. [CrossRef] [PubMed]

32. Leimeister, C.A.; Schellhorn, J.; Dorrer, S.; Gerth, M.; Bleidorn, C.; Morgenstern, B. Prot-SpaM: Fast alignment-free phylogeny reconstruction based on whole-proteome sequences. *GigaScience* **2019**, *8*, giy148. [CrossRef] [PubMed]

33. Zielezinski, A.; Vinga, S.; Almeida, J.; Karlowski, W.M. Alignment-free sequence comparison: Benefits, applications, and tools. *Genome Biol.* **2017**, *18*, 186. [CrossRef] [PubMed]

34. Song, K.; Ren, J.; Reinert, G.; Deng, M.; Waterman, M.S.; Sun, F. New developments of alignment-free sequence comparison: Measures, statistics and next-generation sequencing. *Brief. Bioinform.* **2014**, *15*, 343–353. [CrossRef] [PubMed]

35. Sims, G.E.; Jun, S.R.; Wu, G.A.; Kim, S.H. Alignment-free genome comparison with feature frequency profiles (FFP) and optimal resolutions. *Proc. Natl. Acad. Sci. USA* **2009**, *106*, 2677–2682. [CrossRef] [PubMed]

36. Bankevich, A.; Nurk, S.; Antipov, D.; Gurevich, A.A.; Dvorkin, M.; Kulikov, A.S.; Lesin, V.M.; Nikolenko, S.I.; Pham, S.; Prjibelski, A.D.; et al. SPAdes: A New Genome Assembly Algorithm and Its Applications to Single-Cell Sequencing. *J. Comput. Biol.* **2012**, *19*, 455–477. [CrossRef] [PubMed]

37. NCBI. *BLAST® Command Line Applications User Manual US: Bethesda (MD)*; National Center for Biotechnology Information: Bethesda, MD, USA, 2008.

38. Kearse, M.; Moir, R.; Wilson, A.; Stones-Havas, S.; Cheung, M.; Sturrock, S.; Buxton, S.; Cooper, A.; Markowitz, S.; Duran, C.; et al. Geneious Basic: An integrated and extendable desktop software platform for the organization and analysis of sequence data. *Bioinformatics* **2012**, *28*, 1647–1649. [CrossRef]

39. Carver, T.; Thomson, N.; Bleasby, A.; Berriman, M.; Parkhill, J. DNAPlotter: Circular and linear interactive genome visualization. *Bioinformatics* **2009**, *25*, 119–120. [CrossRef]

40. Shioiri, C.; Takahata, N. Skew of mononucleotide frequencies, relative abundance of dinucleotides, and DNA strand asymmetry. *J. Mol. Evol.* **2001**, *53*, 364–376. [CrossRef]

41. Kurtz, S.; Choudhuri, J.V.; Ohlebusch, E.; Schleiermacher, C.; Stoye, J.; Giegerich, R. REPuter: The manifold applications of repeat analysis on a genomic scale. *Nucleic Acids Res.* **2001**, *29*, 4633–4642. [CrossRef]

42. Benson, G. Tandem repeats finder: A program to analyze DNA sequences. *Nucleic Acids Res.* **1999**, *27*, 573–580. [CrossRef] [PubMed]

43. Gouy, M.; Guindon, S.; Gascuel, O. SeaView version 4: A multiplatform graphical user interface for sequence alignment and phylogenetic tree building. *Mol. Biol. Evol.* **2010**, *27*, 221–224. [CrossRef] [PubMed]

44. Posada, D. jModelTest: Phylogenetic model averaging. *Mol. Biol. Evol.* **2008**, *25*, 1253–1256. [CrossRef] [PubMed]
45. PHYLIP (Phylogeny Inference Package) Version 3.6. Available online: http://evolution.genetics.washington.edu/phylip.html (accessed on 10 September 2019).
46. Barratt, J.; Kaufer, A.; Peters, B.; Craig, D.; Lawrence, A.; Roberts, T.; Lee, R.; McAuliffe, G.; Stark, D.; Ellis, J. Isolation of Novel Trypanosomatid, *Zelonia australiensis* sp. nov. (Kinetoplastida: Trypanosomatidae) Provides Support for a Gondwanan Origin of Dixenous Parasitism in the Leishmaniinae. *PLoS Negl. Trop. Dis.* **2017**, *11*, e0005215. [CrossRef] [PubMed]
47. Harkins, K.M.; Schwartz, R.S.; Cartwright, R.A.; Stone, A.C. Phylogenomic reconstruction supports supercontinent origins for *Leishmania*. *Infect. Genet. Evol.* **2016**, *38*, 101–109. [CrossRef] [PubMed]
48. Felsenstein, J. Confidence-limits on phylogenies—An approach using the bootstrap. *Evolution* **1985**, *39*, 783–791. [CrossRef] [PubMed]
49. Cavalcanti, D.P.; de Souza, W. The Kinetoplast of Trypanosomatids: From Early Studies of Electron Microscopy to Recent Advances in Atomic Force Microscopy. *Scanning* **2018**, *2018*, 9603051. [CrossRef]
50. Flegontov, P.N.; Guo, Q.; Ren, L.; Strelkova, M.V.; Kolesnikov, A.A. Conserved repeats in the kinetoplast maxicircle divergent region of *Leishmania* sp. and *Leptomonas seymouri*. *Mol. Genet. Genom.* **2006**, *276*, 322–333. [CrossRef]
51. Westenberger, S.J.; Cerqueira, G.C.; El-Sayed, N.M.; Zingales, B.; Campbell, D.A.; Sturm, N.R. *Trypanosoma cruzi* mitochondrial maxicircles display species- and strain-specific variation and a conserved element in the non-coding region. *BMC Genom.* **2006**, *7*, 60. [CrossRef]
52. Millan, C.R.; Acosta-Reyes, F.J.; Lagartera, L.; Ebiloma, G.U.; Lemgruber, L.; Nué Martínez, J.J.; Saperas, N.; Dardonville, C.; De Koning, H.P.; Campos, J.L. Functional and structural analysis of AT-specific minor groove binders that disrupt DNA-protein interactions and cause disintegration of the *Trypanosoma brucei* kinetoplast. *Nucleic Acids Res.* **2017**, *45*, 8378–8391. [CrossRef]
53. Asato, Y.; Oshiro, M.; Myint, C.K.; Yamamoto, Y.I.; Kato, H.; Marco, J.D.; Mimori, T.; Gomez, E.A.; Hashiguchi, Y.; Uezato, H. Phylogenic analysis of the genus *Leishmania* by cytochrome b gene sequencing. *Exp. Parasitol.* **2009**, *121*, 352–361. [CrossRef]
54. Horvath, A.; Kingan, T.G.; Maslov, D.A. Detection of the mitochondrially encoded cytochrome c oxidase subunit I in the trypanosomatid protozoan *Leishmania tarentolae*. Evidence for translation of unedited mRNA in the kinetoplast. *J. Biol. Chem.* **2000**, *275*, 17160–17165. [CrossRef] [PubMed]
55. Nawathean, P.; Maslov, D.A. The absence of genes for cytochrome c oxidase and reductase subunits in maxicircle kinetoplast DNA of the respiration-deficient plant trypanosomatid *Phytomonas serpens*. *Curr. Genet.* **2000**, *38*, 95–103. [CrossRef] [PubMed]
56. Maslov, D.A.; Nawathean, P.; Scheel, J. Partial kinetoplast-mitochondrial gene organization and expression in the respiratory deficient plant trypanosomatid *Phytomonas serpens*. *Mol. Biochem. Parasitol.* **1999**, *99*, 207–221. [CrossRef]
57. Jaskowska, E.; Butler, C.; Preston, G.; Kelly, S. *Phytomonas*: Trypanosomatids Adapted to Plant Environments. *PLoS Pathog.* **2015**, *11*, 17. [CrossRef] [PubMed]
58. Porcel, B.M.; Denoeud, F.; Opperdoes, F.; Noel, B.; Madoui, M.A.; Hammarton, T.C.; Field, M.C.; Da Silva, C.; Couloux, A.; Poulain, J.; et al. The streamlined genome of *Phytomonas* spp. relative to human pathogenic kinetoplastids reveals a parasite tailored for plants. *PLoS Genet.* **2014**, *10*, e1004007. [CrossRef] [PubMed]
59. Desquesnes, M.; Holzmuller, P.; Lai, D.H.; Dargantes, A.; Lun, Z.R.; Jittaplapong, S. Trypanosoma evansi and surra: A review and perspectives on origin, history, distribution, taxonomy, morphology, hosts, and pathogenic effects. *BioMed Res. Int.* **2013**, *2013*, 194176. [CrossRef] [PubMed]
60. Lai, D.H.; Hashimi, H.; Lun, Z.R.; Ayala, F.J.; Lukeš, J. Adaptations of *Trypanosoma brucei* to gradual loss of kinetoplast DNA: *Trypanosoma equiperdum* and *Trypanosoma evansi* are petite mutants of *T-brucei*. *Proc. Natl. Acad. Sci. USA* **2008**, *105*, 1999–2004. [CrossRef]
61. Carnes, J.; Anupama, A.; Balmer, O.; Jackson, A.; Lewis, M.; Brown, R.; Cestari, I.; Desquesnes, M.; Gendrin, C.; Hertz-Fowler, C.; et al. Genome and Phylogenetic Analyses of *Trypanosoma evansi* Reveal Extensive Similarity to *T. brucei* and Multiple Independent Origins for Dyskinetoplasty. *PLoS Negl. Trop. Dis.* **2015**, *9*, e3404. [CrossRef]

62. Dewar, C.E.; MacGregor, P.; Cooper, S.; Gould, M.K.; Matthews, K.R.; Savill, N.J.; Schnaufer, A. Mitochondrial DNA is critical for longevity and metabolism of transmission stage *Trypanosoma brucei*. *PLoS Pathog.* **2018**, *14*, e1007195. [CrossRef]

63. Lukes, J.; Butenko, A.; Hashimi, H.; Maslov, D.A.; Votypka, J.; Yurchenko, V. Trypanosomatids Are Much More than Just Trypanosomes: Clues from the Expanded Family Tree. *Trends Parasitol.* **2018**, *34*, 466–480. [CrossRef] [PubMed]

64. Simpson, A.G.B.; Stevens, J.R.; Lukes, J. The evolution and diversity of kinetoplastid flagellates. *Trends Parasitol.* **2006**, *22*, 168–174. [CrossRef] [PubMed]

65. Maslov, D.A.; Votýpka, J.; Yurchenko, V.; Lukeš, J. Diversity and phylogeny of insect trypanosomatids: All that is hidden shall be revealed. *Trends Parasitol.* **2013**, *29*, 43–52. [CrossRef] [PubMed]

66. Podlipaev, S. The more insect trypanosomatids under study-the more diverse Trypanosomatidae appears. *Int. J. Parasitol.* **2001**, *31*, 648–652. [CrossRef]

67. Kostygov, A.Y.; Dobáková, E.; Grybchuk-Ieremenko, A.; Váhala, D.; Maslov, D.A.; Votýpka, J.; Lukeš, J.; Yurchenko, V. Trypanosomatid-Bacterium Association: Evolution of Endosymbiosis in Action. *MBio* **2016**, *7*, e01985-15. [CrossRef] [PubMed]

68. Kraeva, N.; Butenko, A.; Hlaváčová, J.; Kostygov, A.; Myškova, J.; Grybchuk, D.; Leštinová, T.; Votýpka, J.; Volf, P.; Opperdoes, F.; et al. *Leptomonas seymouri*: Adaptations to the Dixenous Life Cycle Analyzed by Genome Sequencing, Transcriptome Profiling and Co-infection with *Leishmania donovani*. *PLoS Pathog.* **2015**, *11*, 23. [CrossRef] [PubMed]

69. Podlipaev, S.A. Insect trypanosomatids: The need to know more. *Mem. Inst. Oswaldo Cruz* **2000**, *95*, 517–522. [CrossRef]

70. Espinosa, O.A.; Serrano, M.G.; Camargo, E.P.; Teixeira, M.M.G.; Shaw, J.J. An appraisal of the taxonomy and nomenclature of trypanosomatids presently classified as *Leishmania* and *Endotrypanum*. *Parasitology* **2018**, *145*, 430–442. [CrossRef] [PubMed]

71. Cox, C.B. Plate tectonics, seaways and climate in the historical biogeography of mammals. *Mem. Inst. Oswaldo Cruz* **2000**, *95*, 509–516. [CrossRef]

72. Haag, J.; O'HUigin, C.; Overath, P. The molecular phylogeny of trypanosomes: Evidence for an early divergence of the Salivaria. *Mol. Biochem. Parasitol.* **1998**, *91*, 37–49. [CrossRef]

73. Hamilton, P.B.; Teixeira, M.M.; Stevens, J.R. The evolution of *Trypanosoma cruzi*: The 'bat seeding' hypothesis. *Trends Parasitol.* **2012**, *28*, 136–141. [CrossRef] [PubMed]

74. Lima, L.; Espinosa-Álvarez, O.; Hamilton, P.B.; Neves, L.; Takata, C.S.; Campaner, M.; Attias, M.; de Souza, W.; Camargo, E.P.; Teixeira, M.M. *Trypanosoma livingstonei*: A new species from African bats supports the bat seeding hypothesis for the Trypanosoma cruzi clade. *Parasites Vectors* **2013**, *6*, 221. [CrossRef] [PubMed]

75. Steverding, D. The history of African trypanosomiasis. *Parasites Vectors* **2008**, *1*, 3. [CrossRef] [PubMed]

76. Lukeš, J.; Mauricio, I.L.; Schönian, G.; Dujardin, J.C.; Soteriadou, K.; Dedet, J.P.; Kuhls, K.; Tintaya, K.W.; Jirků, M.; Chocholová, E.; et al. Evolutionary and geographical history of the *Leishmania donovani* complex with a revision of current taxonomy. *Proc. Natl. Acad. Sci. USA* **2007**, *104*, 9375–9380. [CrossRef] [PubMed]

77. Stevens, J.R.; Noyes, H.; Dover, G.A.; Gibson, W.C. The ancient and divergent origins of the human pathogenic trypanosomes, *Trypanosoma brucei* and *T-cruzi*. *Parasitology* **1999**, *118*, 107–116. [CrossRef] [PubMed]

78. Stevens, J.R.; Noyes, H.A.; Schofield, C.J.; Gibson, W. The molecular evolution of Trypanosomatidae. In *Advances in Parasitology*; Elsevier: Amsterdam, The Netherlands, 2001; Volume 48, pp. 1–56.

79. Espinosa-Álvarez, O.; Ortiz, P.A.; Lima, L.; Costa-Martins, A.G.; Serrano, M.G.; Herder, S.; Buck, G.A.; Camargo, E.P.; Hamilton, P.B.; Stevens, J.R.; et al. *Trypanosoma rangeli* is phylogenetically closer to Old World trypanosomes than to *Trypanosoma cruzi*. *Int. J. Parasitol.* **2018**, *48*, 569–584. [CrossRef] [PubMed]

80. Sauquet, H.; Ho, S.Y.; Gandolfo, M.A.; Jordan, G.J.; Wilf, P.; Cantrill, D.J.; Bayly, M.J.; Bromham, L.; Brown, G.K.; Carpenter, R.J.; et al. Testing the Impact of Calibration on Molecular Divergence Times Using a Fossil-Rich Group: The Case of Nothofagus (Fagales). *Syst. Biol.* **2012**, *61*, 289–313. [CrossRef] [PubMed]

pathogens

MDPI

Article

The Remarkable Metabolism of *Vickermania ingenoplastis*: Genomic Predictions

Fred R. Opperdoes [1,*], Anzhelika Butenko [2,3], Alexandra Zakharova [2], Evgeny S. Gerasimov [4,5],
Sara L. Zimmer [6], Julius Lukeš [3,7] and Vyacheslav Yurchenko [2,5,*]

1   De Duve Institute, Université Catholique de Louvain, 1200 Brussels, Belgium
2   Life Science Research Centre, Faculty of Science, University of Ostrava, 710 00 Ostrava, Czech Republic;
    anzhelika.butenko@paru.cas.cz (A.B.); alexandraz.6946@gmail.com (A.Z.)
3   Biology Centre, Institute of Parasitology, Czech Academy of Sciences,
    370 05 České Budějovice (Budweis), Czech Republic; jula@paru.cas.cz
4   Faculty of Biology, M.V. Lomonosov Moscow State University, 119991 Moscow, Russia; jalgard@gmail.com
5   Martsinovsky Institute of Medical Parasitology, Tropical and Vector Borne Diseases, Sechenov University,
    119435 Moscow, Russia
6   Department of Biomedical Sciences, University of Minnesota Medical School, Duluth Campus,
    Duluth, MN 558812, USA; szimmer3@d.umn.edu
7   Faculty of Science, University of South Bohemia, 370 05 České Budějovice (Budweis), Czech Republic
*   Correspondence: fred.opperdoes@uclouvain.be (F.R.O.); vyacheslav.yurchenko@osu.cz (V.Y.)

Abstract: A recently redescribed two-flagellar trypanosomatid *Vickermania ingenoplastis* is insensitive to the classical inhibitors of respiration and thrives under anaerobic conditions. Using genomic and transcriptomic data, we analyzed its genes of the core metabolism and documented that subunits of the mitochondrial respiratory complexes III and IV are ablated, while those of complexes I, II, and V are all present, along with an alternative oxidase. This explains the previously reported conversion of glucose to acetate and succinate by aerobic fermentation. Glycolytic pyruvate is metabolized to acetate and ethanol by pyruvate dismutation, whereby a unique type of alcohol dehydrogenase (shared only with *Phytomonas* spp.) processes an excess of reducing equivalents formed under anaerobic conditions, leading to the formation of ethanol. Succinate (formed to maintain the glycosomal redox balance) is converted to propionate by a cyclic process involving three enzymes of the mitochondrial methyl-malonyl-CoA pathway, via a cyclic process, which results in the formation of additional ATP. The unusual structure of the *V. ingenoplastis* genome and its similarity with that of *Phytomonas* spp. imply their relatedness or convergent evolution. Nevertheless, a critical difference between these two trypanosomatids is that the former has significantly increased its genome size by gene duplications, while the latter streamlined its genome.

Keywords: *Vickermania ingenoplastis*; *Phytomonas*; metabolism; genome sequencing

check for updates

Citation: Opperdoes, F.R.; Butenko, A.; Zakharova, A.; Gerasimov, E.S.; Zimmer, S.L.; Lukeš, J.; Yurchenko, V. The Remarkable Metabolism of *Vickermania ingenoplastis*: Genomic Predictions. *Pathogens* **2021**, *10*, 68. https://doi.org/10.3390/pathogens 10010068

Received: 5 December 2020
Accepted: 12 January 2021
Published: 14 January 2021

Publisher's Note: MDPI stays neutral with regard to jurisdictional claims in published maps and institutional affiliations.

1. Introduction

Trypanosomatids (Euglenozoa: Kinetoplastea: Trypanosomatidae) are parasites of annelids, arthropods, plants, and vertebrates, with leeches and insects serving as transmission vectors. The best-known trypanosomatids are dixenous (=circulating between two hosts, *Leishmania*, *Phytomonas*, and *Trypanosoma* spp.) and infect vertebrates, including humans and plants [1,2]. Nevertheless, the vast majority of species are monoxenous (=confined to a single host, usually an insect) [3]. Monoxenous Trypanosomatidae are ancestral and significantly more diverse [4–6].

Virtually all trypanosomatids are mono-flagellated (including an "amastigote" stage in some life cycles, which has an extremely short flagellum [7]) and use their flagella for attachment, movement, production of the extracellular vesicles, and environment sensing [8–11]. The only exception to this rule is members of the recently established genus *Vickermania*, *V. ingenoplastis* and *V. spadyakhi* [12]. These flagellates are adapted to the life in the fly midgut, to which they do not attach. Instead, they move constantly, resisting the midgut peristaltic flow within the fly host. To this end, *Vickermania* has disconnected duplication of flagella from the cell cycle and developed a mechanism to join the newly growing flagellum with the old one. As such, these trypanosomatids possess two flagella for a significant period of their life cycle.

In contrast to most other trypanosomatids, the metabolic activity of the *V. ingenoplastis* mitochondrion is strongly reduced. Cytochrome-mediated respiration was found to be missing, and energy metabolism to be based mainly on the fermentative glycolysis with acetate and ethanol as the major end-products and propionate and succinate as minor products. Interestingly, the switch from aerobic to anaerobic conditions had minimal effects [13]. Several enzymes of the Krebs cycle were not detected, and the presence of fumarate reductase activity was interpreted to indicate that CO_2 fixation and reverse flux through part of the Krebs cycle enables growth under anaerobic conditions [14]. This would be a unique feature for trypanosomatid parasites, which are generally considered to be (strictly) aerobic [15]. Respiration was not inhibited by cyanide, while malonate and salicyl-hydroxamic acid strongly reduced succinate oxidation. Taken together, these observations demonstrated that *V. ingenoplastis* lacks the mitochondrial respiratory complex IV, has an inactive complex III, while complex II and an alternative oxidase seem to be functional [16]. All these biochemical observations were experimental. Armed with the recently obtained genomic data for this species [17], we reanalyzed the metabolic potential of *V. ingenoplastis*, explain some earlier contradictory observation and provide a coherent framework for its unique metabolism. In particular, we documented that *V. ingenoplastis* has lost genes coding for the subunits of the respiratory complexes III and IV, increased a number of genes involved in carbohydrate metabolism, and lost the capacity to oxidize fatty acids and a number of aromatic and branched amino acids.

2. Results

2.1. Global Comparison between Vickermania ingenoplastis and Leishmania Major Genomes

The genome of *V. ingenoplastis* was reassembled *de novo* using different programs, resulting in its significant improvement from the previously published version [17]. The obtained assembly is 34.3 Mbp (in contrast to the original 35.3 Mbp, hereafter the data of the previous assembly are given in parentheses) in 241 scaffolds (340) with N_{50} of 591 kb (376 kb) and the longest contig of 2.4 Mb (1.6 Mb). The new assembly is not only more contiguous but also more complete as judged by the higher total number of annotated protein-coding genes (9562 vs. 8619 reported previously) and improved benchmarking universal single-copy orthologs (BUSCO) scores. The percentage of complete BUSCOs increased from 84.6 to 93.1%, and only 4% of BUSCOs are missing compared to 11% in the previous assembly.

The genome of *V. ingenoplastis* contains high copy numbers of the glycolytic pathway genes, such as those encoding hexokinase, as well as glycosomal and cytosolic glyceraldehyde-3-phosphate dehydrogenases. This suggests that the irreversible loss of the mitochondrial oxidative phosphorylation in *V. ingenoplastis* (see below) may have led to severe metabolic stress, by which the expression of a number of the glycolytic enzymes was upregulated by massive amplification of the corresponding genes. This has led, on the one hand, to a reduction in unique metabolic genes (97 *Leishmania major* orthologues were not found, Table S1), and on the other hand, a compensation for this loss by an expansion of the genome from 6500 to 8500 genes typically present in Leishmaniinae to 9562 genes in *V. ingenoplastis*. Most of these genes resulted from gene duplications to form multi-copy gene families (for example, cell surface proteins amastins), the largest of them with copy numbers reaching 200 (Tables S2 and S3). Out of 486 proteins predicted to be involved in the general metabolism of trypanosomatids, 87 have no orthologues in the genome of *V. ingenoplastis*, while it encodes many metabolic genes in high copy numbers. The highest copy numbers were scored for glycosomal glyceraldehyde dehydrogenase (39), cytosolic glyceraldehyde dehydrogenase (38), and hexokinase (19). The genome of *V. ingenoplastis* contains an additional 3612 genes that are not present in the *L. major* genome. Excluding all the genes annotated as coding for hypothetical or viral origin proteins, 620 unique genes were retained (Table S4). Of note, the expression of metabolic genes was further confirmed by whole-transcriptome analysis (Table S2).

2.2. Mitochondrial Enzymes of Oxidative Phosphorylation: Complexes I and II

Genes encoding subunits of complexes I (NADH dehydrogenase) and II (succinate dehydrogenase) are invariably present (Table S5), suggesting that these complexes are fully operational. NADH dehydrogenase is a large complex, the subunits of which are encoded by both the mitochondrial and nuclear genomes [18,19]. *Vickermania ingenoplastis* appears to be endowed with most, if not all, of its essential nuclear-encoded subunits. In the case of succinate dehydrogenase, entirely encoded in the nuclear genome, its two catalytic subunits (the flavoprotein and FeS-containing subunits) are present, along with a large battery of auxiliary proteins identified in *Trypanosoma cruzi* [20].

2.3. Mitochondrial Enzymes of Oxidative Phosphorylation: Complexes III and IV

We identified numerous deletions in the maxicircle kDNA, affecting complexes III and IV. Interestingly, these deletions are accompanied by the corresponding ablation of the complementing nuclear-encoded subunits of these complexes (Table S5). For complex III, not only the mitochondrial-encoded cytochrome *b*, but also the ubiquinol cytochrome *c* reductase, cytochrome c_1, the Rieske FeS protein, and complex III core protein [21] are absent. Of note, two members of the mitochondrial processing peptidase family that have been shown to be core subunits of the complex III (orthologues of the α-MMP or β-MMP) have been retained in *V. ingenoplastis*, likely reflecting their additional functions [22]. For complex IV (cytochrome oxidase), not only all three mitochondrial-encoded subunits (COXI, COXII, and COXIII) but also the nuclear-encoded subunits 4, 5, 6, 7 and 10 are absent, along with the cytochrome oxidase assembly protein COX15, involved in the synthesis of heme *a* [23] and the electron transport protein SCO1/SCO2, a metallochaperone, essential for the assembly of the catalytic core of cytochrome *c* oxidase [24]. Finally, the gene for cytochrome *c*, which transports electrons between complexes III and IV, is absent in the nuclear genome, whereas a gene for the alternative oxidase is prominently present. Interestingly, the same genes have been lost from the genome of *Phytomonas* spp. [25], further supporting the evolutionary relatedness of these genera [12].

We conclude that electrons entering the respiratory chain via NADH or succinate pass through the functional complexes I and II and reach coenzyme Q (ubiquinone), from where they are transferred to molecular oxygen via the alternative oxidase.

2.4. Mitochondrial Enzymes of Oxidative Phosphorylation: Complex V

Although complexes III and IV of the respiratory chain are completely missing, the ATP synthase (complex V) appears to be fully operational. The kinetoplast-encoded subunit of the F_1F_0 ATPase (subunit 6) is present in the maxicircle kDNA, along with all other nuclear-encoded subunits, including 11 auxiliary proteins [26] (Table S5).

2.5. Cytochrome o

The respiration of *V. ingenoplastis* was shown to be cyanide insensitive, with cytochrome (cyt) *o* being suggested as an alternative cyanide-resistant terminal oxidase [16]. However, this was not confirmed experimentally, and no gene for this enzyme has been described in trypanosomatids so far. The spectral characteristics of Cyt *o* inversely correlate with that of reduced Cyt *b*. This allows us to propose that the suspected Cyt *o* was a variant of Cyt *b*, not properly integrated into complex III, and subjected to auto-oxidization. As a consequence, this gene was lost over time. The alternative oxidase, encoded by six copies in *V. ingenoplastis*, might be responsible for the inhibition of respiration by SHAM [27]. Thus, far, an alternative oxidase has been documented in various euglenozoan lineages, including *Bodo*, *Trypanosoma*, *Phytomonas*, and *Angomonas*, as a single-copy gene [25,28,29].

2.6. Krebs Cycle

Our genome analysis indicates that, except for the catabolic NAD-dependent isocitrate dehydrogenase, which has been replaced by an anabolic NADP-dependent isoenzyme, all other Krebs cycle enzymes (citrate synthase, aconitase, isocitrate dehydrogenase, 2-oxoglutarate dehydrogenase, succinyl-CoA ligase, succinate dehydrogenase, fumarate hydratase and malate dehydrogenase) and subunits of the pyruvate dehydrogenase complex are present. The presence of an NADP-dependent isocitrate dehydrogenase, rather than an NAD-dependent enzyme, resembles the situation in other trypanosomatids and predicts that the Krebs cycle in *V. ingenoplastis* functions not as a real cycle [30,31], but CO_2 fixation and reverse flux through a part of the cycle may endow this flagellate with aerobic fermentation (Figure 1).

2.7. Carbohydrate Metabolism

The *Vickermania ingenoplastis* genome contains all genes of the glycolytic pathway, with many of them predicted to possess a peroxisome targeting signal (Table S6). In addition, a whole battery of peroxisome assembly factors is present. Both observations strongly imply the presence of glycosomes, as in all other trypanosomatids [32]. The high copy numbers for several of the glycolytic genes suggest that the corresponding enzymes are abundantly present. This agrees with previous observations of numerous microbodies (likely, glycosomes) in the electron micrographs of *V. ingenoplastis* [14] and the fact that glucose consumption via the glycolytic pathway is *Vickermania*'s principal source of energy [13]. Pyruvate, the end-product of this pathway, is oxidized in the mitochondrion by the pyruvate dehydrogenase complex and the resulting acetyl-CoA is excreted in the form of acetate, the major end-product of *V. ingenoplastis* [13]. We have also documented the presence of Zn-containing alcohol dehydrogenase in the analyzed genome (Figure 1), an enzyme previously reported only from *Phytomonas* sp. [33]. This enzyme is likely responsible for the production of ethanol, since classical alcohol dehydrogenases are absent from all trypanosomatids.

Propionate, the other end-product of carbohydrate metabolism, is normally not produced by trypanosomatids, but only by some anaerobic eukaryotes [34] and *V. ingenoplastis* [13]. Moreover, a ratio of excreted end-products (succinate and propionic acid over acetate, two to one) is reminiscent of that observed in many propionic-acid producing anaerobic eukaryotes, such as the liver fluke *Fasciola hepatica*. A ratio of 2:1 is required to maintain the intracellular redox balance. This kind of fermentation is called "malate dismutation" [35]. Cytosolic malate (or pyruvate) is partly oxidized (*via* pyruvate) to acetate and partly reduced to succinate and propionate in the mitochondrion [34,36]. Succinate production requires the presence of the enzyme fumarate reductase, which, in most anaerobic eukaryotes, is a membrane-bound mitochondrial enzyme, similar or identical to the mitochondrial succinate dehydrogenase, except that it uses rhodoquinone, rather than ubiquinone, as the redox carrier, by which the enzyme is able to run in the reverse direction. In trypanosomatids, the presence of rhodoquinone has never been reported. Thus, the functioning of such a rodoquinone-dependent fumarate reductase was considered unlikely [36]. However, trypanosomatids are unique in that succinate can be readily formed within the highly reduced matrix of the glycosomes [32]. It is formed from phosphoenolpyruvate and CO_2 via oxaloacetate, malate, and fumarate by the glycosomal NADH-dependent fumarate reductase (Figure 1). The two moles of NAD^+ formed in this pathway serve to equilibrate the glycosomal $NAD^+/NADH$ balance. Anaerobically growing *V. ingenoplastis* now has the option to either excrete the succinate, as do all other trypanosomatids or to transport succinate into the mitochondrion, where it is converted to propionate in a cyclic process that contains three enzymes of the mitochondrial methyl-malonyl-CoA pathway (the "propionate cycle"). This way, the formation of each mole of propionate is accompanied by the formation of 1 mole of ATP by substrate-level phosphorylation.

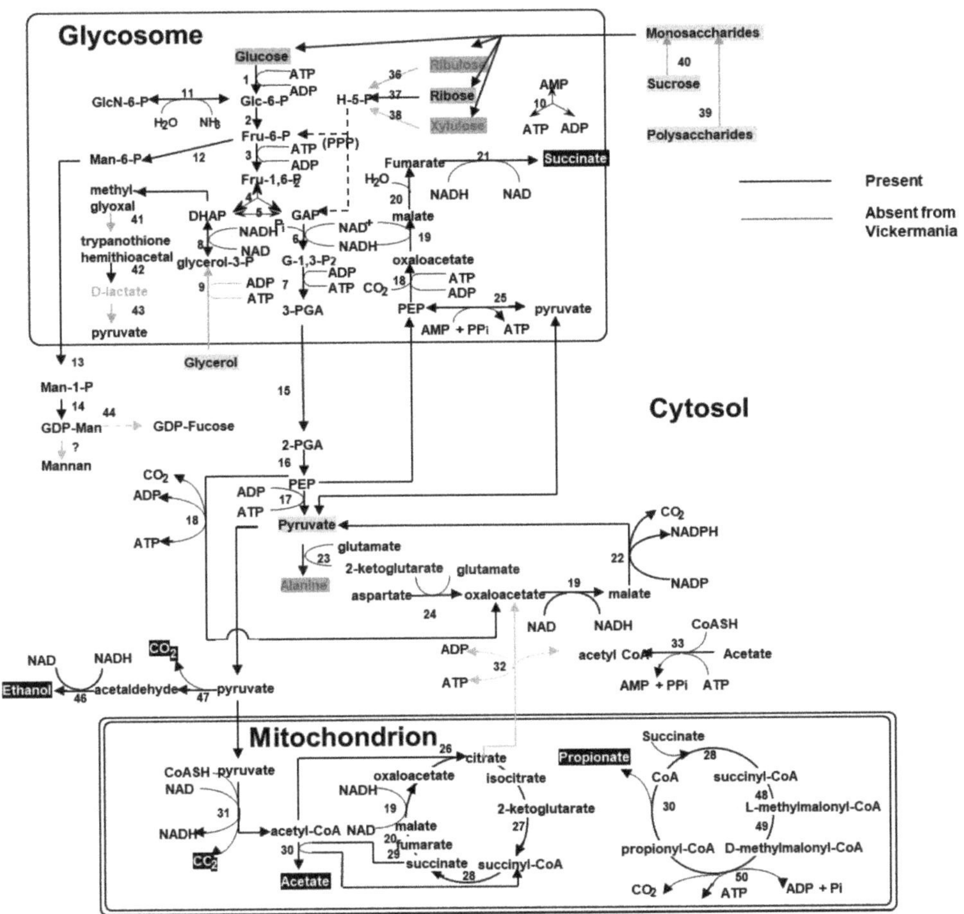

Figure 1. The pathways of core metabolism in *Vickermania ingenoplastis* as compared to that of *Leishmania major*. Boxed metabolites are nutrients (in gray) or end-products (in black). PPP, pentose-phosphate pathway. Enzymes: 1, hexokinase; 2, phosphoglucose isomerase; 3, phosphofructokinase; 4, fructosebisphosphate aldolase; 5, triosephosphate isomerase; 6, glyceraldehyde-3-phosphate dehydrogenase; 7, glycosomal phosphoglycerate kinase; 8, glycerol-3-phosphate dehydrogenase; 9 glycerol kinase; 10, glycosomal adenylate kinase; 11, glucosamine-6-phosphate deaminase; 12, mannose-6-phosphate isomerase; 13, phosphomannomutase; 14, GDP-mannose pyrophosphorylase; 15, phosphoglycerate mutase; 16, enolase; 17, pyruvate kinase; 18, phosphoenolpyruvate carboxykinase; 19, malate dehydrogenase; 20, fumarate hydratase; 21, NADH-dependent fumarate reductase; 22, malic enzyme; 23, alanine aminotransferase; 24, aspartate aminotransferase; 25, pyruvate phosphate di-kinase; 26, citrate synthase; 27, 2-ketoglutarate dehydrogenase; 28, succinyl-CoA ligase; 29, succinate dehydrogenase; 30, acetate:succinate CoA transferase; 31, pyruvate dehydrogenase; 32, citrate lyase; 33, acetyl-CoA synthetase; 36, ribulokinase; 37, ribokinase, 38, xylulokinase; 39, glucoamylase; 40, invertase; 41, glyoxalase I; 42, glyoxalase II; 43, D-lactate dehydrogenase; 44, GDP-mannose 4,6-dehydratase and GDP-L-fucose synthase; 45, phosphoacetylglucosamine mutase and glucosamine-1-phosphate acetyl transferase/UDP-N-acetylglucosamine transferase; 46, alcohol dehydrogenase GroES-like domain/zinc-binding dehydrogenase; 47, pyruvate decarboxylase; 48, methylmalonyl-CoA epimarase; 49, methylmalonyl-CoA mutase; 50, propionyl-CoA carboxylase. Modified from [25].

Thus, under anaerobic conditions, the glucose oxidation in *Vickermania ingenoplastis* leads to the formation of 1 mole of pyruvate in the cytosol and 1 mole of succinate in the glycosome, along with 2 moles of ATP by substrate-level phosphorylation. Half of the pyruvate is transported into the mitochondrion and oxidized by the pyruvate dehydrogenase complex to acetyl-CoA and (*via* the acetate: succinate CoA transferase—succinyl-CoA ligase cycle) acetate with the net synthesis of an additional 1/2 mole of ATP. The other half of the pyruvate is decarboxylated by a cytosolic pyruvate decarboxylase to acetaldehyde and then reduced to 1/2 mole of ethanol. The latter reaction serves to reoxidize the NADH produced in the mitochondrial pyruvate dehydrogenase reaction. Thus, here, redox balance is maintained by a mechanism that should be called "pyruvate dismutation". Glycosomal succinate is either excreted or decarboxylated to propionate in the mitochondrion with the concomitant formation of 1 mole of ATP. According to this scheme, the overall oxidation of 1 mole of glucose leads to the formation of 0.5 mole ethanol, 0.5 mole acetate, 1 mole of (succinate + propionate), and 3.5 moles of ATP. The anaerobic production of 3.5 moles of ATP per 1 mole of glucose consumed must be the explanation of why *V. ingenoplastis* is able to survive and grow under completely anaerobic conditions. In this respect, it is important to note that the bloodstream form of the African trypanosomes produce only 2 moles ATP per 1 mole of consumed glucose aerobically, while under anaerobic conditions (or when oxygen consumption is inhibited by SHAM) when only 1 mole ATP per 1 mole glucose is produced, the trypanosomes die [37]. No L-lactate is formed because a gene for L-lactate dehydrogenase is missing.

2.8. Hexose-Monophosphate Shunt and Gluconeogenesis

The enzymes of the hexose monophosphate pathway, as well as those involved in gluconeogenesis (except for the fructose-1,6-bisphosphatase), are present in the *V. ingenoplastis* genome (Table S2). There is no evidence for the synthesis of glycogen, and no genes for the formation of storage polysaccharides were identified. Nevertheless, the presence of several mannosyl transferases suggests that mannans, rather than glycogen, could serve as polysaccharide storage.

2.9. Sensitivity to Drugs

Vickermania ingenoplastis is sensitive to metronidazole and fexinidazole [14]. Both chemicals require enzyme-mediated reduction by hydrogenases and nitroreductases to generate cytotoxic species [38]. Genes encoding both enzymes, an iron-containing hydrogenase and a nitroreductase, were found in the analyzed genome.

2.10. Beta-Oxidation and Synthesis of Fatty Acids

The oxidation of fatty acids to carbon dioxide and water requires their activation through the linkage to coenzyme A. Subsequently, β-oxidation involves peroxisomes and mitochondria, whereby long-chain fatty acids are shortened first in the glycosomes, after which they are exported to the mitochondrion, where the resulting acetyl-CoA is further oxidized in the Krebs cycle either by acyl-CoA oxidase present in the glycosomes or by an isofunctional acyl-CoA dehydrogenase in the mitochondria [28]. The subsequent reactions are catalyzed by a single trifunctional enzyme in peroxisomes and by two separate enzymes in the mitochondria of most eukaryotes. These two latter enzymes are absent in *Vickermania* and *Phytomonas* spp. Thus, it is unlikely that a complete β-oxidation pathway is operational in *Vickermania*. Similarly, L-Leu cannot be oxidized to acetyl-CoA because isovaleryl-CoA dehydrogenase and 3-methylcrotonoyl-CoA carboxylase are absent.

However, the synthesis of fatty acids is possible due to the presence of genes encoding acetyl-CoA synthetase and acyl carrier protein. Since in all other trypanosomatids, type I fatty acid synthesis is absent, here it proceeds via elongase(s) that act on butyryl-CoA [39]. Several fatty acyl-CoA synthases (ligases) are present (Table S7). Of the four *L. major* fatty acyl dehydrogenases, only one is found in *V. ingenoplastis* (Table S8). Moreover, out of three other β-oxidation enzymes, namely enoyl CoA hydratase, 3-hydroxyacyl-CoA dehydrogenase and 3-ketoacyl-CoA thiolase, only the latter was identified. To oxidize the acetyl-CoA formed by β-oxidation, a functional Krebs cycle is required, as well. Therefore, complexes I and II are of vital importance for *Vickermania*. However, the absence of NAD-isocitrate dehydrogenase prevents the complete oxidation of acetyl-CoA to carbon dioxide and water, and, thus, *Vickermania* is likely unable to oxidize fatty acids to completion.

2.11. Amino Acid Metabolism

In other trypanosomatids, the amino acids Glu, Pro, and Thr may serve as energy sources in the absence of carbohydrates, such as glucose [28]. The presence of mitochondrial Glu and Pro dehydrogenases in the *V. ingenoplastis* genome suggests that it is also able to utilize these two amino acids. In contrast, most trypanosomatids rely on a catabolic Thr dehydrogenase to produce ammonia and 2-ketobutyrate, which is then irreversibly converted to propionyl-CoA and formate, Leishmaniinae and *Phytomonas* spp. use Thr dehydratase [28]. *Vickermania ingenoplastis* apparently uses the first pathway, which includes a mitochondrial Ser hydroxymethyltransferase, or Ser/Thr dehydratase.

The enzyme isovaleryl-CoA dehydrogenase participates in Val, Leu, and Ile degradation in other trypanosomatids [40]. It is absent in *V. ingenoplastis*, as well as kynureninase involved in the oxidation of Trp. Phe cannot be converted into Tyr because two of the three enzymes present in other trypanosomatids have been lost in both *Vickermania* and *Phytomonas* spp. [25] (Table S9). Similarly, a cobalamine-independent methionine synthase was absent from their genomes.

2.12. Catalase and Heme Synthesis

Catalase is present, yet it differs from its counterparts documented in Leishmaniinae [41,42] or *Blastocrithidia* spp. [43]. It remains to be investigated further why a gene encoding this important enzyme was acquired at least three times independently in the evolution of Trypanosomatidae.

All three heme-synthetic enzymes of prokaryotic origin (protoporphyrinogen oxidase, coproporphyrinogen III oxidase, and ferrochelatase) that are present in other Leishmaniinae [44] have been lost (or never acquired) in *V. ingenoplastis*.

3. Discussion

Trypanosomatids are famous for their remarkable adaptability to different environmental conditions. This is held as an explanation for the great variety of hosts (leeches and insects for the monoxenous, and arthropods, vertebrates, and plants for the dixenous parasites) that can be infected with these flagellates [5,6]. With the change of a host or switch between life cycle stages, trypanosomatids display unseen flexibility in metabolism [15,45]. Their insect-dwelling stages have an oxidative metabolism, in which amino and fatty acids serve as energy substrates. This is the case for the amastigotes of *Leishmania* or the epimastigotes and promastigotes of trypanosomes. All these stages require the presence of a fully active mitochondrion [46,47]. The other end of the spectrum is represented by aerobic fermentation of carbohydrates, with the mitochondrial metabolism reduced to a minimum. Examples of the latter are the bloodstream stages of African trypanosomes or their dyskinetoplastic cousins, *T. equiperdum* and *T. evansi* [48–50]. In the vast majority of trypanosomatids, such metabolic changes are reversible, and parasites go back and forth between an active and a less active mitochondrion during the switch of the life cycle stages. An exception to this rule is the plant trypanosomatids, belonging to the genus *Phytomonas* [51]. In these species, the switch back from aerobic fermentation to oxidative metabolism is blocked because several essential mitochondrial and nuclear metabolic proteins have been irreversibly lost. It was proposed that these deletions have occurred in order to make *Phytomonas* spp. insensitive to cyanide, which is present in plant tissues [25], an explanation implausible in the case of *V. ingenoplastis*, though.

Our genome analyses of *V. ingenoplastis* have confirmed the presence of most of the enzymes of carbohydrate metabolism, necessary for the production of all the major end-products under aerobic and anaerobic conditions [13] (Figure 1). Both *Phytomonas* spp. and *V. ingenoplastis* resemble each other in that they metabolize glucose to acetate, succinate and ethanol in varying amounts, depending on the oxygen availability. Ethanol production can be facilitated by isopropanol dehydrogenase (uniquely present in these species) converting acetaldehyde into ethanol [33,52]. However, *Phytomonas* and *Vickermania* differ in that *Phytomonas* excretes pyruvate and glycerol [53], similar to the bloodstream form of the African trypanosomes. In these species, glycerol is produced in the reversal of the glycerol kinase reaction via highly active glycosomal glycerol kinase [54]. *Vickermania ingenoplastis* lacks this enzyme (the annotated glycerol kinase is likely a xylulose kinase rather than a glycerol kinase). Instead, it excretes propionate using enzymes of the methylmalonyl-CoA pathway and the propionate cycle, which is absent in *Phytomonas* spp. (and in most other trypanosomatids).

In this work, we demonstrate that *V. ingenoplastis* is similar to *Phytomonas* spp. also in regard to the gene losses. This suggests that either these species belong to two closely related taxa, or they have been shaped by convergent evolution. *Vickermania ingenoplastis* resembles *Phytomonas* spp. not only because of the parallel loss of genes encoding subunits of complexes III and IV (questioning hypothesis that cyanide was a driving force of this process) but also because they lost numerous genes encoding enzymes involved in β-oxidation of fatty acids, the oxidation of aromatic, long or branched-chain amino acids. As an adaptation to these combined gene losses, *Phytomonas* spp. and *V. ingenoplastis* have drastically augmented their capacity for carbohydrate metabolism either by an increase in the copy number of glycolytic genes and/or the overall number of glycosomes in the cytosol [12,14,55]. Nevertheless, a critical difference between these two trypanosomatids is that *V. ingenoplastis* has increased its genome size by gene duplications, while *Phytomonas* spp. has done exactly the opposite [17,25].

Notably, *V. ingenoplastis* and *Phytomonas* spp. share a unique gene that encodes a Zn-containing NAD(+)-dependent alcohol dehydrogenase/isopropyl alcohol dehydrogenase (iPDH), an enzyme with a broad substrate specificity acting on primary and secondary alcohols [33,56]. In *Phytomonas*, the presence of iPDH facilitates the accumulation of ethanol as an end-product of its glycolysis [53]. Conversely, *V. ingenoplastis* excretes ethanol as one of the end-products of its carbohydrate metabolism [13]. The iPDH gene, not found in the majority of trypanosomatids, has been acquired via horizontal gene transfer from a bacterium and was proposed as a *Phytomonas*-specific marker enzyme [33,56]. The only other trypanosomatid possessing this gene (but as a pseudogene) is *Blechomonas ayalai* [28]. In conclusion, we postulate that *Vickermania* and *Phytomonas* are either closely related or convergently evolved trypanosomatids.

4. Materials and Methods

4.1. Genome Reassembly and Analysis of Its Completeness

The genome of *V. ingenoplastis* was reassembled using sequencing data reported previously [17] with MaSuRCA assembler v. 3.3.9 with the default settings, followed by two rounds of polishing using the assembler-associated Polca software [57]. This pipeline was not tested by us previously [17]. It resulted in a more contiguous assembly, which was used in the current study. For the assembly polishing, trimmed paired-end Illumina reads were mapped onto the genome assembly using the Burrows–Wheeler alignment tool (BWA) v. 0.7.17 with the default settings [58]. The assembly annotation was performed using a Companion server and the genome of *T. brucei* as a reference [59]. The basic assembly statistics and the completeness of the resulting assembly were assessed using QUAST v. 5.0.2 [60] and BUSCO v. 4.0.5 with the eukaryote_odb10 database [61], respectively.

4.2. Gene Expression Analysis

PolyA-enriched RNA library was sequenced on Illumina HiSeq 2500 platform with read length 151 bp, paired-end. Raw data available at NCBI SRA under BioProject accession number PRJNA675748 (SRR13015660). Sequencing reads were trimmed using Trimmomatic v. 0.39 [62] and mapped onto reference genome assembly with Bowtie2 v. 2.3.4.1 [63] with "—end-to-end—no-unal—sensitive" options. Read counts per gene were obtained with (for sorted bam file) and BEDtools [64]. Per gene RPKM, values were calculated using a custom python script.

4.3. Analysis of Metabolic Pathways

Metabolic pathways of *Vickermania ingenoplastis* were analyzed as described previously [28,65], using "all against all" BLASTp searches with an E-value cutoff of 10^{-20} and previously published (but reassembled) genomic data [17]. In some cases, a stricter E-value cutoff of 10^{-50} was used in order to distinguish between true orthologous proteins and more distant homologs, which are not necessarily functional orthologues.

Supplementary Materials: The following are available online at https://www.mdpi.com/2076-0817/10/1/68/s1, Table S1: *Leishmania major* proteins not found in *Vickermania ingenoplastis* genome by BLASTp (hereafter, threshold 10^{-20}). Table S2: Presence/absence for orthologues of 458 genes encoding *L. major* metabolic proteins in *V. ingenoplastis* by BLASTp. Expression (RNA-seq data) is indicated as RPKM values, and Expression level (over genome). The color coding for the expression level is green (below 5%), yellow (5–25%), orange (25–75%), and red (over 75%). Table S3: Copy numbers of genes coding for metabolic enzymes in *V. ingenoplastis*. Table S4: Enzymes unique to *V. ingenoplastis* (absent from *L. major* genome) after exclusion of hypothetical or viral proteins. Table S5: Subunits of the respiratory chain complexes in *L. major*, *T. brucei*, and *V. ingenoplastis*. Table S6: Enzymes of the carbohydrate metabolism in *L. major* and *V. ingenoplastis*. Genes absent in *V. ingenoplastis* are highlighted in red. Table S7: Enzymes of fatty acid oxidation in *B. saltans*, *L. major*, *T. brucei*, and *V. ingenoplastis*. Table S8: Enzymes present in *L. major* but absent in both *V. ingenoplastis* and *Phytomonas* spp. Table S9: Enzymes of other pathways.

Author Contributions: Conceptualization, F.R.O. and V.Y.; methodology, F.R.O., A.B. and E.S.G.; validation, E.S.G., A.B., S.L.Z. and A.Z.; formal analysis, F.R.O., J.L. and V.Y.; investigation, F.R.O., S.L.Z., A.Z. and E.S.G.; resources, V.Y., J.L. and E.S.G.; data curation, F.R.O. and V.Y.; writing—original draft preparation, F.R.O. and V.Y.; writing—review and editing, F.R.O., A.Z., A.B., E.S.G., J.L. and V.Y.; supervision, V.Y. and J.L.; funding acquisition, V.Y., J.L., F.R.O. and E.S.G. All authors have read and agreed to the published version of the manuscript.

Funding: This research was funded by the European Regional Funds (CZ.02.1.01/16_019/0000759), the Grant Agency of Czech Republic (20-07186S and 21-09283S) to V.Y. and J.L., the ERC CZ (LL1601) to J.L., and the University of Ostrava (SGS/PrF/2021) to A.Z. Two specific tasks were funded by the Russian Science Foundation grants 19-15-00054 for analysis of orthologues of *Leishmania major* genes to V.Y. and 19-74-10008 for whole-transcriptome analysis of kinetoplast-encoded gene expression to E.S.G. F.R.O. received support from the de Duve Institute.

Institutional Review Board Statement: Not applicable.

Informed Consent Statement: Not applicable.

Data Availability Statement: The data presented in this study are openly available in NCBI SRA under BioProject accession number PRJNA675748 (SRR13015660).

Acknowledgments: We thank Louis Tielens (Erasmus University, Rotterdam, The Netherlands) for helpful discussions on malate dismutation.

Conflicts of Interest: The authors declare no conflict of interest. The funders had no role in the design of the study; in the collection, analyses, or interpretation of data; in the writing of the manuscript, or in the decision to publish the results.

References

1. Vickerman, K. *Comparative Cell Biology of the Kinetoplastid Flagellates in Biology of Kinetoplastida*; Vickerman, K., Preston, T.M., Eds.; Academic Press: London, UK, 1976; pp. 35–130.
2. Maslov, D.A.; Opperdoes, F.R.; Kostygov, A.Y.; Hashimi, H.; Lukeš, J.; Yurchenko, V. Recent advances in trypanosomatid research: Genome organization, expression, metabolism, taxonomy and evolution. *Parasitology* **2019**, *146*, 1–27. [CrossRef]
3. Maslov, D.A.; Votýpka, J.; Yurchenko, V.; Lukeš, J. Diversity and phylogeny of insect trypanosomatids: All that is hidden shall be revealed. *Trends Parasitol.* **2013**, *29*, 43–52. [CrossRef]

4. D'Avila-Levy, C.M.; Boucinha, C.; Kostygov, A.; Santos, H.L.C.; Morelli, K.A.; Grybchuk-Ieremenko, A.; Duval, L.; Votýpka, J.; Yurchenko, V.; Grellier, P.; et al. Exploring the environmental diversity of kinetoplastid flagellates in the high-throughput DNA sequencing era. *Memórias Inst. Oswaldo Cruz* **2015**, *110*, 956–965. [CrossRef]

5. Lukeš, J.; Butenko, A.; Hashimi, H.; Maslov, D.A.; Votýpka, J.; Yurchenko, V. Trypanosomatids Are Much More than Just Trypanosomes: Clues from the Expanded Family Tree. *Trends Parasitol.* **2018**, *34*, 466–480. [CrossRef]

6. Lukeš, J.; Skalický, T.; Týč, J.; Votýpka, J.; Yurchenko, V. Evolution of parasitism in kinetoplastid flagellates. *Mol. Biochem. Parasitol.* **2014**, *195*, 115–122. [CrossRef]

7. Wheeler, R.J.; Gluenz, E.; Gull, K. The cell cycle of Leishmania: Morphogenetic events and their implications for parasite biology. *Mol. Microbiol.* **2010**, *79*, 647–662. [CrossRef]

8. Broadhead, R.; Dawe, H.R.; Farr, H.; Griffiths, S.; Hart, S.R.; Portman, N.; Shaw, M.K.; Ginger, M.L.; Gaskell, S.J.; Mckean, P.G.; et al. Flagellar motility is required for the viability of the bloodstream trypanosome. *Nat. Cell Biol.* **2006**, *440*, 224–227. [CrossRef]

9. Hughes, L.; Ralston, K.S.; Hill, K.L.; Zhou, Z.H. Three-Dimensional Structure of the Trypanosome Flagellum Suggests that the Paraflagellar Rod Functions as a Biomechanical Spring. *PLoS ONE* **2012**, *7*, e25700. [CrossRef]

10. Beneke, T.; Demay, F.; Hookway, E.; Ashman, N.; Jeffery, H.; Smith, J.; Valli, J.; Bečvář, T.; Myškova, J.; Leštinová, T.; et al. Genetic dissection of a *Leishmania* flagellar proteome demonstrates requirement for directional motility in sand fly infections. *PLoS Pathog.* **2019**, *15*, e1007828. [CrossRef]

11. Szempruch, A.J.; Sykes, S.E.; Kieft, R.; Dennison, L.; Becker, A.C.; Gartrell, A.; Martin, W.J.; Nakayasu, E.S.; Almeida, I.C.; Hajduk, S.L.; et al. Extracellular Vesicles from *Trypanosoma brucei* Mediate Virulence Factor Transfer and Cause Host Anemia. *Cell* **2016**, *164*, 246–257. [CrossRef]

12. Kostygov, A.Y.; Frolov, A.O.; Malysheva, M.N.; Ganyukova, A.I.; Chistyakova, L.V.; Tashyreva, D.; Tesařová, M.; Spodareva, V.V.; Režnarová, J.; Macedo, D.H.; et al. *Vickermania* gen. nov., trypanosomatids that use two joined flagella to resist midgut peristaltic flow within the fly host. *BMC Biol.* **2020**, *18*, 187. [CrossRef]

13. Redman, C.A.; Coombs, G.H. The Products and Pathways of Glucose Catabolism in *Herpetomonas muscarum ingenoplastis* and *Herpetomonas muscarum muscarum*. *J. Eukaryot. Microbiol.* **1997**, *44*, 46–51. [CrossRef]

14. Coombs, G.H. Herpetomonas muscarum ingenoplastis: An anaerobic kinetoplastid flagellate? In *Biochemistry and Molecular Biology of "Anaerobic" Protozoa*; Lloyd, D., Coombs, G.H., Paget, T.A., Eds.; Harwood Academic Publishers: London, UK, 1989; pp. 254–266.

15. Tielens, A.G.M.; Van Hellemond, J.J. Differences in Energy Metabolism between Trypanosomatidae. *Parasitol. Today* **1998**, *14*, 265–272. [CrossRef]

16. Hajduk, S. Studies of Trypanosomatid Flagellates with Special Reference to Antigenic Variation and Kinetoplast DNA. Ph.D. Thesis, Department of Zoology, University of Glasgow, Glasgow, UK, 1980; p. 229.

17. D'Avila-Levy, C.M.; Bearzatto, B.; Ambroise, J.; Helaers, R.; Butenko, A.; Yurchenko, V.; Morelli, K.A.; Santos, H.L.C.; Brouillard, P.; Grellier, P.; et al. First Draft Genome of the Trypanosomatid *Herpetomonas muscarum ingenoplastis* through MinION Oxford Nanopore Technology and Illumina Sequencing. *Trop. Med. Infect. Dis.* **2020**, *5*, 25. [CrossRef]

18. Opperdoes, F.; Michels, P.A. Complex I of Trypanosomatidae: Does it exist? *Trends Parasitol.* **2008**, *24*, 310–317. [CrossRef]

19. Čermáková, P.; Maďarová, A.; Baráth, P.; Bellová, J.; Yurchenko, V.; Horváth, A. Differences in mitochondrial NADH dehydrogenase activities in trypanosomatids. *Parasitology* **2021**, in press.

20. Morales, J.; Mogi, T.; Mineki, S.; Takashima, E.; Mineki, R.; Hirawake, H.; Sakamoto, K.; Ōmura, S.; Kita, K. Novel Mitochondrial Complex II Isolated from *Trypanosoma cruzi*s Composed of 12 Peptides Including a Heterodimeric Ip Subunit. *J. Biol. Chem.* **2009**, *284*, 7255–7263. [CrossRef]

21. Acestor, N.; Zíková, A.; Dalley, R.A.; Anupama, A.; Panigrahi, A.K.; Stuart, K.D. *Trypanosoma brucei* Mitochondrial Respiratome: Composition and Organization in Procyclic Form. *Mol. Cell. Proteom.* **2011**, *10*, 006908. [CrossRef]

22. Peña-Diaz, P.; Mach, J.; Kriegova, E.; Poliak, P.; Tachezy, J.; Lukeš, J. Trypanosomal mitochondrial intermediate peptidase does not behave as a classical mitochondrial processing peptidase. *PLoS ONE* **2018**, *13*, e0196474. [CrossRef]

23. Zíková, A.; Panigrahi, A.K.; Uboldi, A.D.; Dalley, R.A.; Handman, E.; Stuart, K. Structural and Functional Association of *Trypanosoma brucei* MIX Protein with Cytochrome c Oxidase Complex. *Eukaryot. Cell* **2008**, *7*, 1994–2003. [CrossRef]

24. Acestor, N.; Panigrahi, A.K.; Ogata, Y.; Anupama, A.; Stuart, K. Protein composition of *Trypanosoma brucei* mitochondrial membranes. *Proteomics* **2009**, *9*, 5497–5508. [CrossRef] [PubMed]

25. Porcel, B.M.; Denoeud, F.; Opperdoes, F.; Noel, B.; Madoui, M.-A.; Hammarton, T.C.; Field, M.C.; Da Silva, C.; Couloux, A.; Poulain, J.; et al. The Streamlined Genome of *Phytomonas* spp. Relative to Human Pathogenic Kinetoplastids Reveals a Parasite Tailored for Plants. *PLoS Genet.* **2014**, *10*, e1004007. [CrossRef] [PubMed]

26. Zíková, A.; Schnaufer, A.; Dalley, R.A.; Panigrahi, A.K.; Stuart, K. The F_0F_1-ATP Synthase Complex Contains Novel Subunits and Is Essential for Procyclic *Trypanosoma brucei*. *PLoS Pathog.* **2009**, *5*, e1000436. [CrossRef] [PubMed]

27. Opperdoes, F.; Borst, P.; Fonck, K. The potential use of inhibitors of glycerol-3-phosphate oxidase for chemotherapy of African trypanosomiasis. *FEBS Lett.* **1976**, *62*, 169–172. [CrossRef]

28. Opperdoes, F.R.; Butenko, A.; Flegontov, P.; Yurchenko, V.; Lukeš, J. Comparative Metabolism of Free-living *Bodo saltans* and Parasitic Trypanosomatids. *J. Eukaryot. Microbiol.* **2016**, *63*, 657–678. [CrossRef] [PubMed]

29. Butenko, A.; Hammond, M.; Field, M.C.; Ginger, M.L.; Yurchenko, V.; Lukeš, J. Reductionist Pathways for Parasitism in Euglenozoans? Expanded Datasets Provide New Insights. *Trends Parasitol.* **2020**, *37*, 100–116. [CrossRef] [PubMed]

30. Van Weelden, S.W.H.; Van Hellemond, J.; Opperdoes, F.; Tielens, A.G.M. New Functions for Parts of the Krebs Cycle in Procyclic *Trypanosoma brucei*, a Cycle Not Operating as a Cycle. *J. Biol. Chem.* **2005**, *280*, 12451–12460. [CrossRef]

31. Opperdoes, F.; Van Hellemond, J.; Tielens, A. The extraordinary mitochondrion and unusual citric acid cycle in *Trypanosoma brucei*. *Biochem. Soc. Trans.* **2005**, *33*, 967–971. [CrossRef]

32. Opperdoes, F.; Michels, P. The glycosomes of the Kinetoplastida. *Biochimie* **1993**, *75*, 231–234. [CrossRef]

33. Molinas, S.M.; Altabe, S.G.; Opperdoes, F.; Rider, M.H.; Michels, P.A.M.; Uttaro, A.D. The Multifunctional Isopropyl Alcohol Dehydrogenase of *Phytomona* ssp. Could Be the Result of a Horizontal Gene Transfer from a Bacterium to the Trypanosomatid Lineage. *J. Biol. Chem.* **2003**, *278*, 36169–36175. [CrossRef]

34. Muller, M.; Mentel, M.; Van Hellemond, J.; Henze, K.; Woehle, C.; Gould, D.; Yu, R.-Y.; Van Der Giezen, M.; Tielens, A.; Martin, W. Biochemistry and Evolution of Anaerobic Energy Metabolism in Eukaryotes. *Microbiol. Mol. Biol. Rev.* **2012**, *76*, 444–495. [CrossRef]

35. Martin, W.F.; Tielens, A.G.M.; Mentel, M. *Mitochondria and Anaerobic Energy Metabolism in Eukaryotes: Biochemistry and Evolution*; De Gruyter: Düsseldorf, Germany, 2021; p. 252.

36. Van Hellemond, J.J.; Klockiewicz, M.; Gaasenbeek, C.P.H.; Roos, M.H.; Tielens, A.G.M. Rhodoquinone and Complex II of the Electron Transport Chain in Anaerobically Functioning Eukaryotes. *J. Biol. Chem.* **1995**, *270*, 31065–31070. [CrossRef] [PubMed]

37. Fairlamb, A.H.; Opperdoes, F.; Borst, P. New approach to screening drugs for activity against African trypanosomes. *Nat. Cell Biol.* **1977**, *265*, 270–271. [CrossRef] [PubMed]

38. Patterson, S.; Wyllie, S. Nitro drugs for the treatment of trypanosomatid diseases: Past, present, and future prospects. *Trends Parasitol.* **2014**, *30*, 289–298. [CrossRef]

39. Lee, S.H.; Stephens, J.L.; Englund, P.T. A fatty-acid synthesis mechanism specialized for parasitism. *Nat. Rev. Genet.* **2007**, *5*, 287–297. [CrossRef] [PubMed]

40. Millerioux, Y.; Mazet, M.; Bouyssou, G.; Allmann, S.; Kiema, T.-R.; Bertiaux, E.; Fouillen, L.; Thapa, C.; Biran, M.; Plazolles, N.; et al. *De novo* biosynthesis of sterols and fatty acids in the *Trypanosoma brucei* procyclic form: Carbon source preferences and metabolic flux redistributions. *PLoS Pathog.* **2018**, *14*, e1007116. [CrossRef] [PubMed]

41. Kraeva, N.; Horáková, E.; Kostygov, A.; Kořený, L.; Butenko, A.; Yurchenko, V.; Lukeš, J. Catalase in *Leishmaniinae*: With me or against me? *Infect. Genet. Evol.* **2017**, *50*, 121–127. [CrossRef]

42. Škodová-Sveráková, I.; Záhonová, K.; Bučková, B.; Füssy, Z.; Yurchenko, V.; Lukeš, J. Catalase and ascorbate peroxidase in euglenozoan protists. *Pathogens* **2020**, *9*, 317. [CrossRef]

43. Bianchi, C.; Kostygov, A.Y.; Kraeva, N.; Záhonová, K.; Horáková, E.; Sobotka, R.; Lukeš, J.; Yurchenko, V. An enigmatic catalase of *Blastocrithidia*. *Mol. Biochem. Parasitol.* **2019**, *232*, 111199. [CrossRef]

44. Kořený, L.; Oborník, M.; Lukeš, J. Make It, Take It, or Leave It: Heme Metabolism of Parasites. *PLoS Pathog.* **2013**, *9*, e1003088. [CrossRef]

45. Škodová-Sveráková, I.; Verner, Z.; Skalický, T.; Votýpka, J.; Horváth, A.; Lukeš, J. Lineage-specific activities of a multipotent mitochondrion of trypanosomatid flagellates. *Mol. Microbiol.* **2015**, *96*, 55–67. [CrossRef] [PubMed]

46. Rosenzweig, D.; Smith, D.; Opperdoes, F.; Stern, S.; Olafson, R.W.; Zilberstein, D. Retooling *Leishmania* metabolism: From sand fly gut to human macrophage. *FASEB J.* **2008**, *22*, 590–602. [CrossRef] [PubMed]

47. Zikova, A.; Verner, Z.; Nenarokova, A.; Michels, P.A.M.; Lukeš, J. A paradigm shift: The mitoproteomes of procyclic and bloodstream *Trypanosoma brucei* are comparably complex. *PLoS Pathog.* **2017**, *13*, e1006679. [CrossRef] [PubMed]

48. Mazet, M.; Morand, P.; Biran, M.; Bouyssou, G.; Courtois, P.; Daulouède, S.; Millerioux, Y.; Franconi, J.-M.; Vincendeau, P.; Moreau, P.; et al. Revisiting the Central Metabolism of the Bloodstream Forms of *Trypanosoma brucei*: Production of Acetate in the Mitochondrion Is Essential for Parasite Viability. *PLoS Negl. Trop. Dis.* **2013**, *7*, e2587. [CrossRef]

49. Van Hellemond, J.J.; Bakker, B.M.; Tielens, A.G. Energy Metabolism and Its Compartmentation in *Trypanosoma brucei*. *Adv. Microb. Physiol.* **2005**, *50*, 199–226. [CrossRef]

50. Lai, D.-H.; Hashimi, H.; Lun, Z.-R.; Ayala, F.J.; Lukeš, J. Adaptations of *Trypanosoma brucei* to gradual loss of kinetoplast DNA: *Trypanosoma equiperdum* and *Trypanosoma evansi* are petite mutants of *T. brucei*. *Proc. Natl. Acad. Sci. USA* **2008**, *105*, 1999–2004. [CrossRef]

51. Jaskowska, E.; Butler, C.; Preston, G.; Kelly, S. *Phytomonas*: Trypanosomatids adapted to plant environments. *PLoS Pathog.* **2015**, *11*, e1004484. [CrossRef]

52. Uttaro, A.D.; Opperdoes, F. Purification and characterisation of a novel iso-propanol dehydrogenase from *Phytomonas* sp. *Mol. Biochem. Parasitol.* **1997**, *85*, 213–219. [CrossRef]

53. Chaumont, F.; Schanck, A.N.; Blum, J.J.; Opperdoes, F.R. Aerobic and anaerobic glucose metabolism of *Phytomonas* sp. isolated from Euphorbia characias. *Mol. Biochem. Parasitol.* **1994**, *67*, 321–331. [CrossRef]

54. Opperdoes, F.R.; Borst, P. Localization of nine glycolytic enzymes in a microbody-like organelle in *Trypanosoma brucei*: The glycosome. *FEBS Lett.* **1977**, *80*, 360–364. [CrossRef]

55. Sanchez-Moreno, M.; Lasztity, D.; Coppens, I.; Opperdoes, F. Characterization of carbohydrate metabolism and demonstration of glycosomes in a *Phytomonas* sp. isolated from *Euphorbia characias*. *Mol. Biochem. Parasitol.* **1992**, *54*, 185–199. [CrossRef]

56. Uttaro, A.D.; Sanchez-Moreno, M.; Opperdoes, F. Genus-specific biochemical markers for *Phytomonas* spp. *Mol. Biochem. Parasitol.* **1997**, *90*, 337–342. [CrossRef]

57. Zimin, A.V.; Marçais, G.; Puiu, D.; Roberts, M.; Salzberg, S.L.; Yorke, J.A. The MaSuRCA genome assembler. *Bioinformatics* **2013**, *29*, 2669–2677. [CrossRef]
58. Li, H.; Durbin, R. Fast and accurate short read alignment with Burrows-Wheeler transform. *Bioinformatics* **2009**, *25*, 1754–1760. [CrossRef]
59. Steinbiss, S.; Silva-Franco, F.; Brunk, B.; Foth, B.; Hertz-Fowler, C.; Berriman, M.; Otto, T.D. Companion: A web server for annotation and analysis of parasite genomes. *Nucleic Acids Res.* **2016**, *44*, W29–W34. [CrossRef] [PubMed]
60. Gurevich, A.; Saveliev, V.; Vyahhi, N.; Tesler, G. QUAST: Quality assessment tool for genome assemblies. *Bioinformatics* **2013**, *29*, 1072–1075. [CrossRef] [PubMed]
61. Seppey, M.; Manni, M.; Zdobnov, E.M. BUSCO: Assessing genome assembly and annotation completeness. In *Gene Prediction*; Kollmar, M., Ed.; Methods in Molecular Biology; Humana: New York, NY, USA, 2019; Volume 1962, pp. 227–245.
62. Bolger, A.M.; Lohse, M.; Usadel, B. Trimmomatic: A flexible trimmer for Illumina sequence data. *Bioinformatics* **2014**, *30*, 2114–2120. [CrossRef]
63. Langmead, B.; Salzberg, S.L. Fast gapped-read alignment with Bowtie 2. *Nat. Methods* **2012**, *9*, 357–359. [CrossRef]
64. Quinlan, A. BEDTools: The Swiss-Army Tool for Genome Feature Analysis. *Curr. Protoc. Bioinform.* **2014**, *47*, 11.12.1–11.12.34. [CrossRef]
65. Butenko, A.; Kostygov, A.Y.; Sádlová, J.; Kleschenko, Y.; Bečvář, T.; Podešvová, L.; Macedo, D.H.; Žihala, D.; Lukeš, J.; Bates, P.A.; et al. Comparative genomics of *Leishmania* (*Mundinia*). *BMC Genom.* **2019**, *20*, 726. [CrossRef]

 *pathogens*

Concept Paper

Genomics and High-Resolution Typing Confirm Predominant Clonal Evolution down to a Microevolutionary Scale in *Trypanosoma cruzi*

Michel Tibayrenc [1,*] **and Francisco J. Ayala** [2]

1   Maladies Infectieuses et Vecteurs Ecologie, Génétique, Evolution et Contrôle, MIVEGEC (IRD 224-CNRS 5290-UM1-UM2), Institut de Recherche pour le Développement, BP 34394 Montpellier CEDEX 5, France
2   Catedra Francisco Jose Ayala of Science, Technology, and Religion, University of Comillas, 28015 Madrid, Spain; fjayala2018@gmail.com
*   Correspondence: michel.tibayrenc@ird.fr

Received: 23 March 2020; Accepted: 7 May 2020; Published: 8 May 2020

Abstract: *Trypanosoma cruzi*, the agent of Chagas disease, is a paradigmatic case of the predominant clonal evolution (PCE) model, which states that the impact of genetic recombination in pathogens' natural populations is not sufficient to suppress a persistent phylogenetic signal at all evolutionary scales. In spite of indications for occasional recombination and meiosis, recent genomics and high-resolution typing data in *T. cruzi* reject the counterproposal that PCE does not operate at lower evolutionary scales, within the evolutionary units (=near-clades) that subdivide the species. Evolutionary patterns in the agent of Chagas disease at micro- and macroevolutionary scales are strikingly similar ("Russian doll pattern"), suggesting gradual, rather than saltatory evolution.

Keywords: Chagas disease; parasitic protozoa; clonality threshold; genetic recombination; phylogenetic signal; Russian doll pattern

---

1. Preliminary Recalls about the Predominant Clonal Evolution (PCE) Model

The predominant clonal evolution (PCE) pattern does not mean that genetic recombination is either absent, or of little evolutionary significance [1], but rather, that it is not effective enough to erase a persistent and highly detectable phylogenetic signal at all evolutionary scales. The definition of clonality in PCE is therefore based on severe restriction to genetic recombination, a definition that is shared by many authors working on pathogen population genetics (see many references in [2]). The criteria selected for stating that the phylogenetic signal is reliable are the classic, widely accepted, means used in the articles analyzed by us in the present study—(i) mutual corroboration by different markers (see Table 1 in [3]); (ii) posterior probabilities when Bayesian analysis is concerned; (iii) bootstrap, with the limit value of 0.70 considered as significant [4].

PCE is therefore not rejected by the sole detection of genetic exchange, hybridization and meiosis [5,6]. As recalled many times [7], the PCE model is compatible with such traits. Which makes it possible to definitely and specifically challenge the PCE hypothesis is the absence of a stable phylogenetic signal at any evolutionary scale and a population structure that meets panmictic expectations, particularly lack of a statistically significant linkage disequilibrium (nonrandom association of genotypes occurring at different loci) [7].

We have coined the term "near-clades" [8] to designate, within pathogen species, genetic subdivisions that are discrete and stable, but that could be somewhat clouded by occasional genetic exchange. As a matter of fact, "true" clades are supposed to be strictly separated from each other. Now in virtually all pathogen species, even if PCE obtains, as noted above, occasional bouts of genetic exchange are recorded. The term "clade" therefore is not adequate.

2. *Trypanosoma cruzi* and the PCE Model

Trypanosoma cruzi is the parasite responsible for Chagas disease in the New World. It has been the object of early, pioneering studies dealing with its isoenzyme variability, making it possible to characterize its strains [9]. The interpretation of this isoenzyme diversity in population genetic terms has made it possible to propose that this parasite has a predominantly clonal population structure [10]. The evidence for it is as follows—at the level of the whole species, several multilocus genotypes occur at frequencies that are at variance with panmictic expectations, and are widely distributed in various ecosystems and hosts. A highly significant linkage disequilibrium is recorded [10]. The species is subdivided into at least six main "discrete typing units" or DTUs [11,12], namely Tc I to VI. Evolutionary speaking, these DTUs amount to near-clades [8]. More recently, an additional discrete typing unit/near clade has been described under the name of TcBat. It has been isolated exclusively from bats and is widespread over vast geographical areas and time spans [12]. The available data do not make it possible to test our PCE model within TcBat.

3. *T. cruzi* PCE Challengers

Obstacles to genetic recombination and the presence of a ubiquitous, stable, phylogenetic signal at the level of the whole *T. cruzi* species is no longer under debate. However, the PCE model in *T. cruzi* has been challenged with two lines of arguments, namely—(i) it is based on outdated markers that lack resolution [13]. This is not a valid argument—markers that lack resolution should favor the null hypothesis of panmixia (random genetic exchange) through a mechanism of statistical type II error (impossibility to reject the null hypothesis, not because this null hypothesis is true, but because of a lack of resolution of the used means to test it) rather than the working hypothesis of clonality (Figure 1). (ii) The presence of genetic subdivisions (="near-clades") within *T. cruzi* would be "self-evident", which amounts to saying that the outcome of any population genetics and phylogenetic analysis is self-evident. Evidencing obstacles to recombination at the level of the whole species is therefore trivial and vain [14]. However, high-resolution genomic typing will show that similar patterns of obstacles to genetic exchange are not recorded at lower evolutionary scales, under the level of the near-clades [14]. This last argument aims at specifically challenging the "Russian doll model" [15], which states that PCE is verified at all evolutionary scales, and within-near-clade population structure is a miniature form of the population structure of the whole species (Figure 2).

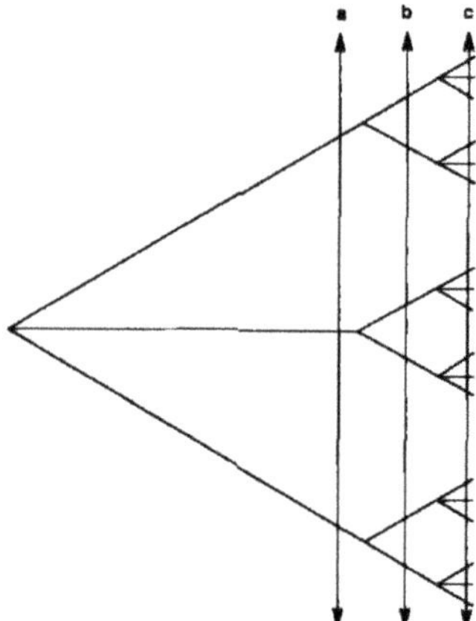

Figure 1. The impact of marker resolution on population genetics and phylogenetic analysis. If a marker with low resolution is used (a), the lesser genetic subdivisions of the species (right part of the figure) will show limited or null genetic variability, which may make it impossible to reject the null hypothesis of panmixia, due to a statistical type II error (after [16]).

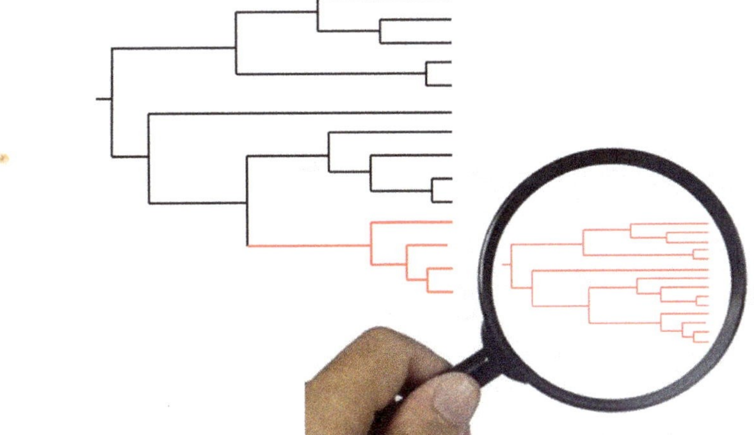

Figure 2. "Russian doll" model. When population genetic analysis is performed with adequate markers (of sufficient resolution) within each of the near-clades that subdivide the species under study (large tree, left), they reveal a miniature picture of the whole species, with the two main predominant clonal evolution features, namely, linkage disequilibrium and lesser near-clades (small tree, right). This is evidence that within the near-clades, predominant clonal evolution also operates (after [2]).

At this microevolutionary level, within each of the main genetic clusters (near-clades) that subdivide the species, two evolutionary models would imply that the Russian doll pattern is not verified. They both deal with lack of restriction to genetic recombination:

(a) Biological speciation—each of the near-clades correspond to cryptic species that are genetically isolated from each other, but within which genetic exchange is random, except for physical obstacles (time and/or space) to this random gene flow (see Figure 3). This hypothesis of speciation has been invoked to claim that the main subdivisions (Savannah, Killifi, Forest) within *Trypanosoma congolense* are not evidence for PCE, because they could correspond to cryptic "species". However, the authors did not clearly refer to a model of biological speciation [17]; and

Figure 3. Cryptic biological speciation: the evolutionary lines that subdivide the species are genetically isolated from each other. However, within each of them, genetic recombination occurs randomly, except when physical obstacles (space and/or time) occur (after [18]).

(b) Progressive clonality—this situation refers to the case where the amount of genetic exchange is inversely proportional to the evolutionary distance between any two given genotypes [16]. If the genotypes are either identical or very similar, genetic exchange is abundant (homogamy, selfing). If they are distantly related, genetic exchange is either severely limited or lacking (Figure 4). Such an evolutionary model is believed to be frequent in bacteria [19].

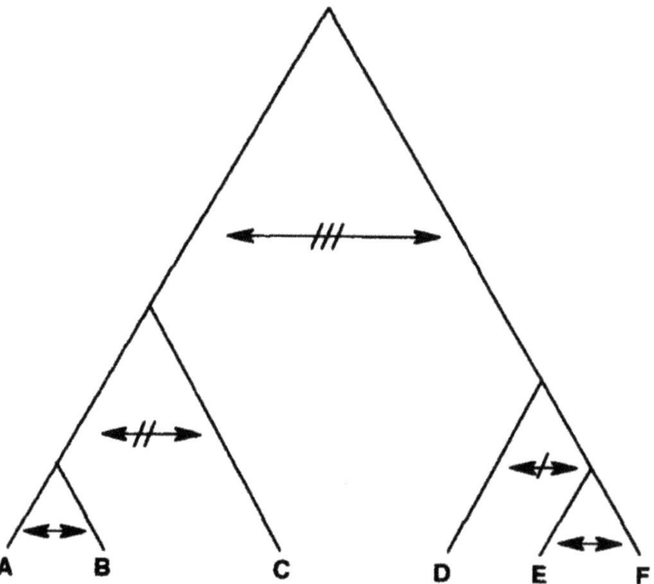

Figure 4. "Progressive clonality". The frequency of genetic exchange is inversely proportional to the evolutionary distance between any two different genotypes. It is virtually random among identical or very closely related genotypes (homogamy, selfing) and is progressively inhibited as genetic distances increase (after [16]).

It is clear that, first, (a) and (b) mean that genetic recombination is not limited or is poorly limited at microevolutionary scales (under the level of the near-clade); second, the means to distinguish the Russian doll model from either (a) or (b) is to give evidence for the presence of PCE traits (linkage disequilibrium and, most of all, constant phylogenetic signal—see Figure 2) within each of the near-clades that subdivide the species under study. However, this demands the use of genetic markers with a sufficient resolution. If this is not the case, lack of resolution of the markers could lead to a wrong hypothesis of panmixia due to a statistical type II error (see Figure 1).

4. New Analyses with High-Resolution Typing Challenge the Challengers

Our previous articles did already include the analysis of studies based on high-resolution markers and genomics data. However, to address the criticisms that (i) our model is based on outdated markers that lack resolution [13]; (ii) our model will not be verified at lower evolutionary scales [14], we have reconsidered the problem of PCE in *T. cruzi* in the light of numerous new published articles. This makes it possible to reliably test the Russian doll model within *T. cruzi* near-clades, and to illustrate some important aspects of the PCE model that are frequently misunderstood.

A wealth of studies show that within the near-clade TcI, in various countries, Russian doll patterns with a highly detectable phylogenetic signal are present. This is against the hypotheses of biological speciation (Figure 3) and progressive clonality (Figure 4).

In the Atlantic forest region of Brazil, the analysis of 107 wild strains, all identified as TcI and isolated from *Didelphis* sp., were analyzed with 27 microsatellite loci (hence coded by nuclear genes), while a subset of this sample was analyzed with 10 maxicircle loci (that are equivalent to mitochondrial genes) [20]. The double tree obtained (Figure 5) shows that this TcI sample is strongly subdivided into various lesser near-clades, with several significant bootstrap values. Some discrepancies are recorded between the two trees, which can be explained by either occasional introgression [20] or different evolutionary patterns, or both. The main fact is that this TcI sample exhibits a highly detectable phylogenetic signal, with a clear Russian doll pattern.

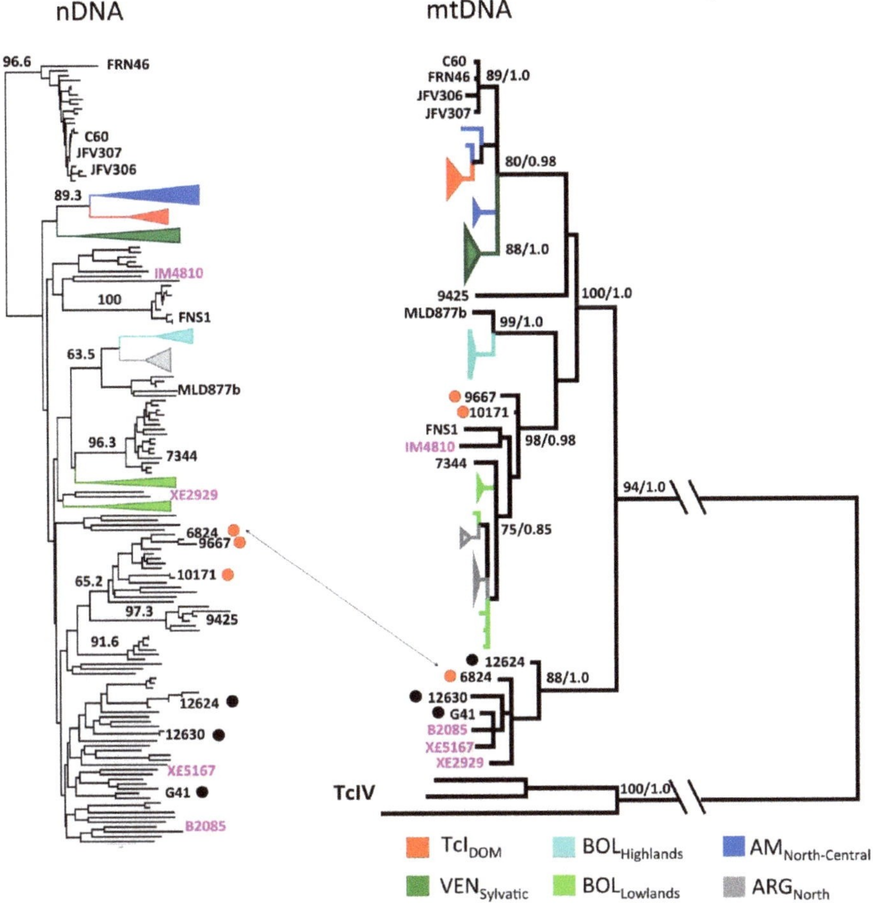

Figure 5. Double phylogenetic tree based on nuclear genes (**left**) and mitochondrial genes (**right**) in a sample of TcI Brazilian strains (after [20]). The TcI discrete typing unit/near clade, itself a discrete subdivision of the species *T. cruzi*, is clustered into various lesser near-clades. Several of these lesser near-clades are supported by significant bootstrap values (numbers along the branches); example—top lesser near-clade—bootstrap 96,6.

In Brazil, 78 TcI strains isolated from various hosts, including *Didelphis* sp., primates, rodents, bats, triatomine bugs, collected over five ecologically diverse biomes, were analyzed with the sequencing of six housekeeping nuclear genes (Multilocus Sequence Typing or MLST), 25 microsatellite loci and one maxicircle gene (*COII*), thus combining slow- and fast-evolving markers [21]. The phylogenies based on individual housekeeping genes exhibit moderate levels of incongruence. However, the concatenated tree shows a clear structuration into several lesser near-clades, many of them being supported by significant bootstrap values (Figure 6). This clustering can be explained by neither geographical repartition nor host specificity.

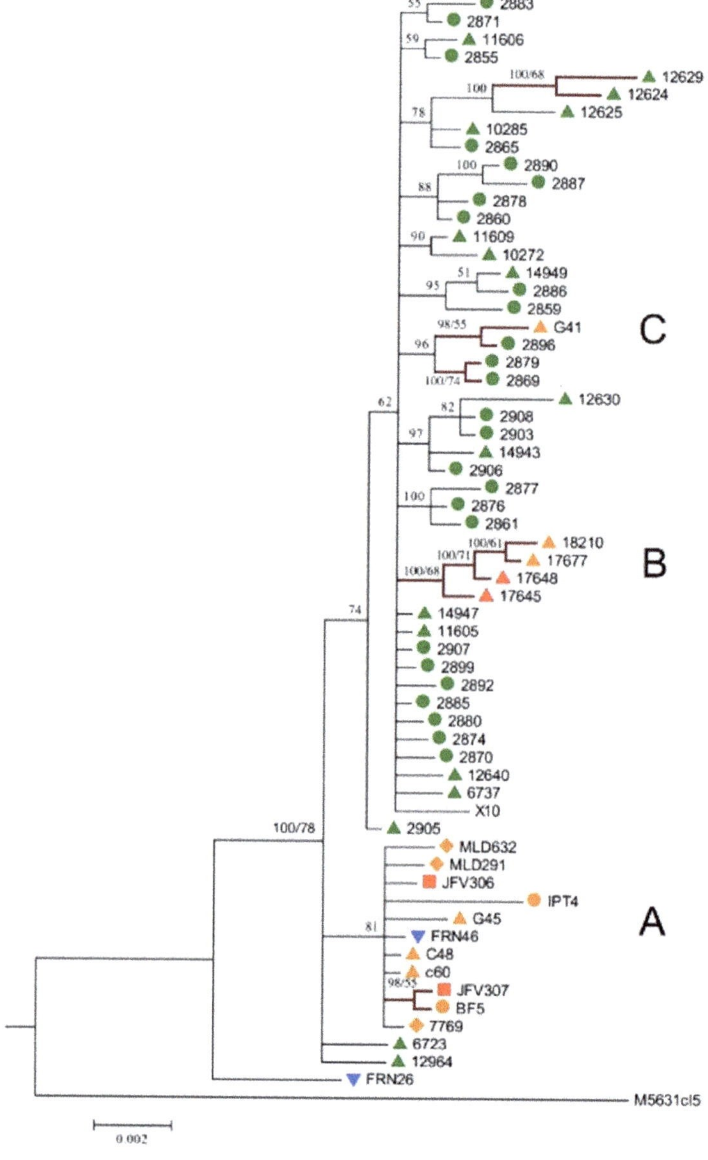

Figure 6. Concatenated Multilocus Sequence Typing (MLST) tree in a sample of TcI Brazilian strains (after [21]). Similarly to Figure 5, a different sampling of Brazilian strains of TcI shows various lesser near-clades within this near-clade. Many of them are supported by bootstrap values that are above the limit of 0.70 used in the present paper [4].

In Venezuela, 246 TcI human strains, some of them being isolated after an outbreak of oral transmission, were typed with 23 microsatellite loci [22]. The tree obtained (Figure 7) again shows the presence of various lesser near-clades with several significant bootstrap values.

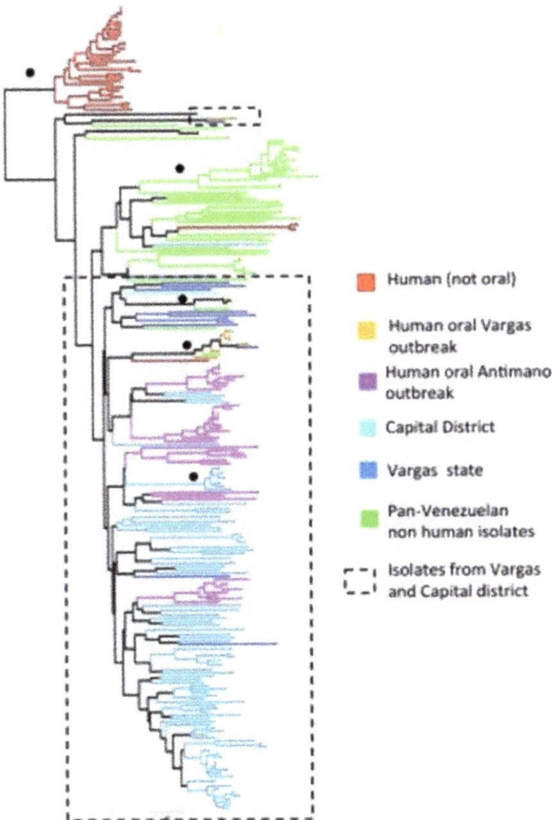

Figure 7. Multilocus microsatellite phylogenetic tree of 246 TcI Venezuelan strains. In Venezuela, the TcI near-clade is also subdivided into many lesser near-clades. Black circles indicate nodes with >60% bootstrap support [22].

In Bolivia, 199 clones isolated from 68 sylvatic TcI strains from both the lowlands and the highlands of the country were typed with 26 microsatellite loci and 10 maxicircle (=mitochondrial) loci [23]. The microsatellite and maxicircle phylogenies show some discrepancies, which the authors explain by introgression events. However, they broadly agree, which shows that these two very different parts of the genome do not evolve independently (linkage disequilibrium). When microsatellite diversity is considered, high levels of linkage disequilibrium are recorded, including within each subpopulation of the sample. The microsatellite phylogeny shows strong clustering patterns (lesser near-clades) that are not explained by either host specificity or geographical separation (Figure 8).

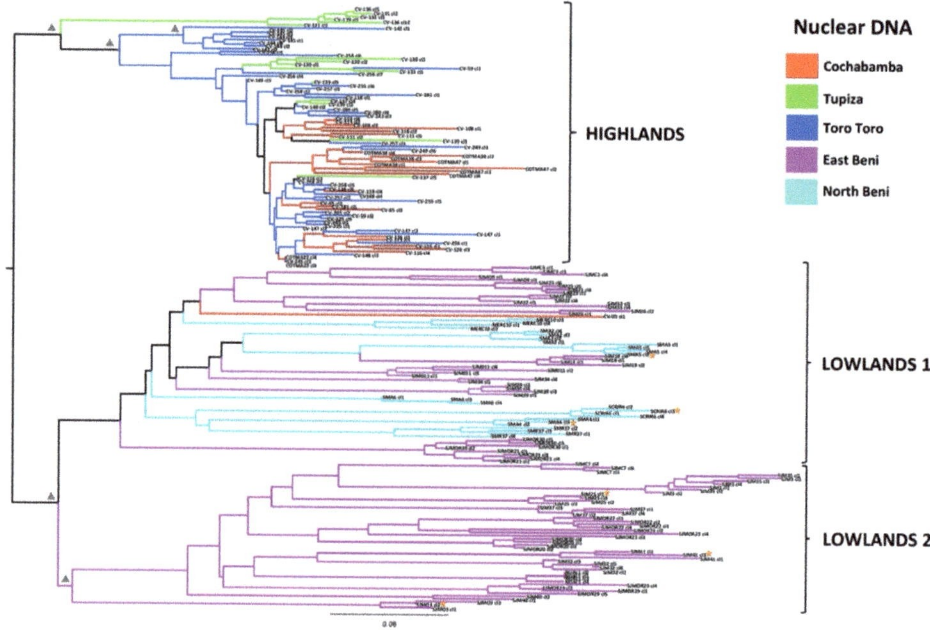

Figure 8. A microsatellite phylogenetic tree of sylvatic TcI strains in Bolivia (after [23]). In Bolivia also, TcI selvatic strains show clustering into many lesser near-clades. Closed grey triangles are adjacent to nodes that receive >60% bootstrap support. Genetic separation accounts only partly for this clustering pattern.

In Ecuador, a population genomics survey has revealed within the near clade TcI two distinct genetic clusters (=lesser near-clades) [6]. One shows clear indications of meiosis, whereas the other one does not. However, as already exposed, the isolated observation of meiosis is not in itself sufficient to conclude a panmictic pattern and to challenge the PCE model. As a matter of fact, in [6], (i) the evidence of two distinct clusters (=near-clades) within the near-clade TcI is in itself a Russian doll pattern; (ii) the occurrence of meiosis proves to be an exceptional event (3 meioses/1000 mitoses [6]); (iii) although the difference in population structure between the two clusters is undisputable, the number of different individuals remains weak—eight individuals, since several samples correspond to laboratory clones of the same isolate. This limited sample size leads to the risk of a statistical type II error with possible erroneous hypothesis of panmixia; (iv) in the first population (Bella Maria locality), even if one considers only the eight isolates that are supposed to exhibit meiosis, in spite of this limited sample size, the phylogenetic signal still is highly detectable—"support is unambiguous for main clusters and high within subclusters, except where last branch lengths are quite short in Cluster 2" (P. Schwabl, personal communication) (see Figure 9). This is evidence that genetic exchange is not frequent and not effective enough to erase a clear phylogenetic signal. This is the very definition of PCE. This is even more evident when including the whole Bella Maria population, which comprises an isolate that pertains to the second cluster and is phylogenetically quite distinct—see Figure 9.

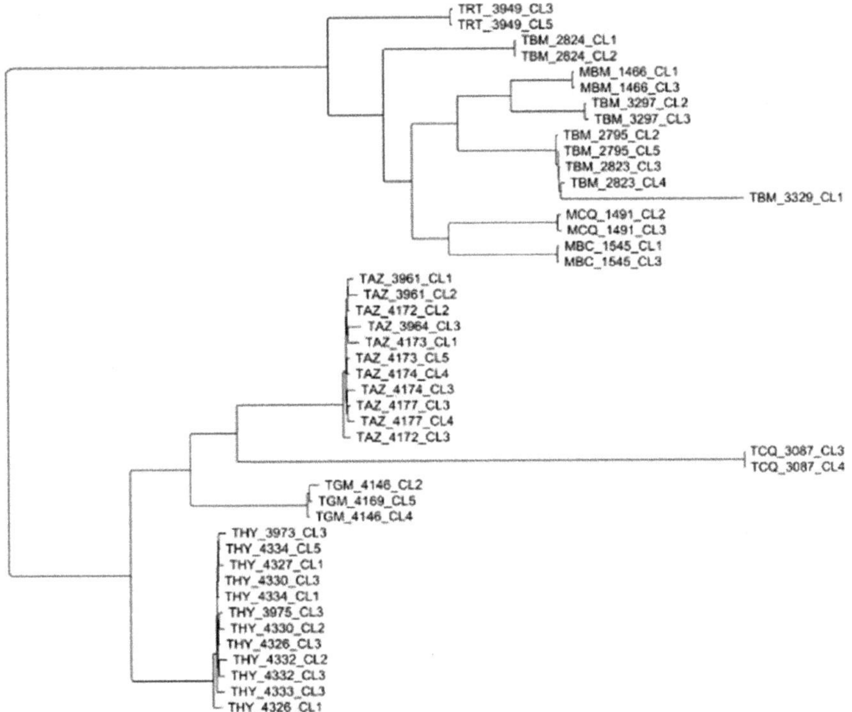

Figure 9. Two lesser near-clades within the TcI near-clade in Ecuador. In spite of clear indications of meiosis in the top cluster, a clear phylogenetic signal is evidenced at the level of the whole sample and within each of the two lesser near-clades (after [6]). "Support is unambiguous for main clusters and high within subclusters, except where last branch lengths are quite short in Cluster 2" (P. Schwabl, personal communication).

When other *T. cruzi* near-clades are considered, within TcII, a phylogenetic signal has been evidenced by genomic data [24]. The study dealt with a limited number (seven) of TCII strains isolated in Minas Gerais (Brazil) and surveyed for both nuclear and mitochondrial genomes. Phylogenies based on the nuclear and mitochondrial genomes show that the majority of branches are shared by both sequences. This gives evidence for the fact that nuclear and mitochondrial genomes do not evolve independently (linkage disequilibrium). The strength of the results is diminished by the limited number of strains. However, clustering (lesser near-clades) is apparent among these strains (Figure 10B in [24]).

Lastly, 19 stocks representative of the 6 *T. cruzi* near-clades (TcI-VI) were analyzed for 335 distinct satellite DNA sequences [25]. The Bayesian phylogeny shows that each of the six near-clades is strongly divided into many lesser near-clades (Figure 10A) with highly significant bootstrap values (Figure 10B).

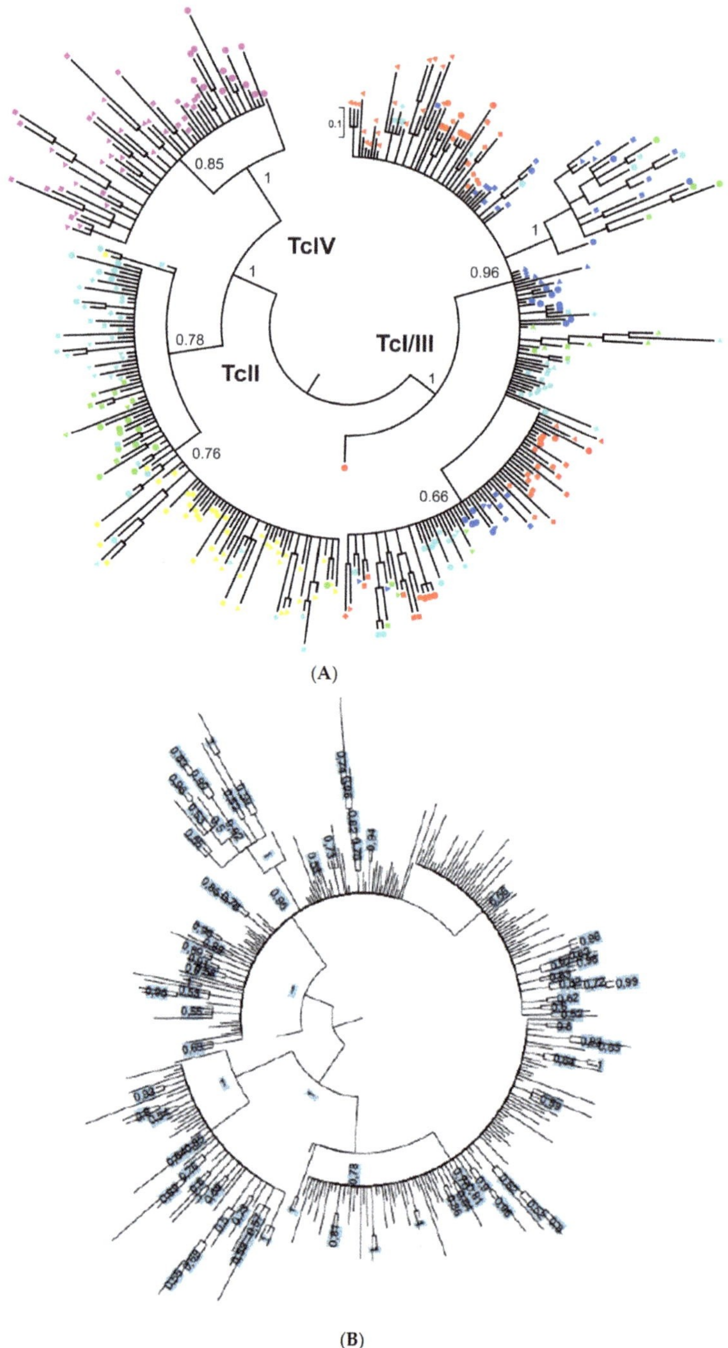

(A)

(B)

Figure 10. (**A**) The analysis by 335 independent satellite DNA sequences of 19 *T. cruzi* strains reveals various lesser near-clades within each of the six *T. cruzi* near-clades (after [25]). (**B**) (Original figure communicated by J.C. Ramírez). The lesser near-clades within each of the six *T. cruzi* near-clades are supported by highly significant bootstrap values (J.C. Ramírez, personal communication).

5. Concluding Remarks

Genomics and high-resolution typing data show that evolutionary patterns at a microevolutionary level (within near-clades) look like a miniature picture of the evolutionary pattern of the full *T. cruzi* species. This is especially well ascertained for the near-clade TcI, for which more data are available. However, data from other near-clades are consistent with this Russian doll pattern [15]. The fact that evolutionary patterns are similar at micro- and macroevolutionary scales suggests that the agent of Chagas disease undergoes progressive, gradual, rather than saltatory, evolution.

The indications for meiosis within TcI in Ecuador [6] undoubtedly constitute a very relevant piece of information about *T. cruzi* evolution. However, this does not challenge the hypothesis of a Russian doll pattern within TcI and the PCE hypothesis in *T. cruzi*. As a matter of fact, the existence of occasional bouts of introgression and hybridization at the level of the whole species [23,26,27] does not challenge PCE in *T. cruzi*, since these occasional events do not break the prevalent PCE pattern (presence of a stable and detectable phylogenetic signal and of near-clades). This maintenance of a detectable phylogenetic signal corroborated by various genetic markers (congruence criterion) corresponds to the "clonality threshold", which is the main trait that specifically gives evidence for PCE [2]. As a matter of fact, beyond this clonality threshold, genetic exchange and recombination are efficiently countered by PCE, and near-clades diverge in an irreversible way. Quite similarly, occasional meiosis events within TcI [6] do not challenge PCE at the within near-clade level, since they do not hamper the persistence of a stable and detectable phylogenetic signal and of lesser near-clades within this near-clade, as clearly evidenced by the many cases exposed in this article (Figures 5–10).

These results show that molecular epidemiology (typing of multilocus genotypes and of lesser near-clades) remains possible within each of the six *T. cruzi* near-clades, since the stability of genotypes is maintained by PCE at this evolutionary level.

It remains to be seen whether genetic clustering and lesser near-clades within each of the six T. *cruzi* near-clades exhibit constant patterns over space and time, in different ecosystems and hosts, and so behave like simili-taxa, a pattern that is observed for example in the yeast *Cryptococcus neoformans* [2].

Author Contributions: Conceptualization, M.T. and F.J.A.; methodology, M.T.; formal analysis, M.T.; investigation, M.T.; resources, M.T.; data curation, M.T.; writing—original draft preparation, M.T.; writing—review and editing, F.J.A. All authors have read and agreed to the published version of the manuscript.

Funding: This research received no external funding.

Acknowledgments: We thank Jenny Telleria (IRD, Montpellier, France) for designing Figure 2.

Conflicts of Interest: The authors declare no conflict of interest.

References

1. Miles, M.A.; Llewellyn, M.S.; Lewis, M.D.; Yeo, M.; Baleela, R.; Fitzpatrick, S.; Gaunt, M.W.; Mauricio, I.L. The molecular epidemiology and phylogeography of *Trypanosoma cruzi* and parallel research on *Leishmania*: Looking back and to the future. *Parasitology* **2009**, *136*, 1509–1528. [CrossRef] [PubMed]
2. Tibayrenc, M.; Ayala, F.J. Is predominant clonal evolution a common evolutionary adaptation to parasitisms in parasitic protozoa, fungi, bacteria and viruses? *Adv. Parasitol.* **2017**, *96*, 243–325.
3. Tibayrenc, M.; Ayala, F.J. Relevant units of analysis for applied and basic research dealing with neglected transmissible diseases: The predominant clonal evolution model of pathogenic microorganisms. *PLoS Neglect. Trop. Dis.* **2017**, *11*, e0005293. [CrossRef] [PubMed]
4. Hillis, D.M.; Bull, J.J. An Empirical Test of Bootstrapping as a Method for Assessing Confidence in Phylogenetic Analysis. *Syst. Biol.* **1993**, *42*, 182–192. [CrossRef]
5. Messenger, L.A.; Llewellyn, M.S.; Bhattacharyya, T.; Franzén, O.; Lewis, M.D.; Ramírez, J.D.; Carrasco, H.J.; Andersson, B.; Miles, M.A. Multiple Mitochondrial Introgression Events and Heteroplasmy in *Trypanosoma cruzi* Revealed by Maxicircle MLST and Next Generation Sequencing. *PLoS Neglect. Trop. Dis.* **2012**, *6*, e1584. [CrossRef]

6. Schwabl, P.; Imamura, H.; Van den Broeck, F.; Costales, J.A.; Maiguashca-Sánchez, J.; Miles, M.A.; Andersson, B.; Grijalva, M.J.; Llewellyn, M.S. Meiotic sex in Chagas disease parasite *Trypanosoma cruzi*. *Nat. Commun.* **2019**, *10*, 3972. [CrossRef]

7. Tibayrenc, M.; Ayala, F.J. A misleading description of the predominant clonal evolution model in *Trypanosoma cruzi*. *Acta Trop.* **2018**, *187*, 13–14. [CrossRef]

8. Tibayrenc, M.; Ayala, F.J. Reproductive clonality of pathogens: A perspective on pathogenic viruses, bacteria, fungi, and parasitic protozoa. *Proc. Nat. Acad. Sci. USA* **2012**, *109*, E3305–E3313. [CrossRef]

9. Miles, M.A.; Souza, A.; Povoa, M.; Shaw, J.J.; Lainson, R.; Toyé, P.J. Isozymic heterogeneity of *Trypanosoma cruzi* in the first autochtonous patients with Chagas'disease in Amazonian Brazil. *Nature* **1978**, *272*, 819–821. [CrossRef]

10. Tibayrenc, M.; Ward, P.; Moya, A.; Ayala, F.J. Natural populations of *Trypanosoma cruzi*, the agent of Chagas'disease, have a complex multiclonal structure. *Proc. Nat. Acad. Sci. USA* **1986**, *83*, 115–119. [CrossRef]

11. Brisse, S.; Barnabé, C.; Tibayrenc, M. Identification of six *Trypanosoma cruzi* phylogenetic lineages by random amplified polymorphic DNA and multilocus enzyme electrophoresis. *Int. J. Parasitol.* **2000**, *30*, 35–44. [CrossRef]

12. Zingales, B.; Miles, M.A.; Campbell, D.A.; Tibayrenc, M.; Macedo, A.M.; Teixeira, M.M.; Schijman, A.G.; Llewellyn, M.S.; Lages-Silva, E.; Machado, C.R.; et al. The revised *Trypanosoma cruzi* subspecific nomenclature: Rationale, epidemiological relevance and research applications. *Infect. Genet. Evol.* **2012**, *12*, 240–253. [CrossRef] [PubMed]

13. Rougeron, V.; De Meeûs, T.; Kako Ouraga, S.; Hide, M.; Bañuls, A.L. "Everything You Always Wanted to Know about Sex (but Were Afraid to Ask)" in *Leishmania* after Two Decades of Laboratory and Field Analyses. *PLoS Pathog.* **2010**, *6*, 1–4. [CrossRef] [PubMed]

14. Ramírez, J.D.; Llewellyn, M.S. Response to Tibayrenc and Ayala: Reproductive clonality in protozoan pathogens–truth or artefact? *Mol. Ecol.* **2015**, *24*, 5782–5784. [CrossRef]

15. Tibayrenc, M.; Ayala, F.J. How clonal are *Trypanosoma* and *Leishmania*? *Trends Parasitol.* **2013**, *29*, 264–269. [CrossRef]

16. Tibayrenc, M. Population Genetics of Parasitic Protozoa and other Microorganisms. *Adv. Parasitol.* **1995**, *36*, 47–115.

17. Van den Broeck, F.; Tavernier, L.J.M.; Vermeiren, L.; Dujardin, J.C.; Van Den Abbeele, J. Mitonuclear genomics challenges the theory of clonality in *Trypanosoma congolense*: Reply to Tibayrenc and Ayala. *Mol. Ecol.* **2018**, *27*, 3425–3431. [CrossRef]

18. Maynard Smith, J.; Smith, N.H.; O'Rourke, M.; Spratt, B.G. How clonal are bacteria? *Proc. Natl. Acad. Sci. USA* **1993**, *90*, 4384–4388. [CrossRef]

19. Hauck, S.; Maiden, M.C. Clonally Evolving Pathogenic Bacteria. In *Molecular Mechanisms of Microbial Evolution*; Rampelotto, P.H., Ed.; Grand Challenges in Biology and Biotechnology Springer International Publishing AG, Part of Springer Nature: Berlin, Germany, 2018; pp. 307–325.

20. Lima, V.S.; Jansen, A.M.; Messenger, L.A.; Miles, M.A.; Llewellyn, M.S. Wild *Trypanosoma cruzi* I genetic diversity in Brazil suggests admixture and disturbance in parasite populations from the Atlantic Forest region. *Parasite Vector* **2014**, *7*, 263. [CrossRef]

21. Roman, F.; das Chagas Xavier, S.; Messenger, L.A.; Pavan, M.G.; Miles, M.A.; Jansen, A.M.; Yeo, M. Dissecting the phyloepidemiology of *Trypanosoma cruzi* I (TcI) in Brazil by the use of high resolution genetic markers. *PLoS Neglect. Trop. D* **2018**, *12*, e0006466. [CrossRef]

22. Segovia, M.; Carrasco, H.J.; Martínez, C.E.; Messenger, L.A.; Nessi, A.; Londoño, J.C.; Espinosa, R.; Martínez, C.; Alfredo, M.; Bonfante-Cabarcas, R.; et al. Molecular Epidemiologic Source Tracking of Orally Transmitted Chagas Disease, Venezuela. *Emerg. Infect. Dis.* **2013**, *19*, 1098–1101. [CrossRef] [PubMed]

23. Messenger, L.A.; Garcia, L.; Vanhove, M.; Huaranca, C.; Bustamante, M.; Torrico, M.; Torrico, F.; Miles, M.A.; Llewellyn, M.S. Ecological host fitting of *Trypanosoma cruzi* TcI in Bolivia: Mosaic population structure, hybridization and a role for humans in Andean parasite dispersal. *Mol. Ecol.* **2015**, *24*, 2406–2422. [CrossRef] [PubMed]

24. Reis-Cunha, J.L.; Baptista, R.P.; Rodrigues-Luiz, G.F.; Coqueiro-dos-Santos, A.; Valdivia, H.O.; de Almeida, L.V.; Cardoso, M.S.; D'Ávila, D.A.; Dias, F.H.C.; Fujiwara, R.T.; et al. Whole genome sequencing of *Trypanosoma cruzi* field isolates reveals extensive genomic variability and complex aneuploidy patterns within TcII DTU. *BMC Genom.* **2018**, *191*, 816. [CrossRef] [PubMed]

25. Ramírez, J.C.; Torres, C.; Curto, M.; Schijman, A.G. New insights into *Trypanosoma cruzi* evolution, genotyping and molecular diagnostics from satellite DNA sequence analysis. *PLoS Neglect. Trop. D* **2017**, *11*, e0006139. [CrossRef] [PubMed]

26. Brisse, S.; Henriksson, J.; Barnabé, C.; Douzery, E.J.; Berkvens, D.; Serrano, M.; De Carvalho, M.R.C.; Buck, G.A.; Dujardin, J.C.; Tibayrenc, M. Evidence for genetic exchange and hybridization in *Trypanosoma cruzi* based on nucleotide sequences and molecular karyotype. *Infect. Genet Evol.* **2003**, *2*, 173–183. [CrossRef]

27. Westenberger, S.J.; Barnabé, C.; Campbell, D.A.; Sturm, N.R. Two Hybridization Events Define the Population Structure of *Trypanosoma cruzi*. *Genetics* **2005**, *171*, 527–543. [CrossRef]

Article

Endosymbiont Capture, a Repeated Process of Endosymbiont Transfer with Replacement in Trypanosomatids *Angomonas* spp.

Tomáš Skalický [1,†], João M. P. Alves [2,†], Anderson C. Morais [2], Jana Režnarová [3], Anzhelika Butenko [1,3], Julius Lukeš [1,4], Myrna G. Serrano [5], Gregory A. Buck [5], Marta M. G. Teixeira [2], Erney P. Camargo [2], Mandy Sanders [6], James A. Cotton [6], Vyacheslav Yurchenko [3,7] and Alexei Y. Kostygov [3,8,*]

1. Institute of Parasitology, Biology Centre, Czech Academy of Sciences, 370 05 České Budějovice (Budweis), Czech Republic; Tomas.Skalicky@seznam.cz (T.S.); rolando24@yandex.ru (A.B.); jula@paru.cas.cz (J.L.)
2. Department of Parasitology, Institute of Biomedical Sciences, University of São Paulo, São Paulo 05508-000, Brazil; jotajj@usp.br (J.M.P.A.); acm2911@usp.br (A.C.M.); mmgteix@icb.usp.br (M.M.G.T.); erney@usp.br (E.P.C.)
3. Life Science Research Centre, Faculty of Science, University of Ostrava, 710 00 Ostrava, Czech Republic; janna.krallova@gmail.com (J.R.); vyacheslav.yurchenko@osu.cz (V.Y.)
4. Faculty of Sciences, University of South Bohemia, 370 05 České Budějovice (Budweis), Czech Republic
5. Department of Microbiology and Immunology, Virginia Commonwealth University, Richmond, VA 23298-0678, USA; myrna.serrano@vcuhealth.org (M.G.S.); gregory.buck@vcuhealth.org (G.A.B.)
6. Wellcome Sanger Institute, Wellcome Genome Campus, Hinxton, Cambridge CB10 1SA, UK; mjs@sanger.ac.uk (M.S.); james.cotton@sanger.ac.uk (J.A.C.)
7. Martsinovsky Institute of Medical Parasitology, Sechenov University, 119435 Moscow, Russia
8. Zoological Institute of the Russian Academy of Sciences, 199034 St. Petersburg, Russia
* Correspondence: kostygov@gmail.com
† These authors contributed equally to this work.

Citation: Skalický, T.; Alves, J.M.P.; Morais, A.C.; Režnarová, J.; Butenko, A.; Lukeš, J.; Serrano, M.G.; Buck, G.A.; Teixeira, M.M.G.; Camargo, E.P.; et al. Endosymbiont Capture, a Repeated Process of Endosymbiont Transfer with Replacement in Trypanosomatids *Angomonas* spp.. *Pathogens* **2021**, *10*, 702. https://doi.org/10.3390/pathogens10060702

Academic Editor: Bruno C. Lemaitre

Received: 10 May 2021
Accepted: 1 June 2021
Published: 4 June 2021

Publisher's Note: MDPI stays neutral with regard to jurisdictional claims in published maps and institutional affiliations.

Abstract: Trypanosomatids of the subfamily Strigomonadinae bear permanent intracellular bacterial symbionts acquired by the common ancestor of these flagellates. However, the cospeciation pattern inherent to such relationships was revealed to be broken upon the description of *Angomonas ambiguus*, which is sister to *A. desouzai*, but bears an endosymbiont genetically close to that of *A. deanei*. Based on phylogenetic inferences, it was proposed that the bacterium from *A. deanei* had been horizontally transferred to *A. ambiguus*. Here, we sequenced the bacterial genomes from two *A. ambiguus* isolates, including a new one from Papua New Guinea, and compared them with the published genome of the *A. deanei* endosymbiont, revealing differences below the interspecific level. Our phylogenetic analyses confirmed that the endosymbionts of *A. ambiguus* were obtained from *A. deanei* and, in addition, demonstrated that this occurred more than once. We propose that coinfection of the same blowfly host and the phylogenetic relatedness of the trypanosomatids facilitate such transitions, whereas the drastic difference in the occurrence of the two trypanosomatid species determines the observed direction of this process. This phenomenon is analogous to organelle (mitochondrion/plastid) capture described in multicellular organisms and, thereafter, we name it endosymbiont capture.

Keywords: genome; bacterial endosymbionts; Trypanosomatidae; *Angomonas*

1. Introduction

The flagellates of the family Trypanosomatidae are well known for human pathogens, such as *Trypanosoma brucei*, *T. cruzi*, and various *Leishmania* spp., yet the majority of trypanosomatid genera are intestinal parasites of insects [1]. In the process of adaptation to this omnipresent and extremely diverse group of hosts, trypanosomatids acquired many peculiar features, the study of which illuminated not only the evolution of parasitism in this group, but also the evolutionary strategies of eukaryotes in general [2,3]. One of the most intriguing phenomena is the presence of bacteria in the cytoplasm of some of these flagellates [4]. Such symbiotic relationships originated in trypanosomatids several times independently and range from recently established and unstable ones to those that demonstrate a high level of integration [5–8]. Mutualistic nature of these endosymbioses is demonstrated by the metabolic cooperation between the bacteria and their trypanosomatid hosts, removing the dependence of the latter on the environmental availability of essential nutrients, such as heme, some amino acids, and vitamins [9–12].

The first discovered and, consequently, most studied group of endosymbiont-bearing trypanosomatids is the subfamily Strigomonadinae, comprising seven described species of the genera *Angomonas*, *Strigomonas*, and *Kentomonas* [7,13]. All of these species have intracytoplasmic bacteria *Candidatus* Kinetoplastibacterium spp. belonging to the family Alcaligenaceae (Betaproteobacteria: Burkholderiales), and, as judged by their respective phylogenies, the origin of the endosymbiosis was a single event followed by a prolonged coevolution [14]. However, the description of *Angomonas ambiguus* revealed a violation of the co-speciation pattern: being a sister species to *A. desouzai*, this flagellate contained an endosymbiont not discernible from that of *Angomonas deanei* by the sequences of the 16S ribosomal RNA gene and the internal transcribed spacer [7]. This discrepancy was reflected in the name of the described trypanosomatid (meaning "ambiguous" in Latin). The endosymbionts of both *A. deanei* and *A. ambiguus* were classified into a single species, *Ca.* Kinetoplastibacterium crithidii [7]. When the same discordance was later shown in the phylogenies of trypanosomatids and their endosymbionts based on the glyceraldehyde 3-phosphate dehydrogenase (GAPDH) gene, it was proposed that *A. ambiguus* obtained its endosymbiont from *A. deanei* by horizontal transfer [15].

In this work, we address these complex evolutionary relationships by analyzing the genomic sequences of two strains of *A. ambiguus* and their respective endosymbionts from geographically distant locations (Brazil and Papua New Guinea) using comparative genomic and phylogenetic tools. Our results not only confirm the transition of bacteria between the two *Angomonas* species, but also demonstrate that this was not a singular event.

2. Results

2.1. Genomic Sequences

The assemblies for the trypanosomatid hosts of the strains TCC2435 and PNG-M02 consisted of 7753 (N50 = 22.5 kb) and 1740 contigs (N50 = 133.9 kb), with total lengths of 21.2 Mb and 23.7 Mb being similar to those of *Angomonas* spp. genomes (21–24 Mb) sequenced previously [16,17].

The genome assembly for the endosymbiont of *Angomonas ambiguus* TCC2435 (hereafter referred to as TCC2435 symbiont) contained 14 contigs with the total size of 803,474 bp and N50 of 126 kb. However, the contigs 8 and 12, comprising the ribosomal operon (~5.6 kb) and the EF-Tu gene (~1.2 kb), respectively, displayed a significantly higher coverage (Table S1) suggesting that they were present in more than one copy. Given that in the genomes of *Ca.* Kinetoplastibacterium spp. the first sequence invariantly has 3 copies and the second one has 2 (except for the very divergent *Ca.* K. sorsogonicusi), we estimate that the actual genome size should be bigger by at least 12.4 kb, i.e., ~816 kb. A similar genome length was obtained for *Ca.* K. crithidii from *A. ambiguus* PNG-M02 (hereafter referred to as PNG-M02 symbiont), the assembly of which contained a single scaffold of 816,901 bp. These values are smaller than that for the genome of the endosymbiont of *A. deanei* TCC036E (821,930 bp; hereafter referred to as TCC036E symbiont) used here as a reference, but are within the known size range for the genomes of bacteria from *Angomonas* spp. and *Strigomonas* spp. (810–830 kb) [16,17].

The GC content of the genomes of the PNG-M02 and ATCC2435 symbionts was 30.33% and 30.65%, respectively. These values are very close to those for the genomes of *Ca.* K. crithidii ATCC036E (30.96%) and the symbionts from *Strigomonas* spp. (31.23–32.55%) [16]. Similarly to other *Ca.* Kinetoplastibacterium spp. [16,18] and bacterial endosymbionts in general [19,20], *Ca.* K. crithidii from *A. deanei* and the two *A. ambiguus* strains showed a very high level of gene order conservation with no detectable rearrangements (Figure 1 and Figure S1).

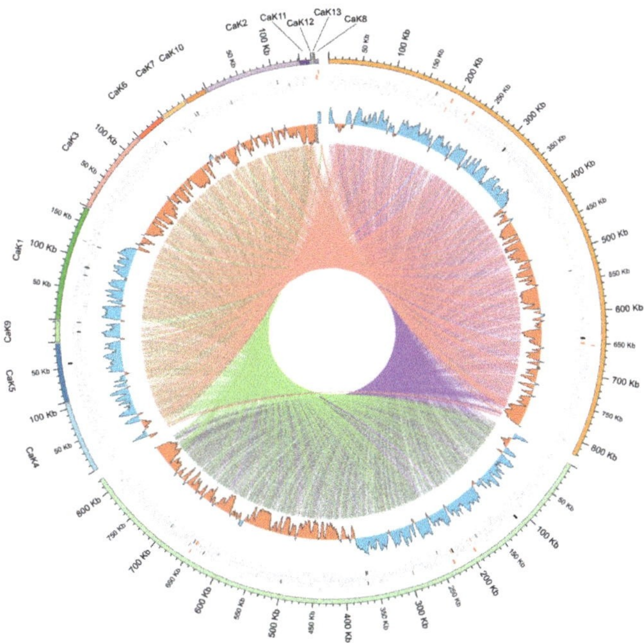

Figure 1. Comparison of the genomes of three *Ca.* K. crithidii strains. The rings in the outside-in direction mean: (i) genomic coordinates of scaffolds; (ii) predicted genes (protein-coding in grey, rRNA in red, tRNA in blue, tmRNA in orange, ncRNA in green, and pseudogenes in black); (iii) GC skew plot (negative values in red and positive ones in blue). The lines in the central area connect orthologous genes between the genomes in a pairwise manner.

The overall genome sequence identity in the TCC2435/TCC036E, TCC2435/PNG-M02, and PNG-M02/TCC036E pairs was 90.8%, 90.4%, and 90.3%, respectively. These values are much higher than the interspecific similarity between the genomes of *Strigomonas* spp. symbionts (83–85%) or *Ca.* K. crithidii and *Ca.* K. desouzaii (73%) [16]. In agreement with the smaller size, the two bacterial genomes studied here were predicted to code for slightly smaller numbers of proteins: 729 and 726 for TCC2435 and PNG-M02 symbionts, respectively, as compared to 733 for the TCC036E symbiont (Table S2). However, the number of annotated pseudogenes in the two newly sequenced genomes was higher, with most of such sequences being frameshifted (Table S2). The distribution of the pseudogenes did not show any hotspots (Figure 1). Only 39 tRNA genes were predicted in the TCC2435 symbiont genome (which may be due to assembly fragmentation), whereas the genomes of PNG-M02 and TCC036E symbionts featured 43 and 44 such genes, respectively. The inspection of the tRNA lists for the three genomes revealed that they all differed from each other, but the differences consisted only in the number of redundant tRNAs, i.e., those with the same anticodon (Table S3).

2.2. Analysis of Orthologous Groups (OGs) of Proteins

Only minor differences in gene content were revealed between the three analyzed endosymbiont genomes (Figure 2). The number of OGs present or absent only in one of the three genomes negatively correlated with the assembly quality, suggesting that at least some of the differences may be artifactual. Thus, the genome of the PNG-M02 symbiont assembled to a single contig based on PacBio and Illumina reads displayed the lowest numbers, whereas those for the fragmented assembly of the TCC2435 symbiont were the highest (Figure 2).

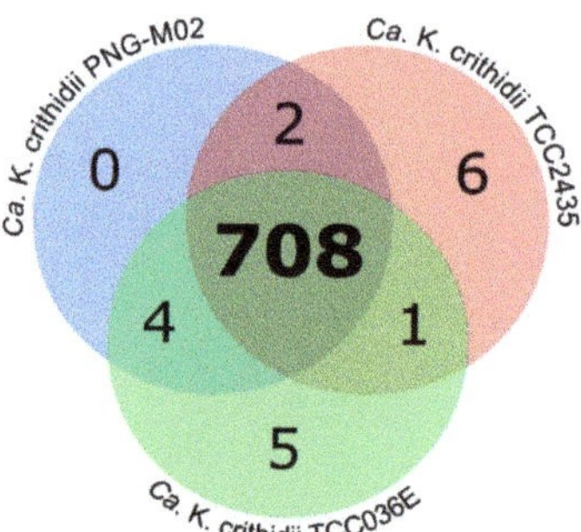

Figure 2. Sharing of orthologous groups of proteins encoded in the genomes of the three *Ca.* K. crithidii strains.

A detailed inspection of the "unique" genes revealed that most of them either represent pseudogenes with a degraded sequence, which leads to clustering them into separate OGs, or potentially spurious short ORFs with no BLAST hits in NCBI nr database (Table S4). After exclusion of annotated or suspected pseudogenes and sequences with no BLAST hits, only two "unique" genes remained, both in the TCC036E symbiont genome: a helix-turn-helix domain-containing protein and tetraacyldisaccharide 4′-kinase. Each of these two genes is present (but not invariably) in other *Ca.* Kinetoplastibacterium spp., suggesting their dispensability. The first one, appearing to be a transcription factor (based on blast results), is absent from the genomes of endosymbionts of all *Strigomonas* spp. The second gene codes for an enzyme phosphorylating a precursor of lipopolysaccharide (component of the outer membrane) and is absent from the genomes of *Ca.* K. galatii and *Ca.* K. oncopelti. This agrees with the previous observation that the functional category "cell wall, membrane, and envelope biogenesis" is overrepresented among lost and pseudogenized genes in the genomes of Strigomonadinae symbionts [16]. Similar results were obtained after the inspection of the OGs missing from one of the three genomes: most of them were associated with the synthesis of the cell wall or lipopolysaccharide (Table S4). In addition, the ribosome-associated translation inhibitor RaiA (also absent from the genomes of endosymbionts of all *Strigomonas* spp.) was not detected in the TCC036E symbiont genome, and a short hypothetical protein was absent from the genome of TCC2435 symbiont, although a potential homolog could be detected with an increased e-value threshold (Table S4).

2.3. Phylogenetic Analyses

For each of the two phylogenomic datasets used (431 and 1549 single-copy genes for bacteria and trypanosomatids, respectively), maximum-likelihood and Bayesian trees showed identical topology with all branches or all but one bearing maximal statistical supports (Figure 3). In accordance with the previous inferences, *Kentomonas sorsogonicus* represents here the earliest branch within the subfamily Strigomonadinae [13], whereas its bacterium, *Ca.* K. sorsogonicusi, occupies the same position among the endosymbionts of this trypanosomatid subfamily [18]. The relationships within the genus *Strigomonas* and their respective endosymbionts are also correlated, suggesting cospeciation of these two groups of organisms. The situation is different for the third genus of Strigomonadinae: although *Angomonas ambiguus* and *A. desouzai* represent sister taxa, the bacteria hosted by the former species are paraphyletic in respect to that of *A. deanei* (Figure 3). This suggests a single horizontal endosymbiont transfer from *A. ambiguus* to *A. deanei*, in contrast to the previous proposal that the transfer had the opposite direction [15]. The alternative explanation of this figure implies two independent endosymbiont switches from *A. deanei* to *A. ambiguus* and is less parsimonious.

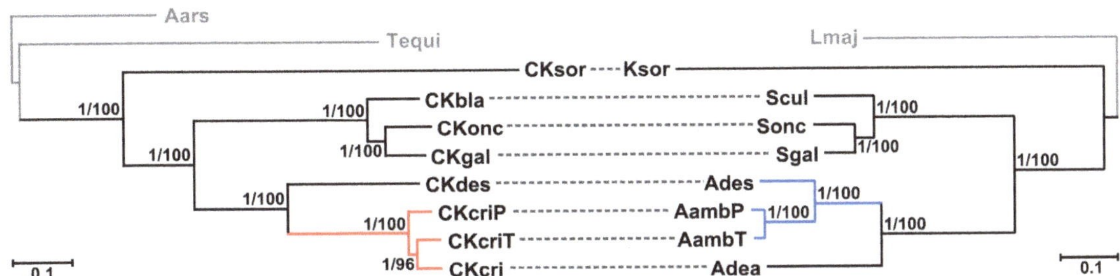

Figure 3. Juxtaposed maximum-likelihood phylogenomic trees of endosymbiotic bacteria and their respective trypanosomatid hosts. Outgroups are shown in grey. Dashed lines connect endosymbionts with their hosts, while colored branches point to the discrepancy between their phylogenies. Numbers at branches indicate bootstrap support and Bayesian posterior probability values, respectively. Scale bars show the number of substitutions per site. Organism codes: Lmaj, *Leishmania major*; Ksor, *Kentomonas sorsogonicus*; Scul, *Strigomonas culicis*; Sonc, *S. oncopelti*; Sgal, *S. galati*; Ades, *Angomonas desouzai*; AambP and AambT, *A. ambiguus* strains PNG-M02 and TCC2535, respectively; Adea, *A. deanei*; Aars, *Achromobacter arsenitoxydans*; Tequi, *Taylorella equigenitalis*; CKsor, *Candidatus* Kinetoplastibacterium sorsogonicusi; CKbla, *Ca.* K. blastocrithidii; CKonc, *Ca.* K. oncopeltii; CKgal, *Ca.* K. galatii; CKdes, *Ca.* K. desouzaii; CKcri, CKcriP, and CKcriT, *Ca.* K crithidii TCC036E, PNG-M02, and TCC2435, respectively.

In order to clarify the situation, we performed an additional phylogenetic analysis using GAPDH gene sequences of *Ca.* Kinetoplastibacterium spp. This allowed investigating the relationships of these bacteria on a much larger set of strains, available from a previous study [15]. The phylogenetic trees inferred using maximum likelihood and Bayesian approaches displayed almost identical topologies differing only in the presence of a single very short branch with a very short length (Figure 4). They were congruent with the previously published GAPDH tree [15] and confirmed the unity of the symbionts from *A. deanei* and *A. ambiguus*, representing the same four subclades (Kcr1–Kcr4). As in the previous inference, all sequences of the endosymbionts from *A. ambiguus* from Brazil (isolates TCC1765, TCC1780, and TCC2435) nested within the Kcr3 subclade and displayed 100% identity to some sequences of the endosymbionts from *A. deanei* originating from the same country. However, *Ca.* K. crithidii from the Papuan *A. ambiguus* isolate PNG-M02 represented a separate lineage, sister to the KCr3+Kcr4 group. The identity of its GAPDH sequence to those of other *Ca.* K. crithidii was only ~91%, almost the same as the observed minimum within this bacterial species (90.8%). Interestingly, the endosymbiont of *A. deanei* PNG-M01, obtained from the same host species and the same locality as PNG-M02, was not related to the latter and nested within the KCr3 + Kcr4 group (Figure 4).

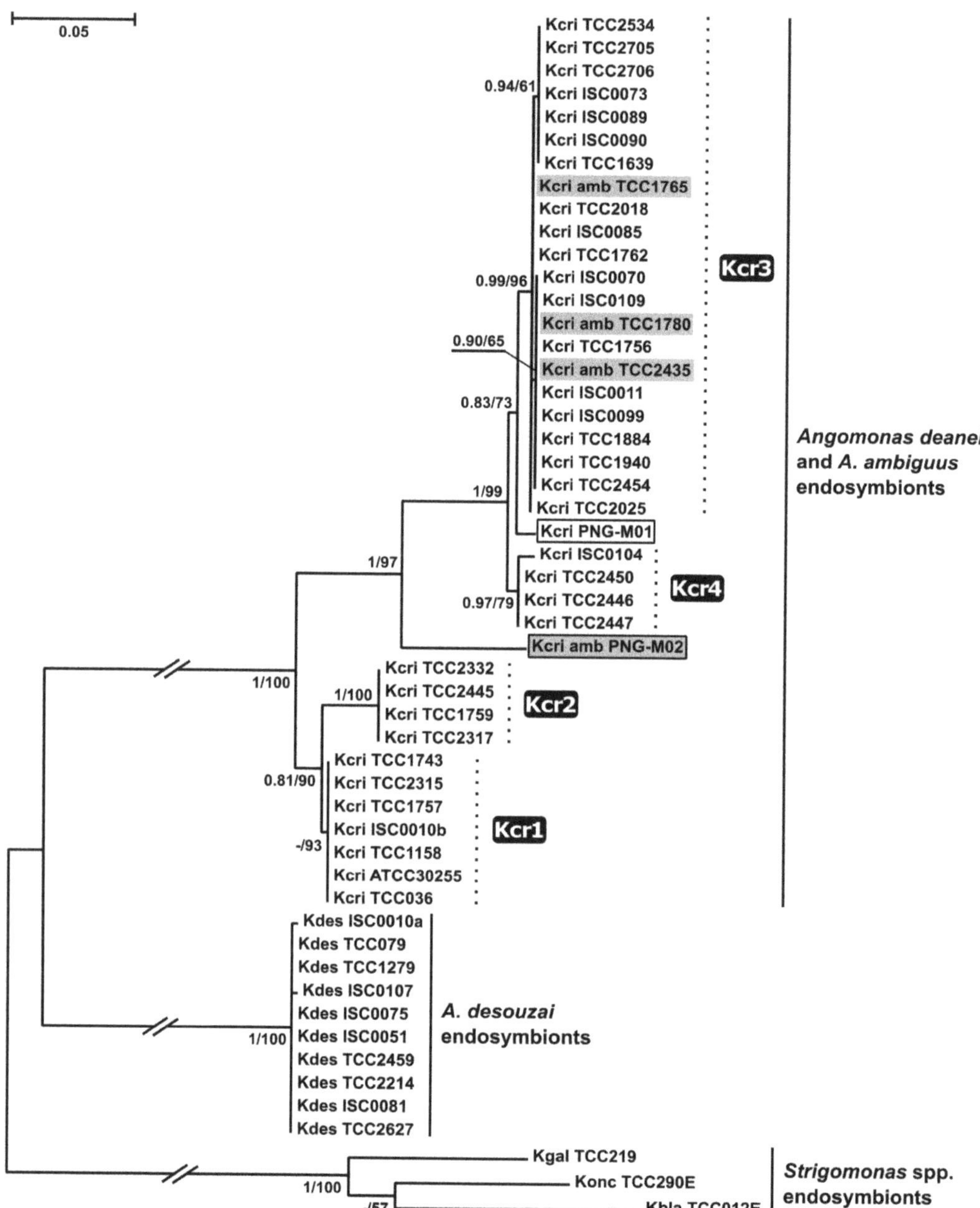

Figure 4. GAPDH-based maximum-likelihood phylogenetic tree of *Ca*. Kinetoplastibacterium spp. The endosymbionts of *Angomonas ambiguus* are highlighted in grey, the isolates from Papua New Guinea are boxed. The labels in black rectangles indicate individual subclades of *Ca*. Kinetoplastibacterium crithidii. Numbers at branches indicate bootstrap supports and Bayesian posterior probabilities, respectively. Scale bar show the number of substitutions per site. The tree is rooted with the sequences of *Strigomonas* spp. endosymbionts.

3. Discussion

Mutualistic endosymbioses of prokaryotes with eukaryotes are quite diverse in terms of involved taxa, time of origin, and level of interdependence, with the latter two factors usually being correlated: evolutionary older relationships demonstrate a higher level of integration [21]. In insects, whose relationships with prokaryotes have been studied quite intensively, symbionts permanently residing in the cytoplasm of the host cells usually display perfect co-evolutionary patterns in contrast to bacteria that do not have such a restriction and, therefore, can switch hosts and/or be replaced by other species [22]. In agreement with this trend, *Ca.* Kinetoplastibacterium spp. also show cospeciation with their trypanosomatid hosts and the only exception concerns the *A. deanei–A. ambiguus* pair, which shares a single endosymbiotic bacterium, *Ca.* K. crithidii. This was first detected using 16S rRNA gene sequences [7] and later confirmed by the analysis of bacterial GAPDH gene sequences [15].

Although being a rare phenomenon, the replacement of permanent endosymbionts is well known in insects [22] and presumably also occurs in ciliates [23,24]. The new bacterium in such a case originates from either a free-living or a facultatively symbiotic species and restores deteriorated functions of the old endosymbiont, whose genome degraded due to Muller's ratchet [25]. The situation with the bacteria of *Angomonas* spp. is drastically different: both of them represent equally ancient endosymbionts and the replacement is combined with horizontal transfer between two related host species. This may appear unprecedented, but only when considering typical bacterial endosymbionts. A remarkable analogy can be found in mitochondria and plastids, the two kinds of organelles with prokaryote ancestry. The organellar capture (replacement of a mitochondrion or plastid of one species by that of another) also known as mitochondrial/plastid introgression (when it refers to genomes) has been described in a wide range of animals and plants and is usually associated with the formation of a hybrid zone between two species [26,27]. These species often have significantly different abundance levels resulting in asymmetrical introgression due to the contrast effects of genetic drift on small and large populations [28]. In general, introgression is driven by the prevalence of interspecific gene flow over the intraspecific one. In the case of mitochondria, this condition is met when dispersal is exerted predominantly by males (in some animals) or pollen (in conifers), not contributing the organelle to the progeny (due to maternal inheritance) and, thus, the intraspecific organellar gene flow for the colonizing species is close to zero [29,30].

Since the outcome of the interspecific interaction between *A. deanei* and *A. ambiguus* is similar to organellar capture, henceforth we will refer to it as endosymbiont capture. In order to understand the mechanism of this phenomenon, we summarize here the available data.

Out of the three *Angomonas* spp. described to date, *A. deanei* has the highest prevalence and the widest (potentially cosmopolitan) distribution. It was documented in various countries of Africa and South America, as well as in Papua New Guinea, Turkey, Czechia, and Russia [15,31,32]. Meanwhile, South America is currently the only known area for *A. desouzai*, whereas *A. ambiguus*, the rarest of the three species, has been also reported from Africa and Papua New Guinea [15,31,32]. All three species occur mostly in blowflies (Calliphoridae), although two clades of *A. deanei* apparently prefer Muscidae [15]. While it is unclear whether the single records of *A. deanei* and *A. desouzai* from Syrphidae [7] represent nonspecific infections, the first isolate of *A. deanei* from the predatory bug *Zelus leucogrammus* [33] undoubtedly is such a case [34].

Here, we sequenced and analyzed the genomes of *Ca.* K. crithidii from two *A. ambiguus* strains and compared them with the previously published genome of the endosymbiont from *A. deanei* TCC036E [16]. The three genomes display very similar sizes and GC content, a high level of nucleotide sequence identity and no significant differences in gene content. Based on these features, the three bacterial endosymbionts can be considered as members of a single species. Previously, the discussion of the discordance in the phylogenies of endosymbionts and their trypanosomatid hosts was based only on data concerning

Brazilian strains, whereas here we also included those from a geographically distant area—Papua New Guinea. Our phylogenomic analysis confirmed the unity of the symbionts from *A. deanei* and *A. ambiguus*, but, due to the small number of included isolates, its results were inconclusive regarding the direction of the endosymbiont transfer. However, the phylogenetic analysis based on the bacterial GAPDH gene sequences allowed taking advantage of a larger *Ca.* K. crithidii sampling. It not only confirmed that the endosymbiont of *A. deanei* was captured by *A. ambiguus* but also demonstrated that this occurred more than once.

With little doubt, the occurrence of *Angomonas* spp. in the same blowfly hosts and the relatedness of the trypanosomatids are the factors that facilitate endosymbiont capture. It was demonstrated that *A. deanei* colonizes the host rectum and forms massive aggregates in the area of rectal papillae [32]. Presumably, upon mixed infections, cells of two different species may come into a close contact and attempt to undergo sexual process. In contrast to multicellular organisms, its successful completion is not required to create a new heritable nucleus-symbiont combination. Of note, a sex-independent (grafting-based) mechanism of chloroplast capture has been proposed for plants [35].

By analogy to organelle capture, the reported very low prevalence of *A. ambiguus* [15] explains the phenomenon to be observed as a unidirectional process with this species being an acceptor. The reason why only *A. deanei* but not *A. desouzai*, being more closely related to *A. ambiguus*, is observed as a donor may be also related to their relative abundance. However, we cannot exclude that this is just due to the small number of *A. ambiguus* strains analyzed to date. Importantly, the endosymbiont capture is a repeated process (there were at least two independent cases) and its incidence may depend on the local demographic situation. The identical GAPDH sequences of *Ca.* K. crithidii of *A. ambiguus* and several *A. deanei* strains from South America indicate a recent event in agreement with the data on the current drastically different prevalence of these two species in that area. Meanwhile, the sequences of this gene in the endosymbionts of the Papuan isolates of both trypanosomatid species obtained from the same population of blowflies were significantly different and positioned distantly on the phylogenetic tree. This might be a result of a relatively ancient endosymbiont capture. Regrettably, for the moment other isolates of these two species from Papua New Guinea and data on their prevalence in that region are not available.

In sum, the replacement of endosymbionts of *Angomonas ambiguus* by those of *Angomonas deanei* is a repeated process analogous to organelle capture described in multicellular organisms and apparently shares with the latter one of the underlying mechanisms.

4. Materials and Methods

4.1. Trypanosomatid Strains: Origin and Cultivation

In this work, two axenically cultivated strains of *Angomonas ambiguus* were used: (i) PNG-M02 from the blowfly *Chrysomya megacephala* collected in Nagada, Papua New Guinea [31]; and (ii) TCC2435 representing a clonal culture of TCC1780 isolated from *C. albiceps* in Campo Grande, Brazil [7]. The cultures were maintained at 27 °C in RPMI 1640 cultivation medium at pH 7.0 supplemented with 10% (*v/v*) fetal calf serum, 10 μg/mL of hemin, 100 units/mL of penicillin, and 100 μg/mL of streptomycin. In addition to cultures, DNA of the non-cultivated *A. deanei* strain PNG-M01 (from the same host species and location as PNG-M02) available from an earlier study [31] was used for PCR amplification of the bacterial GAPDH gene.

4.2. Genome Sequencing, Assembly, and Annotation

DNA extraction from both strains of *A. ambiguus* was performed by the classical phenol-chloroform method, without preceding separation of the endosymbiont and trypanosomatid cells. Sequencing of TCC2435 DNA was performed using Roche 454 GS-FLX Titanium (1.37 mln single-ended reads, 550 Mbp), and Illumina MiSeq (13 mln 2 × 250 bp paired-end reads) platforms. DNA of PNG-M02 strain was sequenced at Wellcome Sanger Institute using Illumina MiSeq and HiSeq 2500 technologies (2,4 mln 2 × 250 bp and

14.7 mln 2 × 125 paired-end reads, respectively) as well as PacBio RS II sequencing system (13,977 long reads, 321 Mbp). The corresponding raw reads are available from Gen-Bank under the following accession numbers: ERS4809514 (PNG-M02) and SRR14208463, SRR14216068, SRR14216074, and SRR14209298 (TCC2435).

After processing the raw reads generated by Illumina and 454 platforms with Trimmomatic V. 0.39 [36] and those from the PacBio system with SMRT Analysis Suite (Pacific BioSciences, Menlo Park, CA, USA), the data quality was assessed using the FastQC v. 0.11.9 software (http://www.bioinformatics.babraham.ac.uk/projects/fastqc, accessed on 30 April 2020). Since the BLAST search against available genomic sequences of trypanosomatids revealed that the PNG-M02 sample was contaminated with DNA from *Crithidia fasciculata*, the data were filtered by mapping the preprocessed reads to the *C. fasciculata* genome Cf-C1 (TritrypDB v. 40) using BBmap v. 38.84 with the settings recommended for contaminant reads removal (http://sourceforge.net/projects/bbmap/, accessed on 6 May 2020). The genomic assembly for the PNG-M02 strain was made with hybridSPAdes v. 3.14.1 [37] using both Illumina and error corrected PacBio reads. The endosymbiont genome was identified using blastn and the closest known endosymbiont genome *Ca.* K. crithidii TCC036E. Two different assemblies were made for the strain TCC2435 with Newbler v. 2.7: (i) trypanosomatid-focused using only Illumina data (with "-large" option); and (ii) endosymbiont-focused using both 454 and Illumina reads. Genes were predicted with Companion [38] and Glimmer v. 3 [39] for the bacteria and their hosts, respectively. Gene annotation for the endosymbionts was performed with PROKKA v. 1.14.5 [40]. The assembled genome sequences have been deposited in GenBank under the Bioproject accession PRJNA673871.

4.3. Synteny Analysis of Bacterial Genomes

The single-scaffold genomic sequences of *Ca.* K. crithidii TCC036E (GCA_000340825.1) and *Ca.* K. crithidii PNG-M02 were circularized using Circlator v. 1.5.5 [41] and the dnaA gene was selected as a start in their linear representation. The scaffolds of *Ca.* K. crithidii TCC2435 genome were reordered and inverted to match the two abovementioned ones, following tripartite genome alignment and synteny analysis using Mauve v. 2015-02-13 [42]. Visualization of genomic alignment was prepared with Circos v. 0.69-9 [43].

4.4. Phylogenomic Analyses

Analyses of protein OGs were performed with OrthoFinder v. 2.3.11 [44] with the default settings. In addition to the sequences obtained in this work, the bacterial dataset (BD) comprised the previously published genomes of all six *Ca.* Kinetoplastibacterium spp. as well as those of *Achromobacter arsenitoxydans* and *Taylorella equigenitalis*, which were used here as outgroups (Table S5). The trypanosomatid dataset (TD) included previously published genomic sequences for Strigomonadinae and *L. major* Friedlin, which served as an outgroup (Table S5), as well as the generated earlier draft sequence of *Kentomonas sorsogonicus* [18]. Out of the total 1645 (BD) and 17,990 (TD) inferred protein OGs, 431 and 1549, respectively, included one protein per species and were used for the subsequent phylogenomic analyses. The amino acid sequences were aligned using Muscle v. 3.8.31 [45], trimmed with Gblocks v. 0.91b [46] and concatenated with FASconCAT-G v.1.04 [47]. The resulting supermatrices contained 133,474 (BD) and 658,788 (ED) positions with respective gap proportions of 0.4% and 5%. Maximum likelihood analyses were performed in RAxML v.8.2.11 [48] with automated selection of the substitution schemes for the partitioned model, linked edge lengths, and 100 bootstrap pseudoreplicates for branch support estimation. Bayesian inference was performed in MrBayes v. 3.2.6 [49] with "mixed" prior for amino acid substitution matrix and rate heterogeneity modelled using 4 discrete Γ-categories. Relative rates, substitution models, and Γ-distribution shape were unlinked across partitions. The analysis was run for 1,000,000 generations with every 100th tree sampled, and other parameters set by default.

4.5. Amplification and Phylogenetic Analysis of GAPDH Gene

The bacterial GAPDH gene of the strain PNG-M01 was amplified and sequenced using the newly designed primers KAGF1 (5′-ATTTTAAGAGCTCATTACGAAGGT-3′) and KAGR1 (5′-GATCTTGCCCTACGCAAATC-3′). The obtained sequence was deposited in GenBank under the accession number MW161049. Other sequences of this gene from the endosymbionts of *Angomonas* and *Strigomonas* available in the GenBank were collected (Table S6) and aligned with MAFFT v. 7.471 using L-INS-I algorithm [50]. Maximum Likelihood analysis was accomplished in IQ-TREE v. 2.0.5 [51] with the best substitution model (TIM2 + F + I) selected by the built-in ModelFinder [52]. The statistical support of branches was estimated by the standard bootstrap method with 1000 pseudoreplicates. Bayesian inference was accomplished in MrBayes v. 3.2.7 under the GTR + I model with run parameters as described above.

Supplementary Materials: The following are available online at https://www.mdpi.com/article/10.3390/pathogens10060702/s1, Figure S1: MAUVE alignment of three genomes of *Ca.* Kinetoplastibacterium spp., Table S1: Read coverage of the genomic contigs of the TCC2435 symbiont, Table S2: Summary statistics on the genes encoded in the three *Ca.* K. crithidii genomes, Table S3: Comparison of lists of tRNA genes in the genomes of endosymbionts from *Angomonas* spp., Table S4: Presence and absence of OGs in the genomes of *Ca.* K. crithidii, Table S5: Previously published genomic sequences used in the phylogenomic analyses, Table S6: GAPDH gene sequences from *Ca.* Kinetoplastibacterium sequences retrieved from GenBank for phylogenetic analysis.

Author Contributions: Conceptualization, J.M.P.A. and A.Y.K.; methodology, A.B., G.A.B., J.A.C., J.M.P.A., and A.Y.K.; validation, A.B., G.A.B., J.A.C., J.M.P.A., and A.Y.K.; formal analysis, T.S., A.B., J.M.P.A., and A.Y.K.; investigation, T.S., A.C.M., J.R., M.G.S., M.S., J.M.P.A., and A.Y.K.; resources, G.A.B., J.A.C., J.L., M.M.G.T., E.P.C., V.Y.; data curation, T.S., A.B., and J.M.P.A.; writing—original draft preparation, J.M.P.A. and A.Y.K.; writing—review and editing, V.Y., J.L., J.M.P.A., and A.Y.K.; visualization, T.S., J.M.P.A., and A.Y.K.; supervision, V.Y., J.L., J.M.P.A., and A.Y.K.; funding acquisition, V.Y., J.L., G.A.B., J.A.C., J.M.P.A., and A.Y.K. All authors have read and agreed to the published version of the manuscript.

Funding: This research was supported by the European Regional Funds project "Centre for Research of Pathogenicity and Virulence of Parasites" CZ.02.1.01/16_019/0000759 to V.Y., A.Y.K., J.R., and J.L., the Grant Agency of the Czech Republic (grant 20-07186S) to V.Y., J.L., and A.Y.K., State Assignment AAAA-A19-119031390116-9 for ZIN RAS to A.Y.K., São Paulo Research Foundation (FAPESP) grants 2016/07487-0 (to M.M.G.T and E.P.C.) and 2013/14622-3 (to J.M.P.A.), CNPq fellowship to A.C.M., Wellcome Trust grant 206194 for M.S. and J.A.C., National Science Foundation's Assembling the Tree of Life program to G.A.B. and M.G.S., and a grant from the University of Ostrava SGS/PrF/2021 for J.R. Computational resources were provided to T.S. by the project "e-Infrastruktura CZ" (e-INFRA LM2018140) within the program Projects of Large Research, Development and Innovations Infrastructures and ELIXIR-CZ project (LM2018131), part of the international ELIXIR infrastructure.

Institutional Review Board Statement: Not applicable.

Informed Consent Statement: Not applicable.

Data Availability Statement: The data used in this study are publicly available from GenBank under the following accession numbers: ERS4809514 (PNG-M02 raw reads), SRR14208463, SRR14216068, SRR14216074, and SRR14209298 (TCC2435 raw reads), Bioproject PRJNA673871 (generated assemblies).

Acknowledgments: We thank the members of our laboratories for the discussion of our results.

Conflicts of Interest: The authors declare no conflict of interest. The funders had no role in the design of the study; in the collection, analyses, or interpretation of data; in the writing of the manuscript, or in the decision to publish the results.

References

1. Kostygov, A.Y.; Karnkowska, A.; Votýpka, J.; Tashyreva, D.; Maciszewski, K.; Yurchenko, V.; Lukeš, J. Euglenozoa: Taxonomy, diversity and ecology, symbioses and viruses. *Open Biol.* **2021**, *11*, 200407. [CrossRef]
2. Lukeš, J.; Butenko, A.; Hashimi, H.; Maslov, D.A.; Votýpka, J.; Yurchenko, V. Trypanosomatids are much more than just trypanosomes: Clues from the expanded family tree. *Trends Parasitol.* **2018**, *34*, 466–480. [CrossRef]
3. Butenko, A.; Hammond, M.; Field, M.C.; Ginger, M.L.; Yurchenko, V.; Lukeš, J. Reductionist pathways for parasitism in euglenozoans? Expanded datasets provide new insights. *Trends Parasitol.* **2021**, *37*, 100–116. [CrossRef] [PubMed]
4. Maslov, D.A.; Opperdoes, F.R.; Kostygov, A.Y.; Hashimi, H.; Lukeš, J.; Yurchenko, V. Recent advances in trypanosomatid research: Genome organization, expression, metabolism, taxonomy and evolution. *Parasitology* **2019**, *146*, 1–27. [CrossRef]
5. Ganyukova, A.I.; Frolov, A.O.; Malysheva, M.N.; Spodareva, V.V.; Yurchenko, V.; Kostygov, A.Y. A novel endosymbiont-containing trypanosomatid *Phytomonas borealis* sp. n. from the predatory bug *Picromerus bidens* (Heteroptera: Pentatomidae). *Folia Parasitol.* **2020**, *67*, 4. [CrossRef] [PubMed]
6. Kostygov, A.Y.; Dobáková, E.; Grybchuk-Ieremenko, A.; Váhala, D.; Maslov, D.A.; Votýpka, J.; Lukeš, J.; Yurchenko, V. Novel trypanosomatid-bacterium association: Evolution of endosymbiosis in action. *mBio* **2016**, *7*, e01985-15. [CrossRef]
7. Teixeira, M.M.; Borghesan, T.C.; Ferreira, R.C.; Santos, M.A.; Takata, C.S.; Campaner, M.; Nunes, V.L.; Milder, R.V.; de Souza, W.; Camargo, E.P. Phylogenetic validation of the genera Angomonas and Strigomonas of trypanosomatids harboring bacterial endosymbionts with the description of new species of trypanosomatids and of proteobacterial symbionts. *Protist* **2011**, *162*, 503–524. [CrossRef] [PubMed]
8. Catta-Preta, C.M.; Brum, F.L.; da Silva, C.C.; Zuma, A.A.; Elias, M.C.; de Souza, W.; Schenkman, S.; Motta, M.C. Endosymbiosis in trypanosomatid protozoa: The bacterium division is controlled during the host cell cycle. *Front. Microbiol.* **2015**, *6*, 520. [CrossRef] [PubMed]
9. Kostygov, A.Y.; Butenko, A.; Nenarokova, A.; Tashyreva, D.; Flegontov, P.; Lukeš, J.; Yurchenko, V. Genome of *Ca. pandoraea novymonadis*, an endosymbiotic bacterium of the trypanosomatid *Novymonas esmeraldas*. *Front. Microbiol.* **2017**, *8*, 1940. [CrossRef] [PubMed]
10. Klein, C.C.; Alves, J.M.; Serrano, M.G.; Buck, G.A.; Vasconcelos, A.T.; Sagot, M.F.; Teixeira, M.M.; Camargo, E.P.; Motta, M.C. Biosynthesis of vitamins and cofactors in bacterium-harbouring trypanosomatids depends on the symbiotic association as revealed by genomic analyses. *PLoS ONE* **2013**, *8*, e79786. [CrossRef]
11. Alves, J.M.; Klein, C.C.; da Silva, F.M.; Costa-Martins, A.G.; Serrano, M.G.; Buck, G.A.; Vasconcelos, A.T.; Sagot, M.F.; Teixeira, M.M.; Motta, M.C.; et al. Endosymbiosis in trypanosomatids: The genomic cooperation between bacterium and host in the synthesis of essential amino acids is heavily influenced by multiple horizontal gene transfers. *BMC Evol. Biol.* **2013**, *13*, 190. [CrossRef]
12. Alves, J.M.; Voegtly, L.; Matveyev, A.V.; Lara, A.M.; da Silva, F.M.; Serrano, M.G.; Buck, G.A.; Teixeira, M.M.; Camargo, E.P. Identification and phylogenetic analysis of heme synthesis genes in trypanosomatids and their bacterial endosymbionts. *PLoS ONE* **2011**, *6*, e23518. [CrossRef] [PubMed]
13. Votýpka, J.; Kostygov, A.Y.; Kraeva, N.; Grybchuk-Ieremenko, A.; Tesařová, M.; Grybchuk, D.; Lukeš, J.; Yurchenko, V. *Kentomonas* gen. n., a new genus of endosymbiont-containing trypanosomatids of Strigomonadinae subfam. n. *Protist* **2014**, *165*, 825–838. [CrossRef] [PubMed]
14. Du, Y.; Maslov, D.A.; Chang, K.P. Monophyletic origin of beta-division proteobacterial endosymbionts and their coevolution with insect trypanosomatid protozoa *Blastocrithidia culicis* and *Crithidia* spp. *Proc. Natl. Acad. Sci. USA* **1994**, *91*, 8437–8441. [CrossRef]
15. Borghesan, T.C.; Campaner, M.; Matsumoto, T.E.; Espinosa, O.A.; Razafindranaivo, V.; Paiva, F.; Carranza, J.C.; Añez, N.; Neves, L.; Teixeira, M.M.G.; et al. Genetic diversity and phylogenetic relationships of coevolving symbiont-harboring insect trypanosomatids, and their Neotropical dispersal by invader African blowflies (Calliphoridae). *Front. Microbiol.* **2018**, *9*, 131. [CrossRef] [PubMed]
16. Alves, J.M.; Serrano, M.G.; Maia da Silva, F.; Voegtly, L.J.; Matveyev, A.V.; Teixeira, M.M.; Camargo, E.P.; Buck, G.A. Genome evolution and phylogenomic analysis of *Candidatus kinetoplastibacterium*, the betaproteobacterial endosymbionts of Strigomonas and Angomonas. *Genome Biol. Evol.* **2013**, *5*, 338–350. [CrossRef]
17. Motta, M.C.; Martins, A.C.; de Souza, S.S.; Catta-Preta, C.M.; Silva, R.; Klein, C.C.; de Almeida, L.G.; de Lima Cunha, O.; Ciapina, L.P.; Brocchi, M.; et al. Predicting the proteins of *Angomonas deanei*, *Strigomonas culicis* and their respective endosymbionts reveals new aspects of the trypanosomatidae family. *PLoS ONE* **2013**, *8*, e60209. [CrossRef]
18. Silva, F.M.; Kostygov, A.Y.; Spodareva, V.V.; Butenko, A.; Tossou, R.; Lukeš, J.; Yurchenko, V.; Alves, J.M.P. The reduced genome of *Candidatus* Kinetoplastibacterium sorsogonicusi, the endosymbiont of *Kentomonas sorsogonicus* (Trypanosomatidae): Loss of the haem-synthesis pathway. *Parasitology* **2018**, *145*, 1287–1293. [CrossRef]
19. Martinez-Cano, D.J.; Reyes-Prieto, M.; Martinez-Romero, E.; Partida-Martinez, L.P.; Latorre, A.; Moya, A.; Delaye, L. Evolution of small prokaryotic genomes. *Front. Microbiol.* **2014**, *5*, 742. [CrossRef]
20. Wernegreen, J.J. Endosymbiont evolution: Predictions from theory and surprises from genomes. *Ann. N. Y. Acad. Sci.* **2015**, *1360*, 16–35. [CrossRef]
21. Wernegreen, J.J. Endosymbiosis. *Curr. Biol.* **2012**, *22*, R555–R561. [CrossRef]
22. McCutcheon, J.P.; Boyd, B.M.; Dale, C. The life of an insect endosymbiont from the cradle to the grave. *Curr. Biol.* **2019**, *29*, R485–R495. [CrossRef] [PubMed]

23. Boscaro, V.; Fokin, S.I.; Petroni, G.; Verni, F.; Keeling, P.J.; Vannini, C. Symbiont replacement between bacteria of different classes reveals additional layers of complexity in the evolution of symbiosis in the ciliate Euplotes. *Protist* **2018**, *169*, 43–52. [CrossRef] [PubMed]

24. Vannini, C.; Ferrantini, F.; Ristori, A.; Verni, F.; Petroni, G. Betaproteobacterial symbionts of the ciliate *Euplotes*: Origin and tangled evolutionary path of an obligate microbial association. *Environ. Microbiol.* **2012**, *14*, 2553–2563. [CrossRef]

25. Bennett, G.M.; Moran, N.A. Heritable symbiosis: The advantages and perils of an evolutionary rabbit hole. *Proc. Natl. Acad. Sci. USA* **2015**, *112*, 10169–10176. [CrossRef]

26. Toews, D.P.; Brelsford, A. The biogeography of mitochondrial and nuclear discordance in animals. *Mol. Ecol.* **2012**, *21*, 3907–3930. [CrossRef]

27. Tsitrone, A.; Kirkpatrick, M.; Levin, D.A. A model for chloroplast capture. *Evolution* **2003**, *57*, 1776–1782. [CrossRef]

28. Harrison, R.G.; Larson, E.L. Hybridization, introgression, and the nature of species boundaries. *J. Hered.* **2014**, *105* (Suppl. 1), 795–809. [CrossRef]

29. Petit, R.J.; Excoffier, L. Gene flow and species delimitation. *Trends Ecol. Evol.* **2009**, *24*, 386–393. [CrossRef]

30. Du, F.K.; Petit, R.J.; Liu, J.Q. More introgression with less gene flow: Chloroplast vs. mitochondrial DNA in the *Picea asperata* complex in China, and comparison with other conifers. *Mol. Ecol.* **2009**, *18*, 1396–1407. [CrossRef]

31. Týč, J.; Votýpka, J.; Klepetková, H.; Šuláková, H.; Jirků, M.; Lukeš, J. Growing diversity of trypanosomatid parasites of flies (Diptera: Brachcera): Frequent cosmopolitism and moderate host specificity. *Mol. Phylogenet. Evol.* **2013**, *69*, 255–264. [CrossRef] [PubMed]

32. Ganyukova, A.I.; Malysheva, M.N.; Frolov, A.O. Life cycle, ultrastructure and host-parasite relationships of *Angomonas deanei* (Kinetoplastea: Trypanosomatidae) in the blowfly *Lucilia sericata* (Diptera: Calliphoridae). *Protistology* **2020**, *14*, 204–218. [CrossRef]

33. Carvalho, A.L.M. Estudos sobre a posição sistemática, a biologia e a transmissão de tripanosomatídeos encontrados em *Zelus leucogrammus* (Perty, 1834) (Hemiptera, Reduviidae). *Rev. Pathol. Trop.* **1973**, *2*, 223–274.

34. Frolov, A.O.; Kostygov, A.Y.; Yurchenko, V. Development of monoxenous trypanosomatids and phytomonads in insects. *Trends Parasitol.* **2021**, *37*, 538–551. [CrossRef]

35. Stegemann, S.; Keuthe, M.; Greiner, S.; Bock, R. Horizontal transfer of chloroplast genomes between plant species. *Proc. Natl. Acad. Sci. USA* **2012**, *109*, 2434–2438. [CrossRef]

36. Bolger, A.M.; Lohse, M.; Usadel, B. Trimmomatic: A flexible trimmer for Illumina sequence data. *Bioinformatics* **2014**, *30*, 2114–2120. [CrossRef]

37. Antipov, D.; Korobeynikov, A.; McLean, J.S.; Pevzner, P.A. hybridSPAdes: An algorithm for hybrid assembly of short and long reads. *Bioinformatics* **2016**, *32*, 1009–1015. [CrossRef]

38. Steinbiss, S.; Silva-Franco, F.; Brunk, B.; Foth, B.; Hertz-Fowler, C.; Berriman, M.; Otto, T.D. Companion: A web server for annotation and analysis of parasite genomes. *Nucleic Acids Res.* **2016**, *44*, W29–W34. [CrossRef]

39. Delcher, A.L.; Bratke, K.A.; Powers, E.C.; Salzberg, S.L. Identifying bacterial genes and endosymbiont DNA with Glimmer. *Bioinformatics* **2007**, *23*, 673–679. [CrossRef]

40. Seemann, T. Prokka: Rapid prokaryotic genome annotation. *Bioinformatics* **2014**, *30*, 2068–2069. [CrossRef]

41. Hunt, M.; De Silva, N.; Otto, T.D.; Parkhill, J.; Keane, J.A.; Harris, S.R. Circlator: Automated circularization of genome assemblies using long sequencing reads. *Genome Biol.* **2015**, *16*, 294. [CrossRef]

42. Darling, A.E.; Mau, B.; Perna, N.T. ProgressiveMauve: Multiple genome alignment with gene gain, loss and rearrangement. *PLoS ONE* **2010**, *5*, e11147. [CrossRef]

43. Krzywinski, M.; Schein, J.; Birol, I.; Connors, J.; Gascoyne, R.; Horsman, D.; Jones, S.J.; Marra, M.A. Circos: An information aesthetic for comparative genomics. *Genome Res.* **2009**, *19*, 1639–1645. [CrossRef]

44. Emms, D.M.; Kelly, S. OrthoFinder: Phylogenetic orthology inference for comparative genomics. *Genome Biol.* **2019**, *20*, 238. [CrossRef]

45. Edgar, R.C. MUSCLE: Multiple sequence alignment with high accuracy and high throughput. *Nucleic Acids Res.* **2004**, *32*, 1792–1797. [CrossRef]

46. Castresana, J. Selection of conserved blocks from multiple alignments for their use in phylogenetic analysis. *Mol. Biol. Evol.* **2000**, *17*, 540–552. [CrossRef]

47. Kuck, P.; Longo, G.C. FASconCAT-G: Extensive functions for multiple sequence alignment preparations concerning phylogenetic studies. *Front. Zool.* **2014**, *11*, 81. [CrossRef]

48. Stamatakis, A. RAxML version 8: A tool for phylogenetic analysis and post-analysis of large phylogenies. *Bioinformatics* **2014**, *30*, 1312–1313. [CrossRef]

49. Ronquist, F.; Teslenko, M.; van der Mark, P.; Ayres, D.L.; Darling, A.; Hohna, S.; Larget, B.; Liu, L.; Suchard, M.A.; Huelsenbeck, J.P. MrBayes 3.2: Efficient Bayesian phylogenetic inference and model choice across a large model space. *Syst. Biol.* **2012**, *61*, 539–542. [CrossRef]

50. Katoh, K.; Standley, D.M. MAFFT multiple sequence alignment software version 7: Improvements in performance and usability. *Mol. Biol. Evol.* **2013**, *30*, 772–780. [CrossRef]

51. Nguyen, L.T.; Schmidt, H.A.; von Haeseler, A.; Minh, B.Q. IQ-TREE: A fast and effective stochastic algorithm for estimating maximum-likelihood phylogenies. *Mol. Biol. Evol.* **2015**, *32*, 268–274. [CrossRef]
52. Kalyaanamoorthy, S.; Minh, B.Q.; Wong, T.K.F.; von Haeseler, A.; Jermiin, L.S. ModelFinder: Fast model selection for accurate phylogenetic estimates. *Nat. Methods* **2017**, *14*, 587–589. [CrossRef] [PubMed]

 *pathogens* 

Article

A Global Analysis of Enzyme Compartmentalization to Glycosomes

Hina Durrani [1], Marshall Hampton [2], Jon N. Rumbley [3] and Sara L. Zimmer [1,*]

[1] Department of Biomedical Sciences, University of Minnesota Medical School, Duluth Campus,
   Duluth, MN 55812, USA; durra018@d.umn.edu
[2] Mathematics & Statistics Department, University of Minnesota Duluth, Duluth, MN 55812, USA;
   mhampton@d.umn.edu
[3] College of Pharmacy, University of Minnesota, Duluth Campus, Duluth, MN 55812, USA;
   jrumbley@d.umn.edu
*   Correspondence: szimmer3@d.umn.edu

Received: 25 March 2020; Accepted: 9 April 2020; Published: 12 April 2020

Abstract: In kinetoplastids, the first seven steps of glycolysis are compartmentalized into a glycosome along with parts of other metabolic pathways. This organelle shares a common ancestor with the better-understood eukaryotic peroxisome. Much of our understanding of the emergence, evolution, and maintenance of glycosomes is limited to explorations of the dixenous parasites, including the enzymatic contents of the organelle. Our objective was to determine the extent that we could leverage existing studies in model kinetoplastids to determine the composition of glycosomes in species lacking evidence of experimental localization. These include diverse monoxenous species and dixenous species with very different hosts. For many of these, genome or transcriptome sequences are available. Our approach initiated with a meta-analysis of existing studies to generate a subset of enzymes with highest evidence of glycosome localization. From this dataset we extracted the best possible glycosome signal peptide identification scheme for in silico identification of glycosomal proteins from any kinetoplastid species. Validation suggested that a high glycosome localization score from our algorithm would be indicative of a glycosomal protein. We found that while metabolic pathways were consistently represented across kinetoplastids, individual proteins within those pathways may not universally exhibit evidence of glycosome localization.

Keywords: evolution; kinetoplastid; organelle; metabolic pathway; glycolysis; gluconeogenesis; meta-analysis; peroxisome targeting sequence; PTS1; PTS2

1. Introduction

An accurate understanding of eukaryotic biology requires representation of studies from the widest possible spectrum of organisms. For instance, within the phylum Euglenozoa, studies of species of the order Trypanosomatida continually reveal new pathways and regulatory mechanisms that change what we believe to be true of the characteristics of eukaryotic organisms (e.g., new post-translational protein modifications [1], polycistronic transcription of eukaryotic nuclear genomes [2,3], genome-scale mRNA trans-splicing [4], and RNA editing [5]). The parasitic protozoan Trypanosomatida species that are insect transmitted and cause diseases in mammalian hosts are the best studied. However, comparisons between these species and their neighbors would be very useful in order to better understand the evolution of novel pathways and events within eukaryotic biology. These comparisons include those between the dixenous, well-studied Trypanosomatida species and others of the same order, many of which are monoxenous and/or have different hosts. Additional useful comparisons are between different orders of the class Kinetoplastea, of which Trypanosomatida belongs, including

Parabodonida and the free-living Eubodonida [6], and between classes of Diplonemea, Euglenoidea, and Kinetoplastea of the phylum Euglenozoa.

One feature common to eukaryotes is the partitioning of enzymatic pathways and material into membrane-bound compartments or organelles. Some of these, such as a nucleus, are common to all eukaryotes. Others are unique to a subset of eukaryotic organisms. In Kinetoplastea and Diplonema, the first seven steps of glycolysis are compartmentalized into an organelle called the glycosome along with parts of other metabolic pathways [7]. The glycosome originates from the same common eukaryotic ancestor as peroxisomes, organelles found in a wider range of eukaryotes with which it shares similar biogenesis and import machinery [8]. Interestingly, while Kinetoplastea and Diplonema possess glycosomes (as defined by their inclusion of glycolytic enzyme content) but not peroxisomes, the opposite is true for the closely related Euglenoidea [9].

Our understanding of the complement of glycosome-localized metabolic enzymes is centered on experiments cataloguing them in *Trypanosoma brucei*, and to a lesser extent *Trypanosoma cruzi* and *Leishmania* spp.—all dixenous organisms capable of causing human and livestock disease (e.g., [10–12]). As the *Trypanosoma* and *Leishmania* genera are phylogenetically distinct, the fact of apparent overlap in their glycosome enzyme composition suggests a high degree of conservation [13]. However, pioneering studies in the kinetoplastid relative Diplonema [14,15] suggest that beyond the seven common glycolytic enzymes, their glycosomes lack other enzymes that are glycosomal in the well-studied kinetoplastid species. Our question is whether evolution and maintenance of glycosome contents may be influenced by organism life cycle and lifestyle as much or more than phylogeny, which would require probing their composition in monoxenous and free-living species, and in dixenous organisms with hosts other than mammals. The answer to this question is confounded by the probability of dual localization of enzymes [12], duplication and separate localization of paralogues (e.g., [16,17]), and the reality of the intracellular connectedness of peroxisomes with other membrane-bound compartments that muddies the waters of subcellular fractionation experiments [18].

Our study objective was to determine the extent that we could leverage existing experimental studies in model kinetoplastids to determine the composition of glycosomes in species for which no experimental evidence is available. This objective required a better in silico identification scheme for signals within predicted proteins that could classify their localization as likely glycosomal. We anticipated that signal-to-noise ratio would still be a challenge in our output. However, enough laboratory-based glycosome composition studies in three species plus enough kinetoplastid sequenced genomes are now available to make our intermediate-throughput analysis possible. Our final work demonstrated that an iterative approach combining automated analysis with manual protein evaluation and annotation can be beneficial in uncovering localization patterns of glycosome proteins among the kinetoplastids.

2. Results

2.1. Meta-Analysis of Existing Studies

In order to establish a starting population of proteins for which glycosomal localization is well supported by experimental evidence, we performed a meta-analysis that included as many different information types as possible. We included studies in species *Trypanosoma cruzi* and *Leishmania donovani* in addition to *T. brucei* where most work is performed (Figure 1, [10–12,19–22]). These three species, all dixenous mammalian parasites, are still similar in lifestyle relative to the full complement of species containing glycosomes. The studies performed in these species utilized a variety of methods. The methodologies for establishing localization or organelle content on a global scale have strengths and weaknesses. Isolation of purified organelles followed by mass spectrometry has revolutionized our understanding of the protein composition of subcellular compartments and structures. However, it is important to remember that proteins will not be equally identifiable by mass spectrometry, as some lack ideal cleavage sites and positions for liberation of peptides of appropriate size and composition, and sample preparation will favor the identification of some proteins over others [23]. Conversely, methods involving fluorescent tagging of proteins followed by microscopy are problematic if tags on the N- or C-terminus effectively block terminal transit peptides and localization signals. There is no perfect single method for this sort of analysis, thus combining results of studies using different methodologies can be useful.

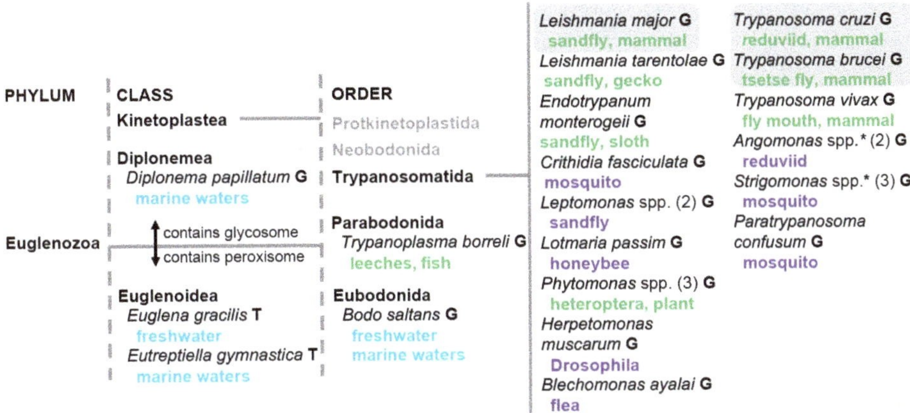

Figure 1. Lifestyle of Euglenozoa species for which glycosome targeting was analyzed. Hosts or environments for these species are shown in color. Blue indicates water environments of free-living species, green indicates hosts of a dixenous lifestyle, and purple indicates insect hosts of a monoxenous lifestyle. For some genera, number of species analyzed is included in brackets. Species included in the initial meta-analysis of glycosome localization are compartmentalized in a grey background box. G, genomic DNA was analyzed. T, translated transcriptomic data was analyzed. Asterisk indicates genus with bacterial symbiont.

We incorporated data of three types. Most studies involved glycosome purification followed by protein composition analysis by mass spectrometry or other methods. When necessary for these studies, we took protein datasets prior to any culling by prediction of peroxisome targeting sequences (PTS1 and PTS2, see below) that the authors may have performed. The second type of data came from the TrypTag project in which *T. brucei* proteins are endogenously tagged with GFP (Green Fluorescent Protein) at one of their native loci [21,24]. Microscopy images of the TrypTag GFP localization pattern within the cell were individually analyzed for each protein for which TrypTag tagging had been successful at its N-terminus, and compared with the distinctive visual pattern of multiple oval-shaped glycosomes shown and described in [24]. The final source of data incorporated was a protein's presence in the resource PeroxisomeDB [19] for any trypanosomatid species. Peroxisome DB is no longer updated, as entries for literature utilized for the resource cease prior to 2010. We treated this as an "archival" source of glycosome localization evidence based off of early individual protein studies in which the protein was determined to be glycosomal.

After inclusion of proteins from all sources, 302 proteins were determined to be glycosomal in at least one study. We subtracted 92 as being part of the import/export machinery or part of the protein composition of the glycosome membrane, as we were instead focused on proteins found in the matrix of the glycosome. We eliminated duplicates when observed. The remaining 209 proteins were then considered as possible glycosome matrix proteins (Table S1). This list consists of 188 that are likely metabolic enzymes and 21 that are hypothetical proteins with no discernable identifying motifs according to TriTrypDB [25]. We also were unable to find any motifs with common motif-finders.

We ranked these enzymes in order of reliability of the data that suggests glycosomal localization. We utilized Bayesian-guided data reliability weight assignments. A glycosomal signal pattern of a tagged protein received a high weight, while a non-glycosomal fluorescence pattern, being less reliable, received a lesser negative weight. Protein studies received a range of individual rankings (Materials and Methods). It is certainly true that another weighting scheme could have resulted in the inclusion or exclusion of lower-ranking proteins from our final group; however, most of these proteins would be retained in our final collection regardless of alternative weight assignment to various studies.

We decided on a cut-off weight (a value of 5) that, except for 1 enzyme (Hypothetical protein Q38C56), exhibited evidence in more than one species. The total number of proteins in our Glycosome Conserved Enzyme Collection (GCEC) is 57 (Table 1). Not surprisingly, the seven first steps of the glycolytic pathway (including triose phosphate isomerase) were clustered within the first 28 highest proteins scored for reliability of glycosome localization. Other known glycosomal pathways were also well-represented such as the pentose phosphate pathway, purine metabolism, and pathways shared with peroxisomes such as fatty acid metabolism. Clearly, as over a hundred proteins are typically identified as potentially glycosomal in any single proteomic study, this list is not comprehensive. Furthermore, individual studies have characterized glycosomal proteins that are not in GCEC. For instance, it is explicitly shown that several superoxide dismutase enzymes are glycosomal from individual *T. brucei* studies, yet this protein does not appear extensively in the global studies of our meta-analysis, so that it is not included by our methods [16,17]. Interestingly, 8 proteins (14%) with sufficient evidence to appear on the list are hypothetical (Table 1). The systematic approach to localization provided by TrypTag has been helpful in bringing these proteins to light, as the fluorescence pattern of their tags all indicated glycosomal localization. TrypTag localization was the only evidence for glycosome localization in six proteins in the master Table S1. In addition to two that are hypothetical, the others are an aspartate carbamoyltransferase that, while found soluble in *Leishmania* in 1981, may be glycosomal in *T. brucei*, as it is part of a pyrimidine biosynthesis pathway with other glycosome components [26]. Others are a ribonucleoside-diphosphate reductase (purine biosynthesis), a mannosyltransferase that likely participates in glycosylphosphatidylinositol-anchor biosynthesis (possibly appearing localized to glycosomes based on the ER's glycosome associations), and a protein with putative phosphatase activity. Other phosphatases have been identified as glycosomal [27].

After establishing the enzymes that would be part of the GCEC, we categorized them into major Kyoto Encyclopedia of Genes and Genomes (KEGG) pathways by assigning them KEGG designations (Table 1). Our goal was to limit the number of pathway categories to allow for patterns to emerge, yet avoid the level of category breadth that erases the essential metabolic depth. Kinetoplastid UniProt accession numbers were used to retrieve primary KEGG reference orthologs and putative pathway associations. Pathways retrieved were organized top to bottom, highest to lowest specificity for any given enzyme. Generally, the top metabolic pathway was retained for our purposes but in some cases more specific subclasses were used, (i.e., urea cycle). Ultimately, nine major KEGG pathways were identified that were represented by two or more enzymes and eight pathways were represented by only one enzyme. Proteins deemed hypothetical could not be categorized due to lack of orthologous sequences.

Table 1. Glycosome Conserved Enzyme Collection (GCEC). Proteins from *Trypanosoma cruzi*, *Trypanosoma brucei*, and *Leishmania donovani* global studies of glycosome composition, evidence of localization of endogenous tagged proteins from the TrypTag project, and historical input of glycosome proteins in PeroxisomeDB culled from individual studies prior to 2010. Proteins are grouped by major KEGG (Kyoto Encyclopedia of Genes and Genomes) pathway inclusion. Presence of a targeting signal within the protein is indicated in the PTS1/PTS2 (Peroxisome Targeting Sequence 1/2) column (glycosome PTS1 or conserved peroxisome targeting signal 2, PTS2). It was not possible to assign a best KEGG reference orthologue for all proteins.

KEGG Reference Ortholog	Major KEGG Pathway	Name of Protein	PTS1/ PTS2
K18561	Glycolysis	NADH-dependent fumarate reductase	PTS1
	Glycolysis	UDP-glc 4'-epimerase	PTS1
K00850	Glycolysis	ATP-dependent 6-phosphofructokinase, glycosomal	PTS1
K00844	Glycolysis/gluconeogenesis	Hexokinase 1	PTS2
K01810	Glycolysis/gluconeogenesis	Glucose-6-phosphate isomerase	PTS1
K00134	Glycolysis/gluconeogenesis	Glyceraldehyde 3-phosphate dehydrogenase	PTS1
K01006	Glycolysis/gluconeogenesis	Pyruvate phosphate dikinase	PTS1
K00927	Glycolysis/gluconeogenesis	Phosphoglycerate kinase	PTS2
K01792	Glycolysis/gluconeogenesis	Aldose 1-epimerase	PTS1
K01623	Glycolysis/gluconeogenesis	Fructose-bisphosphate aldolase	
K01803	Glycolysis/gluconeogenesis	Triose phosphate isomerase	
K03841	Gluconeogenesis	Fructose-1,6-bisphosphatase	PTS1
K01610	Gluconeogenesis	Phosphoenolpyruvate carboxykinase [ATP]	PTS1
K00026	Gluconeogenesis	Glycosomal malate dehydrogenase	PTS1
	Gluconeogenesis	UDP-glucose pyrophosphorylase	
K00760	Purine metabolism	Hypoxanthine-guanine phosphoribosyltransferase 1	PTS1
K00760	Purine metabolism	Hypoxanthine-guanine phosphoribosyltransferase 2	PTS1
K00088	Purine metabolism	Inosine-5'-monophosphate dehydrogenase	PTS1
K00942	Purine metabolism	Guanylate kinase	PTS1
K00759	Purine metabolism	Adenine phosphoribosyltransferase	PTS1
K01490	Purine metabolism	AMP deaminase	PTS1
K00939	Purine metabolism	Adenylate kinase	PTS1
K00088	Purine metabolism	Guanosine monophosphate reductase	PTS1
K00036	Pentose phosphate pathway	Glucose-6-phosphate 1-dehydrogenase (G6PD)	
K00852	Pentose phosphate pathway	Ribokinase	
K01100	Pentose phosphate pathway	Sedoheptulose-1,7-bisphosphatase	PTS1
K01619	Pentose phosphate pathway	Deoxyribose-phosphate aldolase	PTS1
K00615	Pentose phosphate pathway	Transketolase	PTS1
K01057	Pentose phosphate pathway	6-phosphogluconolactonase	
K00864	Glycerophospholipid metabolism	Glycerol kinase	PTS1
K00803	Glycerophospholipid metabolism	Alkyl-dihydroxyacetone phosphate synthase	PTS1
K00649	Glycerophospholipid metabolism	Dihydroxyacetonephosphate acyltransferase	PTS1

Table 1. *Cont.*

KEGG Reference Ortholog	Major KEGG Pathway	Name of Protein	PTS1/ PTS2
K00006	Glycerophospholipid metabolism	Glycerol-3-phosphate dehydrogenase (NAD(+))	PTS1
K00022	Fatty acid metabolism	Enoyl-CoA hydratase/Enoyl-CoA isomerase/3- hydroxyacyl-CoA dehydrogenase	PTS2
K08766	Fatty acid metabolism	Carnitine O-palmitoyltransferase	PTS1
K11262	Fatty acid metabolism	Acetyl-CoA carboxylase	
K11207	Redox maintenance	Trypanothione/tryparedoxin dependent peroxidase 2	
K01833	Redox maintenance	Trypanothione synthetase	
K00103	Redox maintenance	L-galactonolactone oxidase	PTS1
K00869	Terpenoid biosynthesis	Mevalonate kinase	PTS1
K01823	Terpenoid biosynthesis	Isopentenyl-diphosphate delta-isomerase (type II) (idi1)	
K00031	TCA cycle/glutathione metabolism	Isocitrate dehydrogenase	PTS1
K01438	Amino acid biosynthesis (arginine)	Acetylornithine deacetylase	PTS1
K01107	Insositol phosphate metabolism	Inositol polyphosphate 1-phosphatase	
K13421	Pyrimidine metabolism	Orotidine-5-phosphate decarboxylase/Orotate phosphoribosyltransferase	PTS1
K15731	RNA polymerase II C-terminal domain phosphatase	PTP1-interacting protein, 39 kDa/TFIIF-stimulated CTD phosphatase	PTS1
K09829	Steroid biosynthesis (ERG2)	C-8 sterol isomerase-like protein	PTS1
K10703	Long chain fatty acid synthesis	Protein tyrosine phosphatase	
N/A	N/A	Thymine-7-hydroxylase	PTS1
N/A	N/A	Hypothetical protein (Q580K0)	PTS1
N/A	N/A	Hypothetical protein (Q389Y7)	
N/A	N/A	Hypothetical protein (Q38C56)	
N/A	N/A	Hypothetical protein (Q386P8)	PTS1
N/A	N/A	Hypothetical protein(Q388J7)	
N/A	N/A	Hypothetical protein (Q38DM9)	
N/A	N/A	Hypothetical protein (Q383Q3)	
N/A	N/A	Hypothetical protein (Q38AC3)	

2.2. An Adequate Algorithm for Glycosome Localization

It is well understood that many peroxisome proteins possess either a peroxisome targeting signal (PTS1), which is the amino acid triplet [Serine-Lysine-Leucine] on their C-termini, or less commonly, the less-conserved PTS2 very near their N-termini. The predictive power of the PTS1 signal for glycosome localization was analyzed in the most stringent and comprehensive *T. brucei* glycosome proteome study. The conclusion was that the PTS1 signal had a sensitivity of less than 40% and a specificity of less than 50%, making it a remarkably poor predictor of localization [12]. One potential reason for this is that the calculation did not differentiate glycosome membrane and surface proteins from matrix proteins. At least some membrane proteins of peroxisomes originate in the ER [28], and thus utilize an entirely different localization mechanism than the compartmentalized matrix enzymes. A second reason is that presumably only the canonical PTS1 signal, 'SKL', was used in the analysis. It is likely that in kinetoplastids, as in plants [29], certain conserved variations on the canonical sequence are also functional targeting signals.

A major motivation for setting the evidence bar very high for a protein to be included in the GCEC dataset is that we subsequently used it to determine whether there were features of glycosome signaling peptides that differed from the peroxisome targeting signals previously used to interrogate collections of glycosome proteins [13]. The number of proteins utilized is similar to that used to train the plant PTS1 prediction tool PredPlantPTS1 [30], as plant peroxisome proteins can possess PTS1 sequences that are certain conservative substitutions of the consensus 'SKL'. Our approach was similar to that used to generate more recently derived plant PTS1 prediction algorithms [29]. If the kinetoplastid glycosome targeting signal is similar to the PTS1 of plants, any length of input data from three to 40 amino acids will yield similar results. The positive training dataset was the C-terminal 20 amino acids of the GCEC training dataset (using the orthologues from all three species). The negative training dataset consisted of the remainder of the 35,640 proteins in the UniProt database from the same species. After initial comparison, we concluded that only the last three amino acids would be utilized in our glycosome-specific PTS1 predictor, for which amino acid composition preferences were marked (Figure 2A). These amino acid composition preferences were used to establish the algorithm for our PTS1 predictor for glycosomal proteins, which returns a numerical value from 7.35 (classical PTS1 'SKL' sequence; high likelihood of glycosome localization) to −23.9 (File S1). Cumulatively, the PTS1 proteins returned a mean score of 3.45 ± 3.85. For the negative dataset, the mean score is −4.40 ± 4.54. Figure 2B presents the score distributions.

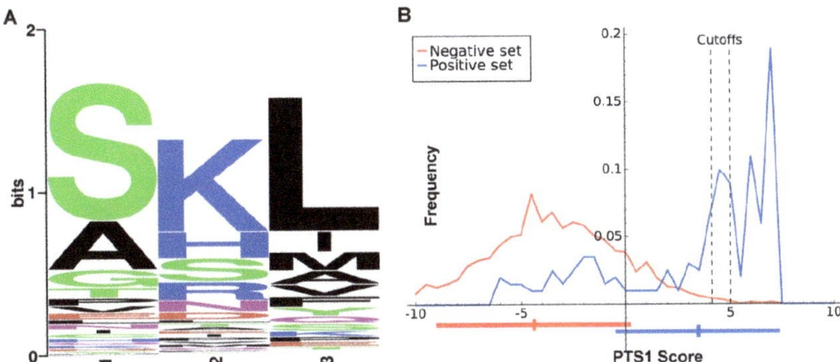

Figure 2. (**A**). The amino acid bias of the C-terminus of proteins in our Glycosome Conserved Enzyme Collection (GCEC) training dataset from *Trypanosoma cruzi*, *Trypanosoma brucei*, and *Leishmania donovani* (*L. major* homologues of *L. donovani* proteins were used in the amino acid bias calculations). The last three amino acids of the proteins could harbor glycosome targeting signals (glycosome PTS1s). The position of the first amino acid of the PTS1 is position 1, with the final amino acid of the protein being position 3. (**B**). Frequencies of PTS1 scores for GCEC proteins (blue, positive set) and a dataset of non-glycosomal proteins (red, negative set). Bars shown underneath the graph indicate mean and standard deviation of each dataset. Cutoff values are shown as vertical dotted lines for pre-selected protein datasets (left) and whole genome prediction (right).

Predication software cutoff values for classifying the last three amino acids of a protein as a glycosome-specific PTS1 are user-selected. The cutoff must account for the tradeoff between specificity and sensitivity, with a higher cutoff value (maximizing specificity) being more appropriate for whole genome prediction, and a lower cutoff value (minimizing false negative rates) more suitable for examining localization of pre-selected protein sets [29]. In this work, we selected a cutoff value of >4.1 for examination of our evidence-based list of proteins (Table S1) and 5 for perusal of complete genome datasets. We based the value of 4.1 on the inflection point of the GCEC training set protein curve after which score frequency sharply increased, and the value of 5 on the point at which the frequency of the negative dataset approached zero in Figure 2B. While a prediction program for a second, glycosome-specific N-terminal signal analogous to PTS2 was desired, only three proteins in the entire dataset had a "classical" PTS2; therefore, this dataset was not deemed appropriate for development of such a tool.

2.3. Glycosome Localization Prediction of Orthologues of Gcec Proteins across Kinetoplastids

Both Leishmania and Trypanosoma species are represented by studies in our global analysis and showed compositional overlap, so similar to earlier conjectures [13], we hypothesized that the contents of the glycosomes are largely conserved between kinetoplastid species. Prediction of localization by presence or absence of a signaling peptide will suggest whether that hypothesis is more or less likely to be true. In order to utilize the glycosome-specific PTS1 prediction algorithm on kinetoplastids that are not part of the major human infectious trypanosomatids, we collected homologues of the 57 GCEC proteins in available kinetoplastid genomes or transcriptomes. When the genome was available in TriTryp [25], we accessed the genomes through the site's BLAST portal as it allowed verification of synteny for the genes. As expected, gene duplications were present in the Leptomonas and Leishmania lineages prone to polyploidy. Some genes were missing entirely from specific genomes. This may result from genome sequencing or assembly issues, or the genes may in fact be missing from these species (Figure 3 blue boxes and Table S2). We assembled predicted homologues from all available kinetoplastid genomes/transcriptomes for each of the GCEC proteins into File S2. We stress that for species not represented in TriTryp, the protein of greatest sequence similarity to the *T. brucei* representative of each GCEC protein may not in fact be a true orthologue, but merely a protein with a conserved domain. This introduces potential but unavoidable noise into our analysis.

Predicted homologues and the training dataset were subject to PTS1 prediction using our previously developed algorithm, and PTS2 prediction using the expanded identification [HKQR][LVIFYA]{5}[HKQR][LVIFYA] [13] that started within the first 20 amino acids. Because of the low specificity of the PTS2 sequence, we only considered a protein with a PTS2 signal likely if the actual PTS2 domain was conserved in both sequence similarity and position, in that protein, in at least than 75% of the interrogated kinetoplastid species. Proteins with a PTS1 or conserved PTS2 are color-coded in pink in Figure 3, while proteins lacking such domains are colored cyan.

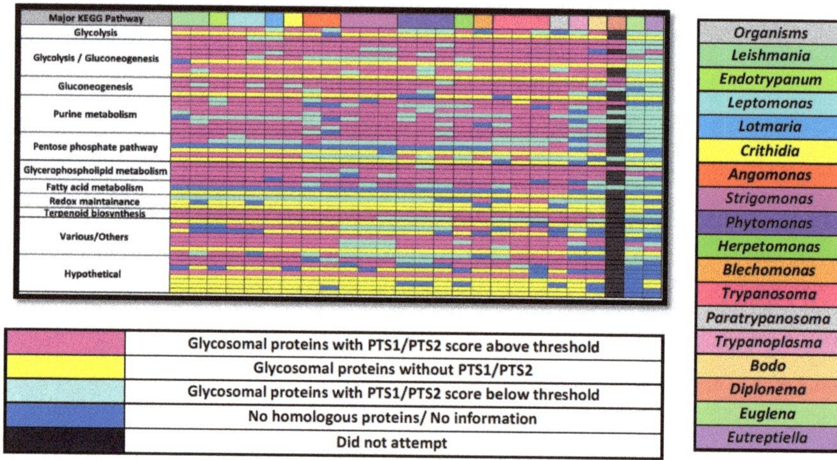

Figure 3. Map of glycosome targeting signal conservation across kinetoplastids, select euglinids and *Diplonema papillatum*. Different species and isolates of a genus are given separate columns. Each row represents one protein. Black rectangles in the *D. papillatum* column represent proteins for which we did not attempt to find an orthologue. The proteins in yellow are those that across all kinetoplastids do not possess a PTS1 or conserved PTS2. The *D. papillatum* entries that were used were those identified in [15]. Specific putative proteins and orthologue Uniprot/TriTryp/NCBI/contig numbers represented by each rectangle are found in Table S2. Organism columns are ordered loosely on phylogeny.

The GCEC dataset also contained 16 proteins that contained no identifiable PTS1 or PTS2 across any of the GCEC organisms, and these were labeled yellow in Figure 2. In only one instance, triose phosphate isomerase, among all interrogated kinetoplastids and diplonema does an orthologue possess evidence of a signaling peptide where none existed in the GCEC organism enzymes. For this enzyme, a PTS1 is apparent in diplonema only. However, a potential cryptic PTS2 is present from approximate amino acids 48 to 55 in 17 of the 23 kinetoplastid species and isolates examined (cryptic because of its downstream position relative to the amino terminus). In the case of peroxisomes, a subset of proteins with very high experimental evidence of localization also lack identifiable targeting signals, and poorly understood or unknown import mechanisms are normally invoked for those proteins. It appears, then, that glycosomal proteins fall into two classes. The ones that have no identifiable signal among our GCEC species universally have no signal among kinetoplastids, while the others primarily do possess a targeting signal but it may be absent in orthologues of some species. We suggest that if an enzyme of the GCEC dataset is lacking a signal normally present, it may be excluded from the glycosomal contents and instead may be present in another cellular location. However, we cannot exclude the possibility that these proteins sporadically acquire the ability to "piggyback" on other proteins, or simultaneously acquire a yet-unidentified alternative localization signal to gain access to the glycosomal matrix. Two enzymes, orotidine-5-phosphate decarboxylase involved in pyrimidine metabolism, and mevalonate kinase of the terpernoid biosynthetic pathway, are the only ones for which evidence of glycosome localization is universally conserved across kinetoplastids.

For perspective on how random the event of a "lost" localization signal is among kinetoplastids, we included multiple species and/or isolates of genera when possible. Reassuringly, often (but not always) when a GCEC enzyme is missing a PTS1 or PTS2 within a genus, the loss is consistent across most or all orthologs of that genus (e.g., compare the loss of proteins across the three Phytomonas, three Strigomonas, or two Angomonas species/isolates in Figure 3). We especially noted the propensity of the PTS1 signal to not be present on homologues of pyruvate phosphate dikinase (glycolysis), ribokinase and 6-phosphoglucolactonase (pentose phosphate pathway), and putative thymine-7-hydrolase in species with a bacterial endosymbiont (Figure 1). This phenomenon suggests that some feature of organismal metabolism within the genus or group may be the driving force for alternative localization. Another validation of our approach to sorting out glycosomal localization by PTS1 or PTS2 is that the localization signals are largely lost in orthologous proteins of the Euglenoidea class that possesses peroxisomes rather than glycosomes (Figures 1 and 3, last columns).

We note that in no KEGG pathway category does a single kinetoplastid species ever entirely lose representation within the glycosome signal-containing enzymes, even given that we are analyzing only a subset of glycosomal proteins. We asked whether proteins of certain metabolic pathways are more likely to conserve glycosome localization across the kinetoplastids than other pathways. Figure 4 tallies all of the orthologues of GCEC dataset enzymes into their respective KEGG pathway categories. One observed trend is some categories such as redox maintenance, pentose phosphate pathway, and the "hypothetical" proteins – those possessing no easily identifiable motifs – possess disproportionally more proteins that lack an identifiable PTS1 or PTS2 among all species. This is likely due to, in the case of the hypothetical proteins, a lack of traditional targeting signal across all orthologues of most hypothetical proteins. It is difficult to conjecture as to what this might mean. We also note that between KEGG pathway categories, there appear to be differences in the relative degree to which a glycosomal PTS1/PTS2 signal is conserved. For instance, a higher proportion of glycophospholipid metabolism enzymes conserve their signaling motifs relative to the proportion of pentose phosphate pathway and purine metabolism proteins that have their signaling motifs conserved. With this limited dataset, however, validation of these trends is likely not statistically possible. More proteins would be required to confirm that enzymes of specific KEGG pathway categories possess different degrees of overall localization with the glycosome. In summary, while we find that the metabolic pathways contributing enzymes to the glycosome are entirely conserved, we have evidence of variability of localization of specific enzymes among kinetoplastids.

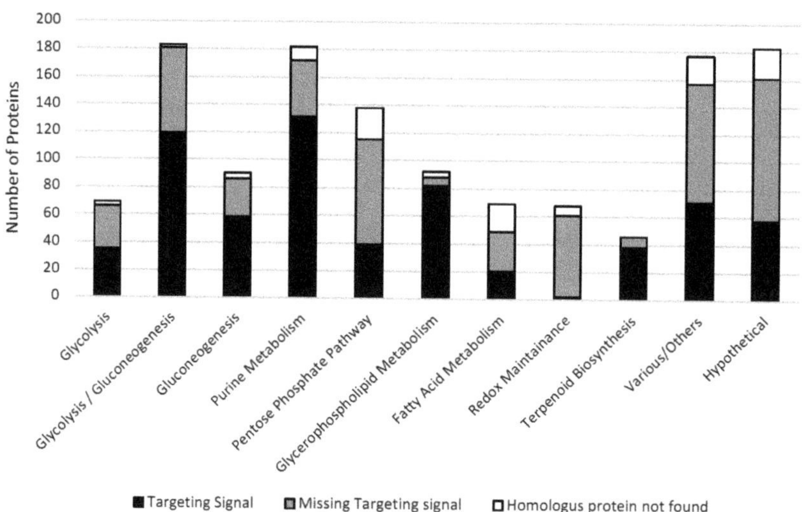

Figure 4. Major metabolic pathways of proteins of the Glycosome Conserved Enzyme Collection (GCEC). All orthologues of the proteins in each metabolic pathway that possess a glycosome-specific targeting signal 1 (PTS1) or a conserved peroxisome targeting signal 2 (PTS2) were classified as "Targeting Signal"; black. Orthologues in which the signal was either not retained, or throughout the orthologous group there was no evidence of either signal were both classified as "Missing Targeting signal"; grey. When an orthologuous protein was not found in a kinetoplastid, it was considered "Homologous protein not found"; white.

2.4. Utilizing the PTS1 Signal Algorithm in Other Datasets

With the results of Figure 3 in hand, we next asked how many enzymes possessing some degree of experimental evidence of glycosome localization scoring below the top GCEC (Table S1) additionally had signal peptide evidence of glycosome localization. An enzyme was considered positive if at least one of the three species (*T. brucei*, *T. cruzi*, and *L. major*) had a glycosome-specific PTS1, or else if all three of the species had a PTS2 in which the motif began within the first 20 amino acids in order to capture potential contributions of sequences that are near neighbors to the final three amino acids. Before applying these standards, we removed twenty-two proteins that, visualized in the TrypTag collection, were obviously targeted to either the nucleus, flagella, mitochondria, or kinetoplastid. While there is a possibility that these proteins could be dual-localized, they also could be part of published glycosome proteomes because cellular compartments physically interact with each other [18] and perfect separation or purification of organelles is not possible.

Of the 126 remaining non-GCEC enzymes, 38 had targeting signals (Table 2), which constitutes 30% of the total. We then asked whether these additional enzymes could be classified with KEGG designations that were either already represented in our GCEC or else specific to peroxisomal pathways. This would constitute further evidence of glycosomal localization for these proteins. We used the same strategy for applying KEGG designations as for GCEC enzymes of Table 1, except that this time some enzymes could not be as precisely defined or lacked orthologues in well-studied systems and could be part of any number of KEGG pathways. Two of the 13 proteins deemed "hypothetical" within the interrogated 126 proteins again also appeared within the signal-peptide containing dataset. In all, 23 enzymes (60%) could be assigned to one of the major KEGG pathways that were represented among the GCEC proteins. Overall then, there is good evidence that the proteins in Table 2 are also localized to the glycosome, at least in the species in which localization signals are present.

Table 2. Additional kinetoplastid proteins with both experimental evidence and a signal sequence indicating localization to the glycosome. Proteins are grouped by major KEGG pathway inclusion. It was not possible to assign a best KEGG orthologue for all proteins.

KEGG Reference Ortholog	Major KEGG Pathway	Name of Protein	PTS1/ PTS2
K00845	Glycolysis	Glucokinase	PTS1
K01809	Glycolysis/gluconeogenesis	Phosphomannose isomerase	PTS1
K17497	Glycolysis/gluconeogenesis	Phosphomannomutase-like protein	PTS1
K00849	Glycolysis/gluconeogenesis	Galactokinase-like protein	PTS1
K02564	Glycolysis/gluconeogenesis	Glucosamine-6-phosphate isomerase	PTS1
K00927	Glycolysis/gluconeogenesis	Pas-domain containing phosphoglycerate kinase	PTS1
K01443	Glycolysis/gluconeogenesis	N-acetylglucosamine-6-phosphate deacetylase-like protein	PTS1
K00863	Glycolysis/gluconeogenesis	Dihydroxyacetone kinase 1-like	PTS1
K01518	Purine metabolism	Kinetoplastid-specific phospho-protein phosphatase	PTS1
K00759	Purine metabolism	Adenine phosphoribosyltransferase	PTS1
K00853	Pentose phosphate pathway	L-ribulokinase	PTS1
K06128	Glycerophospholipid metabolism	Lysophospholipase	PTS1
	Fatty acid metabolism	Acyl-CoA binding protein	PTS1
K00311	Fatty acid metabolism	Electron transfer flavoprotein-ubiquinone oxidoreductase	PTS1
K13356	Fatty acid metabolism	Fatty acyl- CoA reducatase	PTS1
K00632	Fatty acid metabolism	3-ketoacyl- CoA thiolase	PTS2
	Fatty acid epoxide hydrolase	Epoxide hydrolase	PTS1
K04283	Redox maintenance	Trypanothione-disulfide reductase	PTS1
K11185	Redox maintenance	Tryparedoxin peroxidase	PTS1
	Redox maintenance	Dj-1 family protein	PTS1
K04564	Redox maintenance	Iron superoxide dismutase	PTS1
K04564	Redox maintenance	Iron superoxide dismutase	PTS1
	Redox maintenance	2-oxoglutarate (2og) and Fe(II)-dependent oxygenase superfamily protein	PTS1
K01940	Urea cycle	Arginino-succinate synthase	PTS1
K01438	Urea cycle	Acetylornithine deacetylase-like	PTS1
K01745	Amino acid degradation	Histidine ammonia-lyase	PTS1
K02614	Amino acid degradation	Thioesterase-like superfamily	PTS1
	Peptide cleavage	Peptidase T	PTS2
	Protein cleavage	Carboxypeptidase M32	PTS2
K02150	pH regulation	V-ATPase, subunit E	PTS1
	Pyrophosphate and poly phosphate metabolism	Acidocalcisomal exopolyphosphatase	PTS1
K02218	Signal pathway regulation	Casein kinase I, isoform 2	PTS2
K01676	TCA cycle	Fumarate hydratase, class I (FHM)	PTS2
K00972	Amino and nucleotide sugar metabolism	UDP-N-acetylglucosamine pyrophosphorylase	PTS1
	N/A	Hypothetical protein (Q4DBW4)	PTS1
	N/A	Hypothetical protein (Q57TT5)	PTS1
	N/A	Hypothetical protein (Q381V8)	PTS1

Finally, we turned our attention to the question of how extensive the predictive power of the possession of a PTS1 is when interrogating genomes for likely glycosomal proteins. As only 62% of GCEC enzymes have a PTS1, clearly it cannot be used to identify the entire complement of glycosome proteins. However, we wanted to establish how likely it is that a protein found in a genome database with a PTS1 score above a certain level using our glycosome-specific algorithm is indeed targeted to the glycosome. We used the higher-stringency PTS1 cutoff score of 5 or greater to interrogate the dataset of the 81,604 Uniprot Kinetoplastid proteins. This dataset is biased towards organism genomes that previous investigators were interested in capturing, and thus includes the unusual mix of putative proteins of *Leishmania* spp., both lineages of the hybrid *T. cruzi* strain CL Brener, *T. brucei brucei*, *Trypanosoma theileri*, and *Bodo saltans*. However, it is convenient and representative of available datasets.

Of those 81,604 Uniprot Kinetoplastid proteins, only 572 had a PTS1 score of 5 or more (Table S3). As there was no way to further evaluate the proteins that were hypothetical, they were removed. Four hundred twenty-six proteins possessing some identifiable motif remained that were manually evaluated: 202 were actually in our GCEC (Table 1) or were likely orthologues. This number increased to 287 proteins when including all proteins with some experimental evidence of glycosome localization (i.e., found in Table S1). Twenty-two entries had evidence of mitochondrial localization by virtue of the word "mitochondrial" in its name or else are a known mitochondrial protein or a homologue. In other organisms, a subset of mitochondrial protein mRNAs exhibit localization to ribosomes physically associated with mitochondria, and mitochondrial proteins synthesized in the cytosol are most likely highly associated with specific chaperone proteins [31]. It is reasonable to conjecture that in either of these contexts, a PTS1 may exist on a mitochondrial protein that may not be competent to deliver that protein to the glycosome. In conclusion, 76% of the proteins possessing a PTS1 score of 5 or higher had some other evidence of glycosomal localization, or its absence from Table S1 could be explained by a PTS1 overridden by mitochondrial localization (Table S3).

One category of glycosomal proteins that was absent from the set with PTS1 values of 5 or higher were the peroxin (PEX) peroxisome protein import and biogenesis complex proteins. These proteins, best characterized in yeast, mammals, and plants, are loosely conserved in glycosomes and characterized to varying extents [32]. The highest-scoring peroxin of the Uniprot Kinetoplastid proteins was PEX13.1 with a PTS1 score of 4.3. This strengthens our theory that matrix proteins are more likely to require a PTS1 or PTS2 to be properly localized than glycosome membrane proteins or complexes that may derive from the likely glycosome biogenesis involving the ER or even the mitochondrion [28,33]. In the case of PEX13.1, the protein is dual localized, and its PTS1, "TKL", is known to be important for its glucose-dependent glycosome localization [34]. We also analyzed the remaining ~100 proteins to determine how many possessed domains indicating a role in pathways clearly unrelated to glycosome function. (e.g., nucleic acid binding or processing; ribosome protein subunit). We estimate these proteins to comprise less than 25% of this protein list (highlighted in Table S3). This is a considerably better predictive outcome than was anticipated by the previously described assessment that for glycosomes, a PTS1 signal has a sensitivity of less than 40% and a specificity of less than 50% [12].

2.5. Similarities of Protein Compositions of the Glycosome with the Peroxisomes and the Mitochondrion

It is known that some basic metabolic pathways are conserved between peroxisomes and glycosomes, such as elements of fatty acid metabolism. We wished to generate a more specific picture of the enzyme conservation between these two organelles. We compared GCEC proteins and those with lower degrees of experimental evidence but that possess PTS1 or a conserved PTS2 within the first 20 amino acids (proteins of Tables 1 and 2), to peroxisome metabolic enzymes from recent proteomic studies in mammal and plant [35,36]. Figure 5 shows the nine enzymes that are present in both organelles. The data suggest that particular enzymatic steps of fatty acid and glycerophospholipid metabolism and terpenoid biosynthesis, and superoxide dismutase activity may have been compartmentalized fairly early in the evolutionary history of the peroxisome/glycosome. This does not preclude the possibility that these activities are also present at other subcellular locations, as we know to be true of proteins of the pentose phosphate pathway, for example (evidence summarized in [37]).

Figure 5. Overlap in glycosome and peroxisome metabolic pathway and enzyme composition. Proteins with experimental evidence of glycosome and peroxisome localization in the kinetoplastid and mammalian/plant systems, respectively, are arranged by metabolic pathway.

Finally, we looked at the rather obvious overlap between metabolic pathways with components existing in both the glycosome and the mitochondrion. For instance, elements of fatty acid metabolism and enzymes involved in scavenging reactive oxygen species are found in both organelles. Therefore, it is possible that PTS1-containing proteins that we exclude from our analysis because of mitochondrial localization evidence may, in fact, dual-localize. However, many of the 22 proteins in the UniProt PTS1 > 5 population that we deemed mitochondrial are proteins of the electron transport chain localized to the mitochondrion membrane, or are related to electron transport. This pathway is not present in glycosomes, so we believe that other mechanisms override the strong PTS1 that would otherwise localize these proteins to the glycosome. Likewise, all of the proteins that we removed from the analysis resulting in Table 2 are proteins that, when tagged, exhibited mitochondrial localization patterns and/or were known mitochondrial proteins in at least one organism. Also, a possible feature of having a PTS2 is the potential for strong representation of the amino acid arginine. This amino acid is also over-represented in the beginning of many mitochondrial proteins [38,39]. Therefore, the degree of parallel or integrated processing [33] and/or dual localization is unresolved.

3. Discussion

Environmental influences and organism lifestyle profoundly affect energetic flux through anabolic and catabolic metabolic pathways, and regulatory and protective processes. Examples of this can be found in the kinetoplastids, which include free living and monoxenous and dixenous parasites with a wide host range that is likely still incompletely defined [6]. For instance, the absence of the protective enzyme catalase in dixenous but not monoxenous kinetoplastid genomes may be a result of a requirement for low levels of the differentiation signal hydrogen peroxide for the dixenous organisms [40].

Most analysis of glycosomes has occurred in the dixenous *T. brucei*. The importance of glycolysis to *T. brucei* survival is life stage-dependent. Glycosome matrix enzyme composition of cells of the life stage replicating in the amino acid-rich tsetse fly gut differs substantially from those of the replicating life stage in the glucose-rich mammalian bloodstream, a concept nicely illustrated by 2D electrophoresis of purified glycosome lysate in 1990 [20]. This change in response to available nutrients is hypothesized to be the major environmental pressure that maintains a glycosome that contains metabolic as well as oxidative enzymes. Efficiency in establishment of, and changes to, gene product abundances is advantageous. In *T. brucei*, it appears that during changes in nutrient availability and life stage, selective turnover of entire glycosomes occurs. Concurrently, glycosomes containing a different enzyme composition are generated [41,42]. Glycolytic compartmentalization presumably allows *T. brucei* to utilize this process to achieve rapid adaptation.

A complicating factor for this explanation of why glycosomal contents are maintained throughout the kinetoplastid lineage is the fact that the metabolic remodeling of *T. brucei* is very extreme between its replicative stages in its mammalian and insect hosts [43]. Metabolic studies demonstrate that dixenous mammalian kinetoplastids that replicate intracellularly, rather than in the bloodstream as *T. brucei* does, do not have as radical a metabolic transformation, [43]. Even *T. brucei* may experience alternative metabolic states when in fatty tissues rather than in the blood [44]. Furthermore, it is more difficult to envision that the capacity for rapid metabolic adaptation is as much an evolutionary force for organisms that spend their entire existence in a single life stage, a single host, or in an environment such as saltwater. Finally, pathways such as purine metabolism are also represented in the glycosome. What is the reason for retaining compartmentalization of these pathways across kinetoplastids?

Our findings serve as building blocks from which we can begin addressing questions of the origins and maintenance of glycosomes. We have obtained a collection (the consolidation of Tables 1 and 2) of glycosome matrix enzymes with high experimental evidence and/or targeting signal evidence of glycosome localization. We have demonstrated the feasibility of using the presence of PTS1 to identify additional glycosome matrix protein candidates in genomes and transcriptomes of kinetoplastid species, and we expect the number of testable transcriptomes and genomes to increase in number in the coming years [6]. Finally, for many glycosome-containing organisms with publicly available transcriptomes or genomes, we have performed an initial analysis of potential glycosome matrix proteins that may be present or absent in each particular genome (Figure 3). For instance, there are two hypoxanthine-guanine phosphoribosyltransferase paralogues in *T. brucei* and in the *Leishmania* species that we analyzed, both targeted to the glycosome. While we could only find one homologue in *T. cruzi* and some kinetoplastids, in others, only one of the two paralogues retained its PTS1, suggesting an expanded localization of this enzyme beyond the glycosome.

A bias of our approach is that our GCEC is derived from experimental studies in a few specific organisms, especially *T. brucei* proteome studies. If enzymes are compartmentalized into the glycosomes of monoxenous kinetoplastids, but not in the disease causing dixenous organisms, they are absent from our high-confidence lists. In the future it may be possible to interrogate several of the better-annotated monoxenous genomes such as *Leptomonas pyrrhocoris* [45] for PTS1-containing proteins to detect these potential glycosomal enzymes and even additional KEGG pathways that may be partially contained within kinetoplastid glycosomes.

Immediate and long-term future directions emerging from this research are apparent. In the short-term, with Table 2 proteins or future glycosome proteome studies of monoxenous organisms added to our GCEC dataset, we could re-train the PTS1-finding algorithm in an iterative approach to better understand this import signal. In tandem, we could utilize the genetically malleable *T. brucei* to better define appropriate PTS1 cutoff values for prediction of glycosome localization. For this approach, we would genetically tag and modify several PTS1s that score in the intermediate (~4–5.5) range for potential glycosome proteins, and microscopically visualize whether localization changes upon sequence modification. A similar approach was used to better define PTS1 in plants [29].

Longer term, several compelling research directions include characterizing the proteins of unknown function that reliably appear among the proteins purifying with glycosomes (Tables 1 and 2). The fact that so many hypothetical proteins in the GCEC have no PTS1 or conserved PTS2 raises the possibility that perhaps they are membrane proteins that are not part of a glycosome compartmentalized metabolic pathway. Alternately, of course, they could simply possess yet-unidentified signals or piggyback on other glycosome proteins for entry. As a first step, their essentiality could be assessed fairly easily in a variety of kinetoplastids. It would also be intriguing to define a glycosome-specific version of PTS2, although multiple additional glycosome proteomic studies on a variety of kinetoplastids would likely be necessary to acquire the number of proteins needed to pull such a targeting signal out of the noise. Perhaps most globally applicable, one thing that we noted was the continued presence of a PTS1 on the glycolytic enzyme glyceraldehyde-3-phosphate dehydrogenase in one euglenoid species and glucose-6-phosphate isomerase in another, despite a widespread loss of PTS1 on the balance of the glycosomal glycolytic and gluconeogenesis pathway enzymes in the euglenoids. Interestingly, there is increasing evidence, particularly in yeast, of glycolytic proteins including glyceraldehyde-3-phosphate dehydrogenase being at least partially localized to the peroxisome [46]. A better understanding of the protein composition of glycosomes and the composition of the peroxisomes of closely related organisms may be important for understanding the purpose of compartmentalizing glycolytic enzymes among eukaryotes as a whole.

4. Materials and Methods

4.1. Scoring of Meta-Analysis

A repository of potential glycosomal proteins was created using tables and other data from glycosome purification and proteomic studies for three organisms: *T. brucei, T. cruzi,* and *Leishmania donavani*. Tables of likely glycosomal proteins in these studies were mined for those that were parts of enzymatic pathways or had no assigned function. The Peroxisomedb.org site's Protein Families folder [19] was mined for our organisms of interest. Evaluation of protein localization in the TrypTag study initiated from an initial list of potentially glycosomal proteins from Dr. Samuel Dean, TrypTag co-developer, that we independently evaluated for the glycosome tagging pattern of oval-shaped organelles described in [24] using publicly available TrypTag images. Subsequently, TrypTag images of tagged versions of all of the remainder of Table S1 were individually analyzed the same way when available.

Weighting of these studies was as follows: presence in [19] 1 point if yes, 0 points if no; presence in [22] 0.5 points for each life stage—procyclic and/or bloodstream forms—that it was observed in, 0 points if not found; presence in [20] 2 points if yes, 0 points if no; presence in [12] 2 points if yes, 0 points if no; presence in [11] 2 points if yes, 0 points if no, presence in [10] 2 points if yes, 0 points if no; glycosome localization pattern upon N-terminal tagging in TrypTag [21] 3 points if yes, 0 points if information was not available (tagging on the N-terminus was unsuccessful or not attempted), and −2 points if localization of signal from the N-terminus tagged protein was other than glycosomal (we utilized a lesser negative score for non-glycosomal localization because multiple non-relevant factors can lead to non-targeting of a tagged protein while a false positive is rare). As a C-terminal tagging could obscure a terminal peroxisome targeting signal 1, it was ignored. The cut-off score of 5 points was used, as it requires of a protein more than just inclusion in two *T. brucei* proteomic studies to reach this score. Sixty-four proteins made the cut-off value. These proteins were eventually decreased to 57 after combining *T. brucei*-specific duplications and the removal of glycosomal transporters from the list, as they were not deemed metabolic enzymes.

4.2. Categorization of Proteins into Metabolic Pathways

The proteins identified in our current analysis were categorized into groups based on the pathway in which they are involved (e.g., glycolysis/gluconeogenesis proteins). Kinetoplastid UniProt accession numbers were converted to KEGG identifiers using the Convert ID tool in the KEGG Mapper utility. The entries retrieved contained a KEGG reference ortholog and pathways ascribed to the given enzyme. Typically, the top most pathway retrieved was retained as the most broadly descriptive pathway for the search protein. As clusters of enzymes operating in more discrete sub-pathways were identified broad metabolic categories were replaced with specific sub-categories, such as the urea cycle in the degradation of multiple amino acids. The major category "Redox maintenance" is not a KEGG-derived pathway but was used to group the glutathione-based antioxidant cycle (glutathione metabolism) with the enzymes involved in detecting and metabolizing reactive oxygen species. BLASTP (Basic Local Alignment Search Tool for standard protein-protein search) search was used to identify orthologs of the hypothetical sequences but no orthologs were identified.

4.3. Identification of Orthologues in Other Organisms

Outside of the *Trypanosoma* and *Leishmania* genera, we identified additional sequenced genomes of glycosome-containing species and selected 21 for analysis. In addition, we also analyzed *Trypanosoma vivax* because of its presumably simpler lifestyle than *T. brucei*, *T. cruzi*, and *Leishmania tarentolae*, as its non-insect host is non-mammalian. Typically, paralogues were identical or nearly identical, and only one was selected for further analysis. We then performed analysis utilizing automated search functions and manual review to identify orthologues.

For the organisms *Blechomonas ayalai*, *Bodo saltans*, *Crithidia fasciculata*, *Endotrypanum monterogeii*, *Leishmania tarentolae*, *Leptomonas pyrrhocoris*, *Leptomonas seymouri*, *Paratrypanosoma confusum*, and *Trypanosoma vivax*, protein mining and homology searches of the *T. brucei* protein were done using TnBLAST at TriTryp [25] from August through October 2019, as this platform allowed for assurance of synteny. For the organisms *Angomonas deanei*, *Angomonas desouzai*, *Phytomonas serpens*, *Phytomonas* isolate EM1, *Phytomonas* isolate Hart 1, *Lotmaria passim*, *Strigomonas culicis*, *Strigomonas galati*, *Strigomonas oncopelti*, and *Herpetomonas muscarum*, the *T. brucei* orthologue of each GCEC protein was used to TBLASTN whole genome shotgun assembled contigs deposited in NCBI for the presence of Open Reading Frames (ORFs) with high sequence similarity. The best matched contigs were then analyzed using NCBI-ORF Finder to identify the high-similarity ORF. The ORFs were then aligned with the T. brucei standard using CLUSTAL-omega software to evaluate protein start and end positions and sequence similarity of all parts of the putative proteins. If partial and complete ORFs were present, the most complete ORF was selected. The best matched ORF was taken as the most likely orthologous proteins for that species/isolate. If the ORFs were fragmented into more than one contig, individual amino acid sequence fragments were manually joined to produce a single protein sequence. For *Trypanoplasma borreli*, the draft assembled genome was downloaded from the ENA database (accession SAMEA1948381) and searched as was performed for the Euglenoid species below.

In the cases of Diplonema and Euglenoid species, genome assembly has remained challenging [47–49]. However, transcriptomes are available for two Euglenoids. For *Eutreptiella gymnastica* (NCBI SRA accession SRX549022) and *Euglena gracilis* [50], sequences were retrieved from raw or previously assembled transcriptomic data. When necessary, we assembled the transcriptomic data using Trinity [51]. We then searched for homologues of the *T. brucei* 64 consensus glycosome proteins in these assembled genomes with NCBI's BLAST+ [52]. For *Diplonema papillatum*, only the protein sequences identified in a previous study were used [15].

4.4. Development of Glycosome Targeting Signal 1 (PTS1) Algorithm

Our PTS1 score is derived directly from an additive log-odds score in each position. Site-specific frequencies $q_{i,j}$ for each amino acid j at position i were calculated from the GCEC protein set, and corresponding negative frequencies $p_{i,j}$ from the 35,640 other Uniprot proteins from *T. brucei*, *T. cruzi*, and *L. major*. Pseudocounts were added in computing the $q_{i,j}$ to prevent zero valued frequencies. The PTS1 score is then $\sum \ln(q_{i,j}/p_{i,j})$ for the observed amino acids (j) at the last three C-terminal sites.

Supplementary Materials: The following are available online at http://www.mdpi.com/2076-0817/9/4/281/s1, Table S1. Complete list of putative enzymatic proteins with experimental evidence of glycosome localization from *Trypanosoma brucei*, *Trypanosoma cruzi*, and *Leishmania donovani* global studies of glycosome composition. *Leishmania major* is used as the representative Leishmania species in the listed orthologues of the dixenous human-disease causing typanosomatids. Studies represent those that define glycosome proteomes, provide evidence of localization of endogenous tagged proteins from the TypTag project, or that are reflected in the input of glycosome proteins in the Peroxisome database culled from individual studies prior to 2010. Table S2. Specific putative proteins and orthologue identities (Uniprot/TriTryp/NCBI contig numbers) used to determine glycosome targeting signal conservation across kinetoplastids, select euglinids and *Diplonema papillatum*. Contigs listed may be one of several that contain an identical or near-identical amino acid sequence for that specific protein. Different species or isolates of a genus are listed in separate columns. Each row represents one protein of our Glycosome Conserved Enzyme Collection. Black rectangles in the *D. papillatum* column represent proteins for which we did not attempt to find an orthologue. The *D. papillatum* entries that were used were those identified in [15]. A gene with its C-terminus at the extreme end of the contig may not be the complete gene; these are indicated by an asterisk when noted. Table S3. Kinetoplastid UniProt entries with PTS1 scores of 5 or higher. File S1. Code for PTS1 predictor. File S2. Fasta files containing *Trypanoplasma borreli*, *Euglena gracilis*, and *Eutreptiella gymnastica* mRNA coding sequence for orthologues of Glycosome Conserved Enzyme Collection proteins derived from publicly available genomic or transcriptomic data.

Author Contributions: Conceptualization, S.L.Z. and J.N.R.; Methodology, H.D. and M.H. and J.N.R.; Software, M.H.; Validation, S.L.Z. and H.D. and M.H.; Formal Analysis, H.D. and M.H.; Investigation, H.D. and M.H. and J.N.R.; Data Curation, H.D.; Writing—Original Draft Preparation, S.L.Z.; Writing—Review & Editing, M.H. and J.N.R.; Visualization, H.D.; Supervision, S.L.Z.; Project Administration, S.L.Z. All authors have read and agreed to the published version of the manuscript.

Acknowledgments: This work and part of the publication costs were supported by University of Minnesota start-up funds to S.L.Z. H.D. was supported by an Integrated Biosciences Graduate Program Fellowship and a University of Minnesota Medical School Duluth Campus Dean's Fellowship. The University of Minnesota Department of Biomedical Sciences also contributed to publication costs. The authors thank Samuel Dean (University of Oxford) for a TrypTag glycosome protein list, Anzhelika Butenko (Czech Institute of Parasitology) for advice on Euglenozoa genomes, Takahiro Ishikawa (Shimane University) for an *E. gracilis* transcriptome assembly, Marek Eliáš and PhD students at University of Ostrava for helpful discussions, Clara Smoniewski (University of Minnesota) for critical reading of the manuscript, and Rich Melvin (University of Minnesota) for contributions to initial project discussions.

Conflicts of Interest: The authors declare no conflicts of interest.

References

1. Ferguson, M.A.; Low, M.G.; Cross, G.A. Glycosyl-sn-1,2-dimyristylphosphatidylinositol is covalently linked to *Trypanosoma brucei* variant surface glycoprotein. *J. Biol. Chem.* **1985**, *260*, 14547–14555. [PubMed]

2. Imboden, M.A.; Laird, P.W.; Affolter, M.; Seebeck, T. Transcription of the intergenic regions of the tubulin gene cluster of *Trypanosoma brucei*: Evidence for a polycistronic transcription unit in a eukaryote. *Nucleic Acids Res.* **1987**, *15*, 7357–7368. [CrossRef] [PubMed]

3. Muhich, M.L.; Boothroyd, J.C. Polycistronic transcripts in trypanosomes and their accumulation during heat shock: Evidence for a precursor role in mRNA synthesis. *Mol. Cell. Biol.* **1988**, *8*, 3837–3846. [CrossRef] [PubMed]

4. Sutton, R.E.; Boothroyd, J.C. Evidence for Trans splicing in trypanosomes. *Cell* **1986**, *47*, 527–535. [CrossRef]

5. Benne, R.; Van den Burg, J.; Brakenhoff, J.P.; Sloof, P.; Van Boom, J.H.; Tromp, M.C. Major transcript of the frameshifted coxII gene from trypanosome mitochondria contains four nucleotides that are not encoded in the DNA. *Cell* **1986**, *46*, 819–826. [CrossRef]

6. Lukeš, J.; Skalický, T.; Týč, J.; Votýpka, J.; Yurchenko, V. Evolution of parasitism in kinetoplastid flagellates. *Mol. Biochem. Parasitol.* **2014**, *195*, 115–122. [CrossRef]

7. Allmann, S.; Bringaud, F. Glycosomes: A comprehensive view of their metabolic roles in *T. brucei*. *Int. J. Biochem. Cell Biol.* **2017**, *85*, 85–90. [CrossRef]

8. Moyersoen, J.; Choe, J.; Fan, E.; Hol, W.G.J.; Michels, P.A.M. Biogenesis of peroxisomes and glycosomes: Trypanosomatid glycosome assembly is a promising new drug target. *FEMS Microbiol. Rev.* **2004**, *28*, 603–643. [CrossRef]

9. Szöör, B.; Haanstra, J.R.; Gualdrón-López, M.; Michels, P.A. Evolution, dynamics and specialized functions of glycosomes in metabolism and development of trypanosomatids. *Curr. Opin. Microbiol.* **2014**, *22*, 79–87. [CrossRef]

10. Jamdhade, M.D.; Pawar, H.; Chavan, S.; Sathe, G.; Umasankar, P.K.; Mahale, K.N.; Dixit, T.; Madugundu, A.K.; Prasad, T.S.K.; Gowda, H.; et al. Comprehensive proteomics analysis of glycosomes from *Leishmania donovani*. *OMICS* **2015**, *19*, 157–170. [CrossRef]

11. Acosta, H.; Burchmore, R.; Naula, C.; Gualdrón-López, M.; Quintero-Troconis, E.; Cáceres, A.J.; Michels, P.A.M.; Concepción, J.L.; Quiñones, W. Proteomic analysis of glycosomes from *Trypanosoma cruzi* epimastigotes. *Mol. Biochem. Parasitol.* **2019**, *229*, 62–74. [CrossRef] [PubMed]

12. Güther, M.L.S.; Urbaniak, M.D.; Tavendale, A.; Prescott, A.; Ferguson, M.A.J. High-Confidence Glycosome Proteome for Procyclic Form *Trypanosoma brucei* by Epitope-Tag Organelle Enrichment and SILAC Proteomics. *J. Proteome Res.* **2014**, *13*, 2796–2806. [CrossRef] [PubMed]

13. Opperdoes, F.R.; Szikora, J.P. In silico prediction of the glycosomal enzymes of *Leishmania major* and trypanosomes. *Mol. Biochem. Parasitol.* **2006**, *147*, 193–206. [CrossRef] [PubMed]

14. Makiuchi, T.; Annoura, T.; Hashimoto, M.; Hashimoto, T.; Aoki, T.; Nara, T. Compartmentalization of a Glycolytic Enzyme in Diplonema, a Non-kinetoplastid Euglenozoan. *Protist* **2011**, *162*, 482–489. [CrossRef] [PubMed]

15. Morales, J.; Hashimoto, M.; Williams, T.A.; Hirawake-Mogi, H.; Makiuchi, T.; Tsubouchi, A.; Kaga, N.; Taka, H.; Fujimura, T.; Koike, M.; et al. Differential remodelling of peroxisome function underpins the environmental and metabolic adaptability of diplonemids and kinetoplastids. *Proc. R. Soc. B Biol. Sci.* **2016**, *283*, 20160520. [CrossRef] [PubMed]

16. Wilkinson, S.R.; Prathalingam, S.R.; Taylor, M.C.; Ahmed, A.; Horn, D.; Kelly, J.M. Functional characterisation of the iron superoxide dismutase gene repertoire in *Trypanosoma brucei*. *Free Radic. Biol. Med.* **2006**, *40*, 198–209. [CrossRef]

17. Dufernez, F.; Yernaux, C.; Gerbod, D.; Noël, C.; Chauvenet, M.; Wintjens, R.; Edgcomb, V.P.; Capron, M.; Opperdoes, F.R.; Viscogliosi, E. The presence of four iron-containing superoxide dismutase isozymes in Trypanosomatidae: Characterization, subcellular localization, and phylogenetic origin in *Trypanosoma brucei*. *Free Radic. Biol. Med.* **2006**, *40*, 210–225. [CrossRef]

18. Sargsyan, Y.; Thoms, S. Staying in Healthy Contact: How Peroxisomes Interact with Other Cell Organelles. *Trends Mol. Med.* **2020**, *26*, 201–214. [CrossRef]

19. Schlüter, A.; Real-Chicharro, A.; Gabaldón, T.; Sánchez-Jiménez, F.; Pujol, A. PeroxisomeDB 2.0: An integrative view of the global peroxisomal metabolome. *Nucleic Acids Res.* **2010**, *38*, D800-5.

20. Colasante, C.; Ellis, M.; Ruppert, T.; Voncken, F. Comparative proteomics of glycosomes from bloodstream form and procyclic culture form *Trypanosoma brucei brucei*. *Proteomics* **2006**, *6*, 3275–3293. [CrossRef]

21. Parsons, M.; Nielsen, B. *Trypanosoma brucei*: Two-dimensional gel analysis of the major glycosomal proteins during the life cycle. *Exp. Parasitol.* **1990**, *70*, 276–285. [CrossRef]

22. Dean, S.; Sunter, J.D.; Wheeler, R.J. TrypTag.org: A Trypanosome Genome-wide Protein Localisation Resource. *Trends Parasitol.* **2017**, *33*, 80–82. [CrossRef] [PubMed]

23. Rogers, J.C.; Bomgarden, R.D. Sample preparation for mass spectrometry-based proteomics; from proteomes to peptides. In *Advances in Experimental Medicine and Biology*; Springer: New York, NY, USA, 2016; Volume 919, pp. 43–62.

24. Halliday, C.; Billington, K.; Wang, Z.; Madden, R.; Dean, S.; Sunter, J.D.; Wheeler, R.J. Cellular landmarks of *Trypanosoma brucei* and *Leishmania mexicana*. *Mol. Biochem. Parasitol.* **2019**, *230*, 24–36. [CrossRef] [PubMed]

25. Aslett, M.; Aurrecoechea, C.; Berriman, M.; Brestelli, J.; Brunk, B.P.; Carrington, M.; Depledge, D.P.; Fischer, S.; Gajria, B.; Gao, X.; et al. TriTrypDB: A functional genomic resource for the Trypanosomatidae. *Nucleic Acids Res.* **2010**, *38*, D457–D462. [CrossRef] [PubMed]

26. Hammond, D.J.; Gutteridge, W.E.; Opperdoes, F.R. A novel location for two enzymes of de novo pyrimidine biosynthesis in trypanosomes and Leishmania. *FEBS Lett.* **1981**, *128*, 27–29. [CrossRef]

27. Szöor, B.; Ruberto, I.; Burchmore, R.; Matthews, K.R. A novel phosphatase cascade regulates differentiation in *Trypanosoma brucei* via a glycosomal signaling pathway. *Genes Dev.* **2010**, *24*, 1306–1316. [CrossRef]

28. Aranovich, A.; Hua, R.; Rutenberg, A.D.; Kim, P.K. PEX16 contributes to peroxisome maintenance by constantly trafficking PEX3 via the ER. *J. Cell Sci.* **2014**, *127*, 3675–3686. [CrossRef]

29. Wang, J.; Wang, Y.; Gao, C.; Jiang, L.; Guo, D. PPero, a Computational Model for Plant PTS1 Type Peroxisomal Protein Prediction. *PLoS ONE* **2017**, *12*, e0168912. [CrossRef] [PubMed]

30. Neuberger, G.; Maurer-Stroh, S.; Eisenhaber, B.; Hartig, A.; Eisenhaber, F. Prediction of peroxisomal targeting signal 1 containing proteins from amino acid sequence. *J. Mol. Biol.* **2003**, *328*, 581–592. [CrossRef]

31. Fox, T.D. Mitochondrial protein synthesis, import, and assembly. *Genetics* **2012**, *192*, 1203–1234. [CrossRef]

32. Kalel, V.C.; Mäser, P.; Sattler, M.; Erdmann, R.; Popowicz, G.M. Come, sweet death: Targeting glycosomal protein import for antitrypanosomal drug development. *Curr. Opin. Microbiol.* **2018**, *46*, 116–122. [CrossRef] [PubMed]

33. Sugiura, A.; Mattie, S.; Prudent, J.; Mcbride, H.M. Newly born peroxisomes are a hybrid of mitochondrial and ER-derived pre-peroxisomes. *Nature* **2017**, *542*, 251–254. [CrossRef] [PubMed]

34. Bauer, S.T.; McQueeney, K.E.; Patel, T.; Morris, M.T. Localization of a Trypanosome Peroxin to the Endoplasmic Reticulum. *J. Eukaryot. Microbiol.* **2017**, *64*, 97–105. [CrossRef] [PubMed]

35. Pan, R.; Reumann, S.; Lisik, P.; Tietz, S.; Olsen, L.J.; Hu, J. Proteome analysis of peroxisomes from dark-treated senescent *Arabidopsis* leaves. *J. Integr. Plant Biol.* **2018**, *60*, 1028–1050. [CrossRef]

36. Gronemeyer, T.; Wiese, S.; Ofman, R.; Bunse, C.; Pawlas, M.; Hayen, H.; Eisenacher, M.; Stephan, C.; Meyer, H.E.; Waterham, H.R.; et al. The Proteome of Human Liver Peroxisomes: Identification of Five New Peroxisomal Constituents by a Label-Free Quantitative Proteomics Survey. *PLoS ONE* **2013**, *8*, e57395. [CrossRef]

37. Kerkhoven, E.J.; Achcar, F.; Alibu, V.P.; Burchmore, R.J.; Gilbert, I.H.; Trybiło, M.; Driessen, N.N.; Gilbert, D.; Breitling, R.; Bakker, B.M.; et al. Handling Uncertainty in Dynamic Models: The Pentose Phosphate Pathway in *Trypanosoma brucei*. *PLoS Comput. Biol.* **2013**, *9*, e1003371. [CrossRef]

38. Becco, L.; Smircich, P.; Garat, B. Conserved motifs in nuclear genes encoding predicted mitochondrial proteins in *Trypanosoma cruzi*. *PLoS ONE* **2019**, *14*, e0215160. [CrossRef]

39. Zhang, X.; Cui, J.; Nilsson, D.; Gunasekera, K.; Chanfon, A.; Song, X.; Wang, H.; Xu, Y.; Ochsenreiter, T. The *Trypanosoma brucei* MitoCarta and its regulation and splicing pattern during development. *Nucleic Acids Res.* **2010**, *38*, 7378–7387. [CrossRef]

40. Horáková, E.; Faktorová, D.; Kraeva, N.; Kaur, B.; Van Den Abbeele, J.; Yurchenko, V.; Lukeš, J. Catalase compromises the development of the insect and mammalian stages of *Trypanosoma brucei*. *FEBS J.* **2019**, *287*, 964–977. [CrossRef]

41. Bauer, S.; Morris, J.C.; Morris, M.T. Environmentally regulated glycosome protein composition in the African trypanosome. *Eukaryot. Cell* **2013**, *12*, 1072–1079. [CrossRef]

42. Herman, M.; Pérez-Morga, D.; Schtickzelle, N.; Michels, P.A.M. Turnover of glycosomes during life-cycle differentiation of *Trypanosoma brucei*. *Autophagy* **2008**, *4*, 294–308. [CrossRef] [PubMed]

43. Tielens, A.G.; van Hellemond, J.J. Surprising variety in energy metabolism within *Trypanosomatidae*. *Trends Parasitol.* **2009**, *25*, 482–490. [CrossRef] [PubMed]

44. Trindade, S.; Rijo-Ferreira, F.; Carvalho, T.; Pinto-Neves, D.; Guegan, F.; Aresta-Branco, F.; Bento, F.; Young, S.A.; Pinto, A.; Van Den Abbeele, J.; et al. *Trypanosoma brucei* Parasites Occupy and Functionally Adapt to the Adipose Tissue in Mice. *Cell Host Microbe* **2016**, *19*, 837–848. [CrossRef] [PubMed]

45. Flegontov, P.; Butenko, A.; Firsov, S.; Kraeva, N.; Eliáš, M.; Field, M.C.; Filatov, D.; Flegontova, O.; Gerasimov, E.S.; Hlaváčová, J.; et al. Genome of *Leptomonas pyrrhocoris*: A high-quality reference for monoxenous trypanosomatids and new insights into evolution of Leishmania. *Sci. Rep.* **2016**, *6*, 23704. [CrossRef]

46. Freitag, J.; Ast, J.; Bölker, M. Cryptic peroxisomal targeting via alternative splicing and stop codon read-through in fungi. *Nature* **2012**, *485*, 522–525. [CrossRef]

47. Ebenezer, T.G.E.; Carrington, M.; Lebert, M.; Kelly, S.; Field, M.C. Euglena gracilis genome and transcriptome: Organelles, nuclear genome assembly strategies and initial features. In *Advances in Experimental Medicine and Biology*; Springer: New York, NY, USA, 2017; Volume 979, pp. 125–140.

48. Ebenezer, T.E.; Zoltner, M.; Burrell, A.; Nenarokova, A.; Novák Vanclová, A.M.G.; Prasad, B.; Soukal, P.; Santana-Molina, C.; O'Neill, E.; Nankissoor, N.N.; et al. Transcriptome, proteome and draft genome of *Euglena gracilis*. *BMC Biol.* **2019**, *17*, 11. [CrossRef]

49. Kaur, B.; Valach, M.; Peña-Diaz, P.; Moreira, S.; Keeling, P.J.; Burger, G.; Lukeš, J.; Faktorová, D. Transformation of *Diplonema papillatum*, the type species of the highly diverse and abundant marine microeukaryotes Diplonemida (Euglenozoa). *Environ. Microbiol.* **2018**, *20*, 1030–1040. [CrossRef]

50. Yoshida, Y.; Tomiyama, T.; Maruta, T.; Tomita, M.; Ishikawa, T.; Arakawa, K. De novo assembly and comparative transcriptome analysis of *Euglena gracilis* in response to anaerobic conditions. *BMC Genom.* **2016**, *17*, 182. [CrossRef]

51. Grabherr, M.G.; Haas, B.J.; Yassour, M.; Levin, J.Z.; Thompson, D.A.; Amit, I.; Adiconis, X.; Fan, L.; Raychowdhury, R.; Zeng, Q.; et al. Full-length transcriptome assembly from RNA-Seq data without a reference genome. *Nat. Biotechnol.* **2011**, *29*, 644–652. [CrossRef]

52. Camacho, C.; Coulouris, G.; Avagyan, V.; Ma, N.; Papadopoulos, J.; Bealer, K.; Madden, T.L. BLAST+: Architecture and applications. *BMC Bioinform.* **2009**, *10*, 421. [CrossRef]

 *pathogens*

Article

Catalase and Ascorbate Peroxidase in Euglenozoan Protists

Ingrid Škodová-Sveráková [1,2,*,†] , Kristína Záhonová [1,3,†] , Barbora Bučková [2], Zoltán Füssy [3] , Vyacheslav Yurchenko [4,5] and Julius Lukeš [1,6,*]

1   Institute of Parasitology, Biology Centre, Czech Academy of Sciences, 370 05 České Budějovice (Budweis), Czech Republic; kika.zahonova@gmail.com
2   Faculty of Natural Sciences, Comenius University, 841 04 Bratislava, Slovakia; barbora@bucko.sk
3   Faculty of Science, Charles University, BIOCEV, 128 00 Prague, Czech Republic; zoltan.fussy@gmail.com
4   Life Science Research Centre, Faculty of Science, University of Ostrava, 710 00 Ostrava, Czech Republic; vyacheslav.yurchenko@osu.cz
5   Martsinovsky Institute of Medical Parasitology, Tropical and Vector Borne Diseases, Sechenov University, 119435 Moscow, Russia
6   Faculty of Sciences, University of South Bohemia, 370 05 České Budějovice (Budweis), Czech Republic
*   Correspondence: skodovaister@gmail.com (I.Š.-S.); jula@paru.cas.cz (J.L.)
†   These authors contributed equally to this work.

Received: 27 March 2020; Accepted: 22 April 2020; Published: 24 April 2020

Abstract: In this work, we studied the biochemical properties and evolutionary histories of catalase (CAT) and ascorbate peroxidase (APX), two central enzymes of reactive oxygen species detoxification, across the highly diverse clade Eugenozoa. This clade encompasses free-living phototrophic and heterotrophic flagellates, as well as obligate parasites of insects, vertebrates, and plants. We present evidence of several independent acquisitions of CAT by horizontal gene transfers and evolutionary novelties associated with the APX presence. We posit that Euglenozoa recruit these detoxifying enzymes for specific molecular tasks, such as photosynthesis in euglenids and membrane-bound peroxidase activity in kinetoplastids and some diplonemids.

Keywords: Euglenozoa; ascorbate peroxidase; catalase; enzymatic activity; phylogeny

---

1. Introduction

Aerobic metabolism is associated with the undesirable production of reactive oxygen species (ROS) due to the leakage of electrons to molecular oxygen [1]. The ROS molecules include free radicals, such as superoxide anion ($O_2^{\bullet-}$) or hydroxyl radical ($^{\bullet}OH$), and non-radical molecules, e.g., hydrogen peroxide (H_2O_2) [2]. In the eukaryotic cell, respiration in mitochondria and photosynthesis in plastids are the main ROS producers. However, ROS may accumulate as a by-product in any cell compartment where aerobic metabolism occurs. While molecular oxygen shows relatively low reactivity towards most cellular components, its partially reduced forms are much more reactive. Oxidative damage of proteins, lipids, and nucleic acids happens when the level of ROS exceeds the physiological threshold under oxidative stress conditions [3]. Hydrogen peroxide is often used as a second messenger in cell signaling pathways and its accumulation has long been documented to play an important role in mediating programmed cell death—apoptosis or (at very high concentrations) necrosis [4]. Hence, both the production and removal of ROS must be strictly controlled. This is why enzymatic and non-enzymatic mechanisms for ROS detoxification are employed by virtually every cell [5].

The family of enzymatic antioxidants comprises catalase (CAT), ascorbate peroxidase (APX), superoxide dismutase (SOD), glutathione peroxidase (GPX), and peroxiredoxin (PrxR) [6–8]. Both CAT and APX are heme-containing enzymes widely distributed among aerobes. As such, the APX activity was documented in higher plants and a range of protist lineages, including chlorophytes, rhodophytes, stramenopiles, and euglenozoans [9–11]. The structure and function of APX have been well described in plants, which usually carry several isoforms [12]. While APX requires ascorbate for ROS detoxification, with four reactions of the ascorbate–glutathione cycle necessary for ascorbate regeneration, CAT represents a system for direct ROS dissociation by conversion of H_2O_2 molecules into water and molecular oxygen. This enzyme requires one molecule of H_2O_2 to bind at the CAT active site in order to generate a reaction intermediate that binds the second molecule of H_2O_2 [13]. The affinities of APX and CAT for H_2O_2 are in μM and mM ranges, respectively, reflecting that they belong to two fundamentally different classes of peroxidases [14]. Having a low affinity to H_2O_2, CAT is most effective at high concentrations of peroxide [15].

ROS scavenging systems are classified according to their subcellular localization, which is primarily determined by the organelle-specific targeting signals found in the N- and C-termini of the corresponding proteins. Soluble forms are found in the cytosol, mitochondria, and plastids, while membrane-bound isoforms are found in microbodies (including peroxisomes and glyoxysomes) and plastid thylakoids [16–18]. Usually, multiple systems are present in a given cellular compartment and cooperate to scavenge for ROS [19]. CAT occurs in either soluble or membrane-bound forms, and is typically localized in the peroxisomes and mitochondria, where H_2O_2 production is most significant [20].

Kinetoplastids and euglenids possess yet another mechanism for scavenging ROS centered on the glutathione analog trypanothione, that is unique to these protists [21]. The trypanothione system reduces H_2O_2 via tryparedoxin and tryparedoxin peroxidase. Since CAT and the selenium-containing GPX are absent in the kinetoplastid flagellates (for exceptions, see below), trypanothione is particularly important in this group of organisms [22]. For the same reasons, APX appears to be a key enzyme for redox homeostasis in these flagellates [23]. In addition, trypanosomatids also lack the whole thioredoxin/thioredoxin reductase pathway, and it was proposed that trypanothione actually substitutes it. However, in the absence of CAT, trypanothione and thioredoxin systems co-exist in *Euglena gracilis*, indicating that these systems are not entirely redundant [24–26].

It was shown that, under stress, plants boost the activity of all of their ROS scavenging enzymatic systems, namely CAT, APX, SOD, GPX, and PrxR [16]. Interestingly, an increase in the APX activity can compensate for the loss of the CAT activity in plants and prevent the accumulation of intracellular ROS, suggesting the functional overlap of both systems [5]. Indeed, in tobacco, the CAT and APX machineries are (to some extent) functionally redundant, as one can compensate for the lack of another [27]. Given the complex pattern of antioxidant systems in the euglenozoan protists, we performed detailed phylogenetic and functional analyses of their APX proteins, dissected their co-occurrence with CAT, and identified multiple acquisitions of these nearly ubiquitous enzymes from unrelated sources.

Euglenozoa are as diverse as can be, and they comprise free-living, parasitic, and photosynthetic species, having clinical, economical, and environmental importance. Yet, how euglenozoans actually cope with ROS is still poorly understood. Given the complex pattern of antioxidant systems in euglenozoan protists, we performed phylogenetic and biochemical analyses of their APX and CAT to shed more light on the importance of these nearly ubiquitous enzymes.

2. Results

2.1. Euglenozoans Encode Unique HPXs and CAT of Different Origins

An extensive phylogenetic analysis of the APX domain-containing sequences revealed the subdivision of sequences derived from Euglenozoa into several clades (Figure 1; File S1). They are well-separated from heme (HPX) and cytochrome *c* peroxidases (CCP), suggesting a sub-functionalization of the HPX superfamily early in the eukaryotic evolution. All studied sequences can be divided into the following clades: (i) diplonemid peroxidases, which are invariably predicted as mitochondrion-localized (in turquoise) (Supplementary Table S1), form a moderately supported sister lineage to mitochondrial CCPs; (ii) a sequence from the diplonemid *Lacrimia lanifica* forms a sister branch to the kinetoplastid-specific hybrid APX-CCP proteins (hAPX-CCPs, in blue). While the kinetoplastid hAPX-CCP orthologues carry a mitochondrial targeting signal, the *L. lanifica* sequence possesses a peroxisomal targeting signal (Supplementary Table S1), which should navigate the corresponding protein to the glycosomes [28]; (iii) APXs of the diplonemids *Artemidia motanka*, *Namystynia karyoxenos*, and *Sulcionema specki* are nested inside the plastid/cytosolic APX clade along with various algae, parasitic chytridiomycota, and filter-feeding choanoflagellates, all of which carry a PTS2 signal, strongly indicating their glycosomal localization (in bright orange); (iv) a clade, consisting solely of sequences from the diplonemids *A. motanka* and *N. karyoxenos*, branches inside the plastid hAPX-CCPs cluster, close to the plastid-targeted euglenid APX (in dark green). Consistent with the absence of plastid in diplonemids, their APXs are predicted to be cytosolic (Supplementary Table S1); (v) a small clade contains an additional APX homolog from the plastid-carrying euglenids, otherwise specific for Chloroplastida (in light green); (vi) all remaining 29 diplonemid and six euglenid sequences constitute five independent clades that are unrelated to known APX sequences (in black). In the absence of an appropriate reference ortholog, we cannot infer their function, and thus we denoted them as Euglenozoa-specific HPX. We predict these proteins to be targeted to a range of cellular compartments, including mitochondria, glycosomes, and the secretory organelles (Supplementary Table S1).

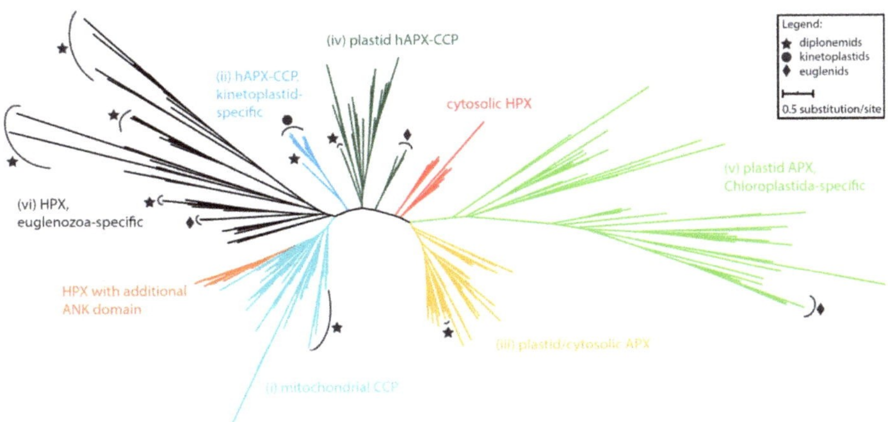

Figure 1. Maximum-likelihood phylogeny of heme peroxidases possessing APX domains in Euglenozoa. Taxa representing euglenozoan sequences are marked by symbols according to the graphical legend. Full tree in Newick format can be found in File S1.

It was previously shown that kinetoplastid flagellates acquired their CAT enzyme at least two times independently from bacteria. As revealed by the phylogenetic analysis, CAT of Leishmaniinae and the *Blastocrithidia*/"*jaculum*" clade derive from different bacterial groups [29,30], and this was confirmed here using a larger dataset (Figure 2A; File S2). While euglenids, studied so far, do not encode CAT [11,31], we wondered whether the same pattern applies to diplonemids, which constitute a sister clade to kinetoplastids [32,33]. For this purpose, we took advantage of the transcriptomic data derived from the axenic cultures of several diplonemid species [26]. As supported by maximum likelihood, the CAT sequences of *Diplonema japonicum*, *N. karyoxenos*, *Rhynchopus humris*, *A. motanka*, and *S. specki* are nested within eukaryotes with maximum support, consistent with the ancestral origin of their CAT (Figure 2B). However, yet another diplonemid, *Diplonema papillatum*, has apparently gained its CAT by horizontal gene transfer from an alpha-proteobacterium (Figure 2A,C). Hence, the inspected euglenozoans have acquired CAT from at least three distinct bacterial sources, while the genes of the majority of studied diplonemids are clearly of eukaryotic origin. This shows an unusual propensity of this group of protists to functionally replace CAT with homologs from bacteria that they likely prey upon.

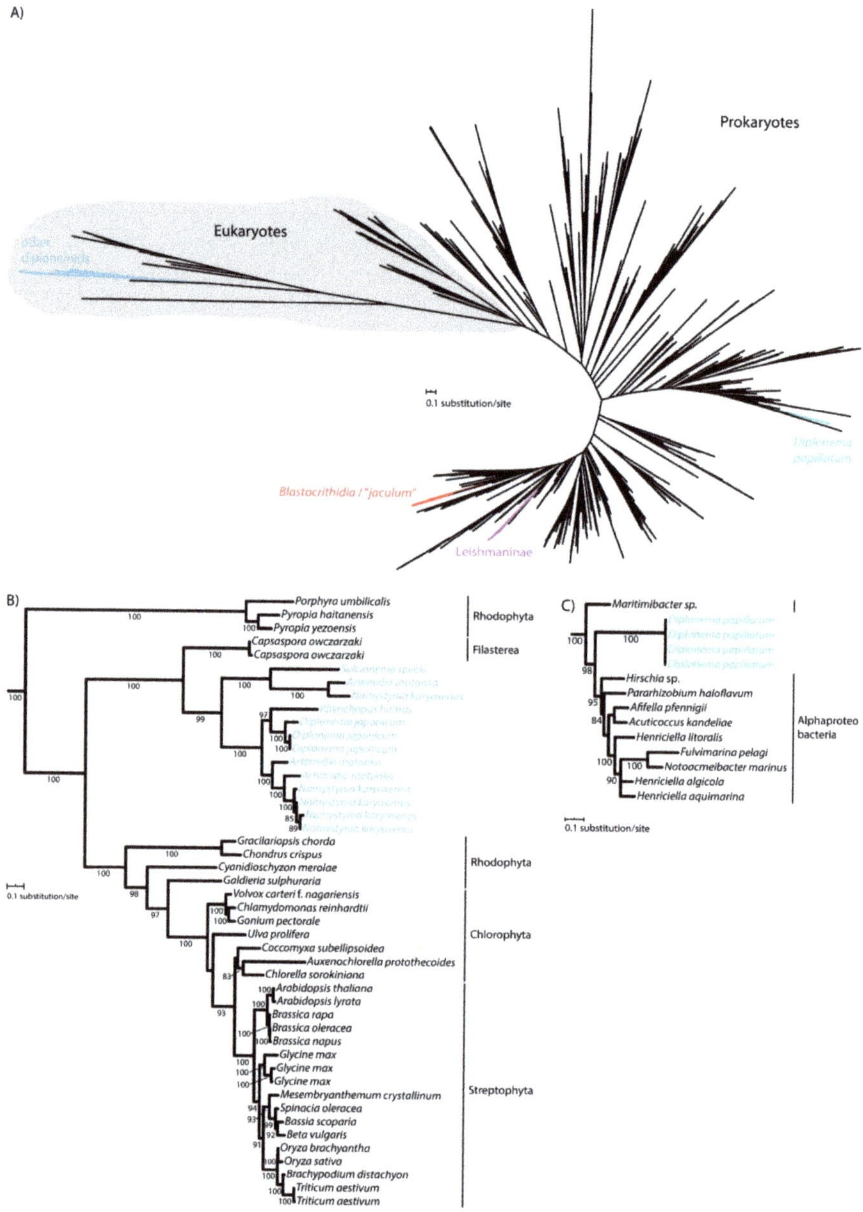

Figure 2. Maximum-likelihood phylogeny of catalases in Euglenozoa. (**A**) Unrooted tree showing ancestral origin of diplonemid CAT branching within eukaryotes and three horizontal gene transfer events in *Blastocrithidia*/"*jaculum*", Leishmaninae and *D. papillatum*. Full tree in Newick format can be found in File S2. (**B**) Subtree showing the ancestral diplonemid CAT. (**C**) Subtree showing the *D. papillatum* CAT related to alpha-proteobacteria. UFBoot support values are shown when ≥ 80.

2.2. APX Activity Widely Varies Among Species

To corroborate computational results by biochemical evidence, we measured the CAT and APX activities separately in total cell lysates. In a good correlation, *D. papillatum* lacks both APX genes and the corresponding biochemical activity (Figure 3).

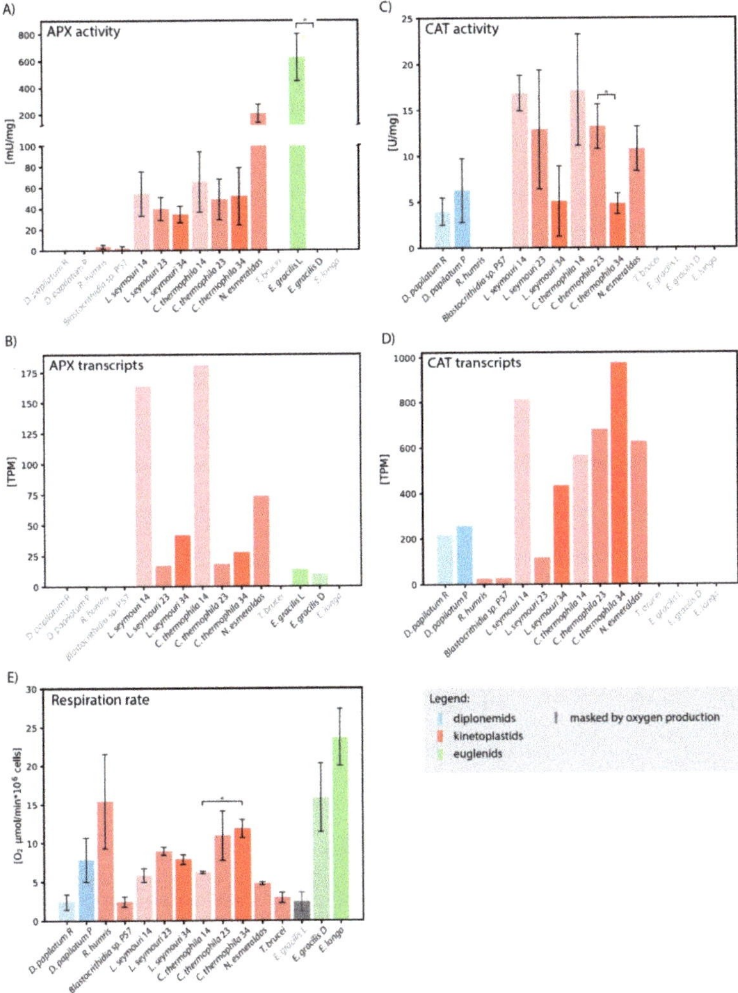

Figure 3. Biochemical and transcriptomic analysis. Comparison of (**A**) APX and (**C**) CAT activities, (**B**) APX and (**D**) CAT expression levels, and (**E**) oxygen uptake in *Diplonema papillatum* cultivated in nutrient-rich (R) and nutrient-poor (P) medium, *Rhynchopus humris*, *Blastocrithidia* sp. P57, *Leptomonas seymouri* cultivated at 14 °C (14), 23 °C (23) and 34 °C (34), *Crithidia thermophila* cultivated at 14 °C (14), 23 °C (23) and 34 °C (34), *Novymonas esmeraldas*, *Trypanosoma brucei*, *Euglena gracilis* cultivated in light (L) and dark (D), and *Euglena longa*. Species names in grey denote organisms, in which corresponding enzyme was not identified. Activity U is defined as the amount of the enzyme which catalyzes the conversion of 1 µmol of ascorbate (APX) or H_2O_2 (CAT) per 1 min. Each experiment was performed in two biological replicates. Statistical significance of differences between organisms was evaluated by an unpaired *t*-test. * statistically significant ($p < 0.05$). Note that respiration value in light-grown *E. gracilis* is masked by photosynthetic oxygen consumption (grey bar).

Surprisingly, in *R. humris* and *Blastocrithidia* sp. P57 very low APX activity was detected, despite the fact that both species apparently lack the corresponding gene. *Leptomonas seymouri*, *Crithidia thermophila* (both cultivated at a standard temperature of 23 °C), and *Novymonas esmeraldas* possess the kinetoplastid-specific hAPX-CCP, and consistently, specific activities of 39 ± 10, 48 ± 19, and 206 ± 69 mU/mg, respectively, were documented in these species. (Figure 3A; Supplementary Table S2). It should be noted that the activity of *N. esmeraldas* APX is comparable to that of the full-length APX of *Leishmania major* [34]. Although the APX activity changed when *L. seymouri* and *C. thermophila* were shifted to 14 °C and 34 °C, the limit temperatures at which both organisms are able to grow, the change was not statistically significant (Figure 3A). Interestingly, gene expression followed the same pattern in both species with the highest number of transcripts at 14 °C and their decrease with elevated temperature (Figure 3B; Supplementary Table S2).

When *E. gracilis* is grown under light conditions, promoting photosynthesis, it exhibits high APX activity of 625 ± 176 mU/mg (Supplementary Table S2), which is still 4-times lower than the activity reported for the pea plastids [35]. When the *E. gracilis* culture was transferred from light to dark, its color changed from green to pale reddish and the photosynthetic activity ceased. Although a low amount of APX transcripts were still present in these conditions, the enzymatic activity dropped below the limit of detection (Figure 3B; Supplementary Table S2). Consistent with the localization of APX in the plastid, neither the dark-grown *E. gracilis* nor the non-photosynthetic *Euglena longa* exhibit any APX activity (Figure 3A,B; Supplementary Table S2).

2.3. The Catalytic Center of APX is Altered in Euglenozoans

The primary APX sequences from selected representatives, namely the plastid-bearing *Arabidopsis thaliana*, *Glycine max*, and *E. gracilis*, and the plastid-lacking *L. major*, *L. seymouri*, *C. fasciculata*, and *N. esmeraldas* possess the domains required for the H_2O_2-reducing APX activity (Figure 4; Supplementary Figure S1).

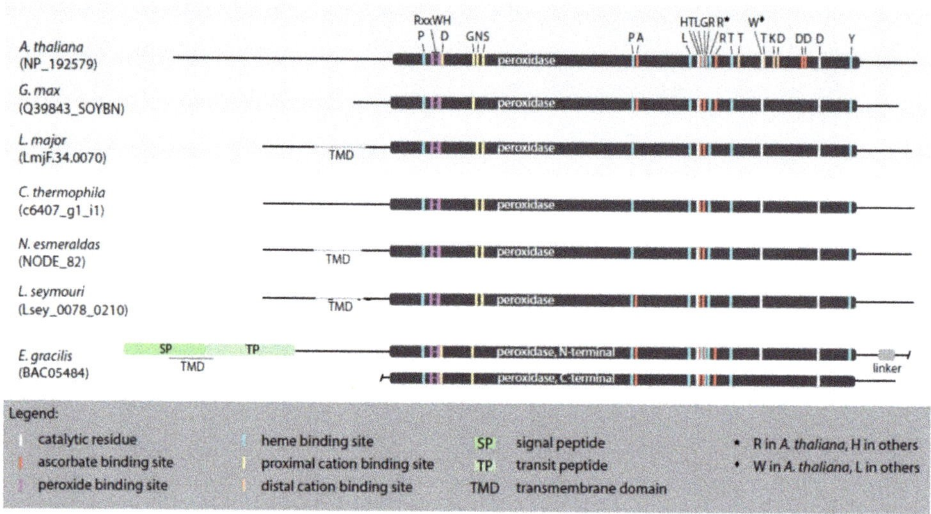

Figure 4. Schematic alignment of APX sequences from selected taxa. APXs from organisms studied previously are boxed in grey. Important amino acid residues are highlighted by different colors explained in the graphical legend.

The critical residues that coordinate binding of H_2O_2 by APX are R^{158}, W^{161}, and H^{162} (numbering according to the alignment in Supplementary Figure S1) [36], with the active site composed of the catalytic triad H^{293}, W^{325}, and D^{354} [37]. However, in the catalytic center of the studied APX sequences W^{325} is invariably replaced with L^{325}, except for the APX of *A. thaliana* (Supplementary Figure S1). Furthermore, we have identified residues critical for the heme and ascorbate binding, the number of which being species-specific (Figure 4; Supplementary Figure S1). Regarding the heme binding residues, in all analyzed sequences (except for *A. thaliana*), H^{299} substitutes R^{299}, similarly to the multiple isoforms of cytosolic APX [36]. Surprisingly, when compared to the model *A. thaliana* APX sequence, the proximal and distal cation-binding sites have been significantly changed or lost altogether, respectively, in all inspected euglenozoans and *G. max* (Figure 4; Supplementary Figure S1).

2.4. CAT Activity is Temperature-Dependent

In consonance with the absence of the CAT-encoding gene in euglenids [25] and *T. brucei* [38], we did not detect any CAT activity in these species (Figure 3C; Supplementary Table S2). *N. Esmeraldas* exhibited the CAT activity of 10 ± 2.4 U/mg, but with the employed methodology we failed to detect any activity in *Blastocrithidia* sp. P57 and *R. humris*, despite the fact that corresponding transcripts were expressed (Figure 3C,D; Supplementary Table S2). While the rich and poor cultivation medium has a significant impact on the metabolism of *D. papillatum*, the CAT activity remained stable under different conditions (Figure 3C; Supplementary Table S2).

Recently, it has been demonstrated that the human CAT and its orthologue from a monoxenous (insect-hosts only) trypanosomatid *C. fasciculata* have very different temperature optima [39]. In order to investigate the thermal properties of euglenozoan CAT in more detail, we examined its activity in *C. thermophila* and *L. seymouri*, both at their optimal cultivation temperature of 23 °C, as well as at 14 °C and 34 °C. These monoxenous species were selected because they have the highest expression and activity of CAT among the studied euglenozoans (Figure 3C,D; Supplementary Table S2), and are also thermo-tolerant, being able to withstand temperature changes [40,41]. In both species, the elevated temperature lowered CAT activity (Figure 3C). Nevertheless, the increased temperature caused a mild increase (1.5- and 3.7-times in *C. thermophila* and *L. seymouri*, respectively) in the mRNA level of CAT. The resulting pattern of CAT activity and transcription is rather complex. In both species, lower temperature generally resulted in increased CAT activity (Figure 3C), although the correlation between temperature and transcription levels was different for *C. thermophila* and *L. seymouri* (Figure 3D).

2.5. Respiration Rate Does Not Correlate with CAT and APX Activities

We assumed that organisms with high oxygen uptake require highly active systems for effective ROS detoxification. If so, CAT was the best candidate for this role, since its activity is not linked to any other system and it can provide direct and rapid reduction of H_2O_2. However, we did not find any correlation between the presence or activity of the CAT or APX detoxification systems and the rate of respiration (Figure 3E; Supplementary Table S2). Although it has the highest oxygen uptake, *E. longa* encodes neither CAT nor APX in its transcriptome (presumed to be highly representative [42]), and, correspondingly, their activities were lacking (Figure 3). Consistently, the dark-grown *E. gracilis* does not display any APX activity, with its respiration rate only slightly lower than that of *E. longa* (Figure 3E). Since oxygen consumption was masked by photosynthesis in the illuminated *E. gracilis* with fully developed plastids, we could not properly evaluate respiration in this case. The CAT activity of *D. papillatum* was not significantly influenced by the cultivation conditions (rich versus poor medium), even with increased respiration rate in the latter (Figure 3E). The respiration rate of thermostable *C. thermophila* and *L. seymouri* was lowest at 14 °C, however the enzymatic activities were highest (Figure 3A,C,E).

3. Discussion

The complex phylogenetic pattern, diversity, and distribution of CAT and APX in euglenozoans testify to their importance for these protists. Indeed, *E. gracilis* contains a photosynthesis-specific APX shared with other phototrophic euglenophytes, along with a putative plastidial APX acquired from and limited to Chloroplastida (Figure 1). Moreover, both diplonemids and euglenids encode a novel clade of peroxidases with an unknown function, while most kinetoplastids share a unique hAPX-CCP enzyme exhibiting both the APX and CCP activities [43]. Surprisingly, the kinetoplastid *Blastocrithidia* sp. P57 and the diplonemid *R. humris* apparently lack hAPX-CCP, so the low APX activity in both species must be assigned to another, possibly horizontally transferred, yet unidentified enzyme. Another such example is CAT in *D. papillatum* that was gained horizontally from an alpha-proteobacterium (Figure 2C). The distribution of APX and CAT in diplonemids is best explained by a scenario, in which the predecessor of these marine protists lacked both enzymes, which were reacquired by horizontal gene transfer from either prokaryotic or eukaryotic sources (Figures 1 and 2). The importance of possessing such detoxifying systems is highlighted by the fact that this has occurred independently in several euglenozoan lineages.

For a long time, APX was considered characteristic for photosynthetic organisms, while CAT was thought to be ubiquitous in the aerobic systems [44]. However, the distribution of these enzymes in euglenozoans challenges both claims. Evidence for the presence of APX and prominent absence of CAT in *Trypanosoma cruzi* and *Leishmania major* suggest that the former contributes to the ROS scavenging also in these parasites [34]. The absence of CAT in certain parasitic trypanosomatids is likely due to an adaptation to their dixenous (two-hosts) lifestyle, as the transition in the development from the insect to the mammalian stages of *T. brucei* and *Leishmania* spp. seem to rely on H_2O_2 production [39,45,46]. However, the lack of CAT in dixenous kinetoplastids is not universal, as exemplified by *Trypanoplasma borelli* harboring a glycosomal CAT [47].

CAT and APX complement each other's function in tobacco [27], so to clarify the evolutionary context of their functional differentiation in euglenozoans, we searched for the CAT and APX sequences in genomes and/or transcriptomes of representatives, for which such data are available. From these organisms, only *T. brucei*, *E. gracilis*, and *E. longa* lack both the CAT genes and corresponding activity (Figure 3C,D). However, despite the presence of CAT transcripts in *R. humris* and *Blastocrithidia* sp. P57, respective enzymatic activity was below our detection limit (Figure 3C), which is not consistent with the detection of low CAT activity via heme-dependent oxygen production [29]. Our methodology is based on a spectrophotometric measurement of a decrease in H_2O_2 (for CAT) and ascorbate (for APX) levels. Although the sensitivity of oxygen detection appears to be significantly higher than that for the spectrophotometric detection of H_2O_2 reduction, we used the latter method in order to have consistent comparison for the CAT and APX enzymes. The kinetic parameters of CAT suggest yet another explanation for the lack of measurable activity in *R. humris* and *Blastocrithidia* sp. P57. The low affinity of CAT to H_2O_2 implies that it is responsible for the removal of excessive ROS when their concentration is high, while high-affinity APX modulates low concentration of ROS, necessary for cell signaling [14]. Thus, it is plausible that under the applied experimental setup *R. humris* and *Blastocrithidia* sp. P57 were not exposed to the conditions in which ROS would exceed a threshold and upregulate CAT activity.

Since at least *E. longa* and *T. brucei* have neither CAT nor APX activity, the plant-like pattern with APX complementing the lack of CAT does not apply to the euglenozoan protists. In *E. gracilis*, a high APX activity was limited to the phototrophic growth conditions and, supposedly, the plastid. The corresponding APX contains two homologous catalytic domains, forming an intramolecular dimeric structure and a class II plastid-targeting bipartite sequence [48]. Previously, the APX activity was demonstrated to be cytosolic [48], which is most likely an artifact of the procedure, given the presence of the encoded targeting sequence and recent proteomic evidence [49]. We propose that the high APX activity in *E. gracilis*, comparable with that in plant plastids [35], mainly mitigates photosynthetic ROS production in plastids, rather than amends the absence of CAT. Furthermore, the APX activity in plastid-lacking protists, assayed herein, is 3–18-times lower when compared to the phototrophic *E. gracilis*, suggesting that its expression in these species does not meet conditions where high amounts of ROS need to be combatted.

The catalytic properties of APX depend on the architecture of its domains, substrate binding and orienting sites. Our results show that not all previously described catalytic amino acids are conserved.

For instance, a substitution of W for L within the catalytic triad at position 325 can apparently be tolerated and it does not affect the enzymatic activity, since *G. max* APX (with L^{325}) shows activity comparable to the *A. thaliana* orthologue (with W^{325}) [50,51]. Uniquely, APXs of *L. major*, *L. seymouri*, and *N. esmeraldas* (and, probably, other representatives of Leishmaniinae [52]) possess the N-terminal anchoring trans-membrane domain that modulates their catalytic activity. While full-length proteins exhibit specific APX activities comparable to the cytosolic counterparts, deletion of this domain from *L. major* caused a 5-fold decrease in activity [34]. This highlights the importance of membrane tethering for enzyme architecture or substrate accessibility for this type of APX. On the other hand, the absence of several cation-binding residues has no effect on enzymatic activity, suggesting that they may be either redundant or not critical.

ROS production is a phenomenon, accompanying aerobic life, as both photosynthesis and respiration are major sources of ROS. However, we did not document a direct link between the rate of oxygen consumption and the activity of studied peroxidases. High ROS production via respiration is considered to recruit the CAT and APX systems to limit ROS reactivity [53]. However, we posit that this correlation may not be as straightforward. An increase in oxygen uptake was documented in *D. papillatum*, cultivated in the nutrient-poor medium, yet this had no effect on its CAT activity. Surprisingly, in thermostable *C. thermophila* and *L. seymouri*, elevated temperature triggered an increase in respiration, but a decrease in the activities of both CAT and APX. That transcript abundance did not follow the same pattern as activity can be explained by the key role of post-transcriptional regulation in the gene expression of euglenids and trypanosomatids [54–58]. Indeed, weak-to-moderate correlation has been observed between transcript abundance and protein levels in both trypanosomatids and *Euglena*. This suggests an important role of mRNA turnover, translational efficiency and protein degradation in modulating biological responses in these protists [57]. All euglenozoan protists are known for polycistronic transcription and the important role of post-transcriptional processes. Hence, the rather weak correlation between transcript levels and enzymatic activities of CAT and APX is not unexpected.

To conclude, despite the fact that the ROS detoxification systems in plants are upregulated under different conditions, we documented a similar activity pattern only for CAT in thermostable kinetoplastids, but not in the other studied euglenozoans. When present, these peroxidases appear to be constitutively transcribed, yet the extent of their enzymatic activity varies widely across the examined species. CAT and APX are retained (or horizontally acquired) only in a subset of studied protists, which is reflected in their complex phylogenies.

4. Materials and Methods

4.1. Sequence Searches and Phylogenetic Analyses

CAT sequences from *D. papillatum* were found by tBLASTn [59] search in an unpublished genome and transcriptome assembly using kinetoplastid sequences, identified previously [29], as queries. The *D. papillatum* sequences served as queries for searches in the transcriptomes of other diplonemid species – *Diplonema japonicum*, *Rhynchopus humris*, *Lacrimia lanifica*, *Sulcionema specki*, *Artemidia motanka*, and *Namystynia karyoxenos*.

APX sequences were downloaded from RedOxiBase [60] and GenBank [61] databases. These served as queries and references for APX identification in all studied euglenozoans and other protists. To distinguish between APX and cytochrome *c* peroxidases, with which they share the common phylogenetic origin [43], all euglenozoan sequences were submitted to the InterProScan [62]. Only those with identified APX domains or strong affiliation to the reference APX clades were retained for further phylogenetic analysis.

Sequences were clustered (50% identity and 80% coverage) using MMseqs2 [63]. Datasets were aligned by MAFFT [64] and poorly aligned positions were discarded by trimAl [65] using -gt 0.5 option. Maximum likelihood trees were inferred from the alignments using the LG + C20 + F + G model and the posterior mean site frequency method [66], LG + F + G guide tree in the IQ-TREE software [67] and employing the strategy of rapid bootstrapping followed by a "thorough" maximum likelihood search with 1000 bootstrap replicates.

4.2. Localization Predictions

To assess a putative subcellular localization of all studied euglenozoan proteins, PrediSi [68], NommPred (in kinetoplastid setting; [69]), TargetP v.2.0 [70], and MultiLoc2 (in fungal and animal settings; [71]) tools were employed. Glycosomal predictions were determined by an in-house python script based on previously identified targeting signals in kinetoplastids [28]. The number of transmembrane domains was predicted by TMHMM [72] or Phobius [73], implemented in the Geneious Prime software [74].

4.3. Transcript Expression Levels

Trimmed RNA-Seq reads were mapped onto the assembled transcriptomes using BBMap (part of the BBTools suite; https://jgi.doe.gov/data-and-tools/bbtools/). The expression values for each transcript were calculated as Fragments Per Kilobase of transcript per Million mapped reads (FPKM). From these, TPM (Transcripts Per Million) values were calculated as FPKM / sum(FPKMs) $\times$ 10^6. For proteins with several transcript models, mean TPM was calculated.

4.4. Strains and Culture Conditions

D. papillatum cells were inoculated into the nutrient-rich and nutrient-poor media, to a final concentration 5 $\times$ 10^5 cells/mL. The nutrient-rich medium contained 36 g/L sea salts (Red Sea), and 1% (v/v) horse serum (Sigma-Aldrich, St. Louise, USA), and 1 g/L tryptone (Duchefa Biochemie, Amsterdam, Netherlands), while the nutrient-poor medium consisted of 36 g/L sea salts, 1% (v/v) horse serum, and 0.01 g/L tryptone [75].

R. humris was cultivated in artificial sea water containing 3.6% sea salts (Sigma-Aldrich), enriched with 1% (v/v) heat-inactivated horse serum (Sigma-Aldrich), and 0.025 g/L LB broth powder (Amresco, Solon, USA) [76].

Blastocrithidia sp. P57 was cultivated in a medium composed of 40% (v/v) Schneider medium, 40% (v/v) RPMI medium, and 20% (v/v) inactivated fetal bovine serum (all Sigma-Aldrich).

L. seymouri and *C. thermophila* were cultivated at 23 °C in the Brain Heart Infusion medium (Sigma-Aldrich) supplemented with 10 µg/mL of hemin (Jena Bioscience, Jena, Germany), 10% fetal bovine serum, 100 units/mL of penicillin, and 100 µg/mL of streptomycin (all Sigma-Aldrich) [77]. For experiments at different temperatures (14 °C and 34 °C), cells were seeded at a concentration of 3 $\times$ 10^5 cells/mL and cultured for 72 h as described previously [40].

T. brucei (strain 29-13) was cultured at 27 °C in SDM79 medium (GE Healthcare, Chicago, USA) containing 10% (v/v) heat-inactivated fetal bovine serum (Biosera, Nuaillé, France) and 2.5 mg/mL hemin [78].

E. gracilis cells were cultivated statically at 27 °C under constant illumination (10 µm/m²s¹) and dark in liquid Hutner medium [79]. *E. longa* cells were cultivated statically under constant illumination (10 µm/m²s¹) at 27 °C in Cramer-Myers medium supplemented with ethanol (0.8% v/v) [80].

4.5. RNA Isolation, Sequencing and Read Processing

Total RNA of 5 $\times$ 10^7 cells from each *L. seymouri* and *C. thermophila* shifted to 14 °C was isolated using the RNeasy Mini kit (Qiagen, Hilden, Germany) according to the manufacturer's instruction for three independent biological replicates. Paired-end strand-specific cDNA libraries were sequenced on Illumina NovaSeq platform (Macrogen Inc., Seoul, Korea). RNA-Seq reads were adapter and quality trimmed using BBDuk (part of the BBTools suite). As described above, trimmed reads were mapped onto previously assembled transcriptomes [40,41]. The raw sequencing data for *L. seymouri* and *C. thermophila* are available at NCBI (https://www.ncbi.nlm.nih.gov/) as BioProjects PRJNA611003 and PRJNA611063, respectively.

4.6. Protein Isolation

Five million cells of *D. papillatum* and *R. humris* were incubated with 2 mg of digitonin (AppliChem, Darmstadt, Germany) at room temperature for 5 min. Aliquot of 10^6 lysed cells were used for activity measurements. 5×10^8 *T. brucei*, *L. seymouri*, *C. thermophila*, *N. esmeraldas*, and *Blastocrithidia* sp. P57 cells were lysed on ice with 2% (w/v) dodecyl maltoside in 0.5 M aminocaproic acid (both AppliChem) for 1 h with subsequent 30 min centrifugation at 21,300× *g* at 4 °C. 10^5 *E. gracilis* and *E. longa* cells were disrupted using 200 mg of 1 mm silica spheres (Silica matrix C; MP Biomedicals, Irvine, USA), and cycle of 3 × 15 s, 4.0 M/S on FastPrep-24 (MP Biomedicals) with cooling on ice between cycles. Residual intact cells were separated from lysate by centrifugation at 1800× *g* for 10 min at 4 °C. Supernatant was used for activity measurements.

Protein concentration in cell lysates was determined using Bradford assay [81].

4.7. Measurements of Activities

Both CAT and APX activity were measured at 25 °C in whole-cell lysates. CAT activity was measured in total volume 1.3 mL that comprised 50 mM KPi pH 7.2, 0.005% (v/v) H_2O_2, and 10 μL of kinetoplastid lysate (*T. brucei*, *L. seymouri*, *C. thermophila*, *N. esmeraldas*, or *Blastocrithidia* sp. P57) or 100 μL of diplonemid lysate (*D. papillatum* or *R. humris*) or euglenid lysate (*E. gracilis* or *E. longa*). Activity of CAT was monitored for 2–4 min as a decrease in absorbance at 240 nm in the cuvette for UV-VIS spectra (Varian-Agilent Quartz Semi-Micro Cuvette Cell, Agilent, Santa Clara, USA). Activity U was calculated as the amount of enzyme that reduces 1 μmol of H_2O_2 ($\varepsilon_{240} = 43.6$ $M^{-1}cm^{-1}$) per 1 min. Specific activity was calculated as U per mg of cell proteins.

APX activity was measured analogously to CAT activity with the addition of freshly prepared 15 μM ascorbate (AppliChem). The activity was monitored for 2–4 min at 274 nm as a decrease in absorbance in the cuvette for UV-VIS spectra (Agilent). Activity U was calculated as the amount of enzyme required for a reduction of 1 μmol of ascorbic acid ($\varepsilon_{274} = 14,900$ $M^{-1}cm^{-1}$) per 1 min. Specific activity was calculated as U per mg of cell proteins.

In order to distinguish between H_2O_2 and ascorbate absorption spectra, we scanned absorbance of each substrate from 200 to 350 nm. Both molecules had absorption peaks at 300 nm, the wavelength at which CAT and APX activities are routinely detected [10,35]. After reducing the concentration of both substrates compared to the original protocols, 0.005% (v/v) H_2O_2 displayed an additional peak at 240 nm. Thus, we picked this wavelength (240 nm) for CAT measurements, because ascorbate did not absorb in this area. The control for CAT activity was the formation of oxygen that was visible in the cuvette in the form of bubbles. Since ascorbate displayed only one absorption peak around 300 nm, we adjusted the wavelength to 274 nm where 0.005% (v/v) H_2O_2 was not detected and only an increase in reduced ascorbate was monitored. Our optimized assay for APX activity measurement was closest to the protocol published previously [82], in which the concentration of ascorbate varied from 10 to 80 μM and oxidation of substrate was monitored at 265 nm with extinction coefficient for ascorbate $\varepsilon_{274} = 14,900$ $M^{-1}cm^{-1}$.

Each experiment was performed in two biological replicates. Each biological replicate consisted of 2–6 technical replicates. Statistical significance of differences between organisms was evaluated by unpaired *t*-test.

4.8. Oxygen Uptake Analysis

Clark oxygen electrode (Oxytherm System; Hansatech Instruments, Norfolk, UK) was used for measuring the oxygen consumption by intact cells. Each culture in the logarithmic growth phase was diluted to a concentration of 10^6 cells/mL. The electrode chamber was filled with 1 mL of culture and oxygen consumption was recorded for 4–6 min. Final values were calculated as the difference in oxygen consumption per 1 min caused by 10^6 cells.

Each experiment was performed in two biological replicates. Each biological replicate consisted of two to six technical replicates. Statistical significance of differences between organisms was evaluated by unpaired *t*-test.

Supplementary Materials: The following are available online at http://www.mdpi.com/2076-0817/9/4/317/s1, Figure S1: Full alignment of APX sequences from selected species. APX domains, signal and transit peptide, and linker are boxed by a colored background corresponding to domain colors in Figure 4. Important amino acid residues are highlighted in different colors corresponding to colors in Figure 4. Transmembrane domains are underlined. Table S1: Predicted subcellular localization of euglenozoan sequences used in this study. Table S2: Inferred expression levels and measurements of enzyme activities and oxygen uptake in studied euglenozoans. File S1: Maximum-likelihood phylogenetic tree of Euglenozoa HPX possessing APX domains in Newick format. File S2: Maximum-likelihood phylogenetic tree of Euglenozoa CAT in Newick format.

Author Contributions: Conceptualization, I.Š.-S. and K.Z.; methodology, I.Š.-S.; software, K.Z. and Z.F.; validation, I.Š.-S., K.Z., and Z.F.; formal analysis, I.Š.-S., K.Z., and B.B.; investigation, I.Š.-S. and K.Z.; resources, I.Š.-S. and V.Y.; data curation, I.Š.-S., K.Z., B.B. and Z.F.; writing—original draft preparation, I.Š.-S., K.Z., and Z.F.; writing—review and editing, I.Š.-S., K.Z., Z.F., V.Y., and J.L.; visualization, I.Š.-S., K.Z. and Z.F.; supervision, V.Y. and J.L.; project administration, I.Š.-S. and K.Z.; funding acquisition, I.Š.-S., V.Y. and J.L. All authors have read and agreed to the published version of the manuscript.

Funding: This research was funded by Czech Ministry of Education (project LM2015042), CERIT-Scientific Cloud (project LM2015085), ERC CZ (project LL1601), the Grant Agency of Czech Republic (projects 18-15962S and 20-07186S), the European Regional Funds (project OPVVV/0000759), the Grant Agency of the Slovak Ministry of Education and the Academy of Sciences (projects 1/0781/19 and 1/0387/17), and the Slovak Research and Development Agency (project APVV-0286-12).

Acknowledgments: We thank members of our laboratories for stimulating discussions.

Conflicts of Interest: V.Y. is an Academic Editor of Pathogens. Other authors declare no conflict of interest. The funders had no role in the design of the study; in the collection, analyses, or interpretation of data; in the writing of the manuscript, or in the decision to publish the results.

References

1. Bilinski, T. Oxygen toxicity and microbial evolution. *Biosystems* **1991**, *24*, 305–312. [CrossRef]
2. Lü, J.M.; Lin, P.H.; Yao, Q.; Chen, C. Chemical and molecular mechanisms of antioxidants: Experimental approaches and model systems. *J. Cell Mol. Med.* **2010**, *14*, 840–860. [CrossRef] [PubMed]
3. Gill, S.S.; Tuteja, N. Reactive oxygen species and antioxidant machinery in abiotic stress tolerance in crop plants. *Plant Physiol. Biochem.* **2010**, *48*, 909–930. [CrossRef]
4. Avery, S.V. Molecular targets of oxidative stress. *Biochem. J.* **2011**, *434*, 201–210. [CrossRef] [PubMed]
5. Apel, K.; Hirt, H. Reactive oxygen species: Metabolism, oxidative stress, and signal transduction. *Annu. Rev. Plant Biol.* **2004**, *55*, 373–399. [CrossRef]
6. Asada, K.; Takahashi, M. *Production and Scavenging of Active Oxygen in Chloroplasts*; Photoinhibition, D.J., Kyle, C.B., Arntzen, C.J., Eds.; Elsevier: Amsterdam, The Netherlands, 1987; pp. 227–287.
7. Bowler, C.; Montagu, M.V.; Inze, D. Superoxide dismutase and stress tolerance. *Annu. Rev. Plant Physiol. Plant Mol. Biol.* **1992**, *43*, 83–116. [CrossRef]
8. Willekens, H.; Chamnongpol, S.; Davey, M.; Schraudner, M.; Langebartels, C.; Van Montagu, M.; Inze, D.; Van Camp, W. Catalase is a sink for H_2O_2 and is indispensable for stress defence in C3 plants. *EMBO J.* **1997**, *16*, 4806–4816. [CrossRef]
9. Takeda, T.; Yoshimura, K.; Yoshii, M.; Kanahoshi, H.; Miyasaka, H.; Shigeoka, S. Molecular characterization and physiological role of ascorbate peroxidase from halotolerant *Chlamydomonas* sp. W80 strain. *Arch. Biochem. Biophys.* **2000**, *376*, 82–90. [CrossRef]
10. Wilkinson, S.R.; Obado, S.O.; Mauricio, I.L.; Kelly, J.M. *Trypanosoma cruzi* expresses a plant-like ascorbate-dependent hemoperoxidase localized to the endoplasmic reticulum. *Proc. Natl. Acad. Sci. USA* **2002**, *99*, 13453–13458. [CrossRef]
11. Shigeoka, S.; Nakano, Y.; Kitaoka, S. Metabolism of hydrogen peroxide in *Euglena gracilis* Z by L-ascorbic acid peroxidase. *Biochem. J.* **1980**, *186*, 377–380. [CrossRef]

12. Caverzan, A.; Passaia, G.; Rosa, S.B.; Ribeiro, C.W.; Lazzarotto, F.; Margis-Pinheiro, M. Plant responses to stresses: Role of ascorbate peroxidase in the antioxidant protection. *Genet. Mol. Biol.* **2012**, *35* (Suppl. 4), 1011–1019. [CrossRef] [PubMed]

13. Vlasits, J.; Jakopitsch, C.; Schwanninger, M.; Holubar, P.; Obinger, C. Hydrogen peroxide oxidation by catalase-peroxidase follows a non-scrambling mechanism. *FEBS Lett.* **2007**, *581*, 320–324. [CrossRef] [PubMed]

14. Mittler, R. Oxidative stress, antioxidants and stress tolerance. *Trends Plant Sci.* **2002**, *7*, 405–410. [CrossRef]

15. De Marco, A.; Roubelakis-Angelakis, K.A. The complexity of enzymic control of hydrogen peroxide concentration may affect the regeneration potential of plant protoplasts. *Plant Physiol.* **1996**, *110*, 137–145. [CrossRef]

16. Shigeoka, S.; Ishikawa, T.; Tamoi, M.; Miyagawa, Y.; Takeda, T.; Yabuta, Y.; Yoshimura, K. Regulation and function of ascorbate peroxidase isoenzymes. *J. Exp. Bot.* **2002**, *53*, 1305–1319. [CrossRef]

17. Teixeira, F.K.; Menezes-Benavente, L.; Margis, R.; Margis-Pinheiro, M. Analysis of the molecular evolutionary history of the ascorbate peroxidase gene family: Inferences from the rice genome. *J. Mol. Evol.* **2004**, *59*, 761–770. [CrossRef]

18. Teixeira, F.K.; Menezes-Benavente, L.; Galvao, V.C.; Margis, R.; Margis-Pinheiro, M. Rice ascorbate peroxidase gene family encodes functionally diverse isoforms localized in different subcellular compartments. *Planta* **2006**, *224*, 300–314. [CrossRef]

19. Mittler, R.; Vanderauwera, S.; Gollery, M.; Van Breusegem, F. Reactive oxygen gene network of plants. *Trends Plant Sci.* **2004**, *9*, 490–498. [CrossRef]

20. Glorieux, C.; Calderon, P.B. Catalase, a remarkable enzyme: Targeting the oldest antioxidant enzyme to find a new cancer treatment approach. *Biol. Chem.* **2017**, *398*, 1095–1108. [CrossRef]

21. Fairlamb, A.H.; Blackburn, P.; Ulrich, P.; Chait, B.T.; Cerami, A. Trypanothione: A novel bis(glutathionyl)spermidine cofactor for glutathione reductase in trypanosomatids. *Science* **1985**, *227*, 1485–1487. [CrossRef]

22. Carnieri, E.G.; Moreno, S.N.; Docampo, R. Trypanothione-dependent peroxide metabolism in *Trypanosoma cruzi* different stages. *Mol. Biochem. Parasitol.* **1993**, *61*, 79–86. [CrossRef]

23. Adak, S.; Pal, S. Ascorbate peroxidase acts as a novel determiner of redox homeostasis in *Leishmania*. *Antioxid. Redox Signal.* **2013**, *19*, 746–754. [CrossRef] [PubMed]

24. Montrichard, F.; Le Guen, F.; Laval-Martin, D.L.; Davioud-Charvet, E. Evidence for the co-existence of glutathione reductase and trypanothione reductase in the non-trypanosomatid Euglenozoa: *Euglena gracilis* Z. *FEBS Lett.* **1999**, *442*, 29–33. [CrossRef]

25. Zimorski, V.; Rauch, C.; van Hellemond, J.J.; Tielens, A.G.M.; Martin, W.F. The mitochondrion of *Euglena gracilis*. *Adv. Exp. Med. Biol.* **2017**, *979*, 19–37.

26. Butenko, A.; Opperdoes, F.R.; Flegontova, O.; Horak, A.; Hampl, V.; Keeling, P.; Gawryluk, R.M.R.; Tikhonenkov, D.; Flegontov, P.; Lukeš, J. Evolution of metabolic capabilities and molecular features of diplonemids, kinetoplastids, and euglenids. *BMC Biol.* **2020**, *18*, 23. [CrossRef]

27. Mittler, R.; Herr, E.H.; Orvar, B.L.; van Camp, W.; Willekens, H.; Inze, D.; Ellis, B.E. Transgenic tobacco plants with reduced capability to detoxify reactive oxygen intermediates are hyperresponsive to pathogen infection. *Proc. Natl. Acad. Sci. USA* **1999**, *96*, 14165–14170. [CrossRef]

28. Opperdoes, F.R.; Szikora, J.P. In silico prediction of the glycosomal enzymes of *Leishmania major* and trypanosomes. *Mol. Biochem. Parasitol.* **2006**, *147*, 193–206. [CrossRef]

29. Bianchi, C.; Kostygov, A.Y.; Kraeva, N.; Záhonová, K.; Horáková, E.; Sobotka, R.; Lukeš, J.; Yurchenko, V. An enigmatic catalase of *Blastocrithidia*. *Mol. Biochem. Parasitol.* **2019**, *232*, 111199. [CrossRef]

30. Kraeva, N.; Horáková, E.; Kostygov, A.; Kořený, L.; Butenko, A.; Yurchenko, V.; Lukeš, J. Catalase in Leishmaniinae: With me or against me? *Infect. Genet. Evol.* **2017**, *50*, 121–127. [CrossRef]

31. Schott, E.J.; Di Lella, S.; Bachvaroff, T.R.; Amzel, L.M.; Vasta, G.R. Lacking catalase, a protistan parasite draws on its photosynthetic ancestry to complete an antioxidant repertoire with ascorbate peroxidase. *BMC Evol. Biol.* **2019**, *19*, 146. [CrossRef]

32. Adl, S.M.; Bass, D.; Lane, C.E.; Lukeš, J.; Schoch, C.L.; Smirnov, A.; Agatha, S.; Berney, C.; Brown, M.W.; Burki, F.; et al. Revisions to the classification, nomenclature, and diversity of eukaryotes. *J. Eukaryot. Microbiol.* **2019**, *66*, 4–119. [CrossRef] [PubMed]

33. Lukeš, J.; Skalický, T.; Týč, J.; Votýpka, J.; Yurchenko, V. Evolution of parasitism in kinetoplastid flagellates. *Mol. Biochem. Parasitol.* **2014**, *195*, 115–122. [CrossRef] [PubMed]

34. Adak, S.; Datta, A.K. *Leishmania major* encodes an unusual peroxidase that is a close homologue of plant ascorbate peroxidase: A novel role of the transmembrane domain. *Biochem. J.* **2005**, *390*, 465–474. [CrossRef] [PubMed]

35. Mittler, R.; Zilinskas, B.A. Purification and characterization of pea cytosolic ascorbate peroxidase. *Plant Physiol.* **1991**, *97*, 962–968. [CrossRef] [PubMed]

36. Wada, K.; Tada, T.; Nakamura, Y.; Ishikawa, T.; Yabuta, Y.; Yoshimura, K.; Shigeoka, S.; Nishimura, K. Crystal structure of chloroplastic ascorbate peroxidase from tobacco plants and structural insights into its instability. *J. Biochem.* **2003**, *134*, 239–244. [CrossRef] [PubMed]

37. Pitsch, N.T.; Witsch, B.; Baier, M. Comparison of the chloroplast peroxidase system in the chlorophyte *Chlamydomonas reinhardtii*, the bryophyte *Physcomitrella patens*, the lycophyte *Selaginella moellendorffii* and the seed plant *Arabidopsis thaliana*. *BMC Plant Biol.* **2010**, *10*, 133. [CrossRef] [PubMed]

38. Opperdoes, F.R.; Butenko, A.; Flegontov, P.; Yurchenko, V.; Lukeš, J. Comparative metabolism of free-living *Bodo saltans* and parasitic trypanosomatids. *J. Eukaryot. Microbiol.* **2016**, *63*, 657–678. [CrossRef]

39. Horáková, E.; Faktorová, D.; Kraeva, N.; Kaur, B.; Van Den Abbeele, J.; Yurchenko, V.; Lukeš, J. Catalase compromises the development of the insect and mammalian stages of *Trypanosoma brucei*. *FEBS J.* **2020**, *287*, 964–977. [CrossRef]

40. Ishemgulova, A.; Butenko, A.; Kortišová, L.; Boucinha, C.; Grybchuk-Ieremenko, A.; Morelli, K.A.; Tesařová, M.; Kraeva, N.; Grybchuk, D.; Pánek, T.; et al. Molecular mechanisms of thermal resistance of the insect trypanosomatid *Crithidia thermophila*. *PLoS ONE* **2017**, *12*, e0174165. [CrossRef]

41. Kraeva, N.; Butenko, A.; Hlaváčová, J.; Kostygov, A.; Myškova, J.; Grybchuk, D.; Leštinová, T.; Votýpka, J.; Volf, P.; Opperdoes, F.; et al. *Leptomonas seymouri*: Adaptations to the dixenous life cycle analyzed by genome sequencing, transcriptome profiling and co-infection with *Leishmania donovani*. *PLoS Pathog.* **2015**, *11*, e1005127. [CrossRef]

42. Záhonová, K.; Füssy, Z.; Birčák, E.; Novák-Vanclová, A.M.G.; Klimeš, V.; Vesteg, M.; Krajčovič, J.; Oborník, M.; Eliáš, M. Peculiar features of the plastids of the colourless alga *Euglena longa* and photosynthetic euglenophytes unveiled by transcriptome analyses. *Sci. Rep.* **2018**, *8*, 17012. [CrossRef] [PubMed]

43. Zámocký, M.; Gasselhuber, B.; Furtmüller, P.G.; Obinger, C. Turning points in the evolution of peroxidase-catalase superfamily: Molecular phylogeny of hybrid heme peroxidases. *Cell Mol. Life Sci.* **2014**, *71*, 4681–4696. [CrossRef] [PubMed]

44. Asada, K. Ascorbate peroxidase—A hydrogen peroxide-scavenging enzyme in plants. *Physiol. Plant* **1992**, *85*, 235–241. [CrossRef]

45. Mittra, B.; Cortez, M.; Haydock, A.; Ramasamy, G.; Myler, P.J.; Andrews, N.W. Iron uptake controls the generation of *Leishmania* infective forms through regulation of ROS levels. *J. Exp. Med.* **2013**, *210*, 401–416. [CrossRef]

46. Khan, Y.A.; Andrews, N.W.; Mittra, B. ROS regulate differentiation of visceralizing *Leishmania* species into the virulent amastigote form. *Parasitol. Open* **2018**, *4*, e19. [CrossRef]

47. Opperdoes, F.R.; Nohýnková, E.; Van Schaftingen, E.; Lambeir, A.M.; Veenhuis, M.; Van Roy, J. Demonstration of glycosomes (microbodies) in the Bodonid flagellate *Trypanoplasma borelli* (Protozoa, Kinetoplastida). *Mol. Biochem. Parasitol.* **1988**, *30*, 155–163. [CrossRef]

48. Ishikawa, T.; Tajima, N.; Nishikawa, H.; Gao, Y.; Rapolu, M.; Shibata, H.; Sawa, Y.; Shigeoka, S. *Euglena gracilis* ascorbate peroxidase forms an intramolecular dimeric structure: Its unique molecular characterization. *Biochem. J.* **2010**, *426*, 125–134. [CrossRef]

49. Novák Vanclová, A.M.G.; Zoltner, M.; Kelly, S.; Soukal, P.; Záhonová, K.; Füssy, Z.; Ebenezer, T.E.; Lacová Dobáková, E.; Eliáš, M.; Lukeš, J.; et al. Metabolic quirks and the colourful history of the *Euglena gracilis* secondary plastid. *New Phytol.* **2020**, *225*, 1578–1592. [CrossRef]

50. Lazzarotto, F.; Teixeira, F.K.; Rosa, S.B.; Dunand, C.; Fernandes, C.L.; Fontenele Ade, V.; Silveira, J.A.; Verli, H.; Margis, R.; Margis-Pinheiro, M. Ascorbate peroxidase-related (APx-R) is a new heme-containing protein functionally associated with ascorbate peroxidase but evolutionarily divergent. *New Phytol.* **2011**, *191*, 234–250. [CrossRef]

51. Koussevitzky, S.; Suzuki, N.; Huntington, S.; Armijo, L.; Sha, W.; Cortes, D.; Shulaev, V.; Mittler, R. Ascorbate peroxidase 1 plays a key role in the response of *Arabidopsis thaliana* to stress combination. *J. Biol. Chem.* **2008**, *283*, 34197–34203. [CrossRef]

52. Kostygov, A.Y.; Yurchenko, V. Revised classification of the subfamily Leishmaniinae (Trypanosomatidae). *Folia Parasitol.* **2017**, *64*, 020. [CrossRef] [PubMed]

53. Ajithkumar, I.P.; Panneerselvam, R. ROS scavenging system, osmotic maintenance, pigment and growth status of *Panicum sumatrense* roth under drought stress. *Cell Biochem. Biophys.* **2014**, *68*, 587–595. [CrossRef]

54. Maslov, D.A.; Opperdoes, F.R.; Kostygov, A.Y.; Hashimi, H.; Lukeš, J.; Yurchenko, V. Recent advances in trypanosomatid research: genome organization, expression, metabolism, taxonomy and evolution. *Parasitology* **2019**, *146*, 1–27. [CrossRef] [PubMed]

55. Záhonová, K.; Füssy, Z.; Oborník, M.; Eliáš, M.; Yurchenko, V. RuBisCO in non-photosynthetic alga *Euglena longa*: Divergent features, transcriptomic analysis and regulation of complex formation. *PLoS ONE* **2016**, *11*, e0158790.

56. Ebenezer, T.E.; Zoltner, M.; Burrell, A.; Nenarokova, A.; Novák Vanclová, A.M.G.; Prasad, B.; Soukal, P.; Santana-Molina, C.; O'Neill, E.; Nankissoor, N.N.; et al. Transcriptome, proteome and draft genome of *Euglena gracilis*. *BMC Biol.* **2019**, *17*, 11. [CrossRef]

57. Smircich, P.; Eastman, G.; Bispo, S.; Duhagon, M.A.; Guerra-Slompo, E.P.; Garat, B.; Goldenberg, S.; Munroe, D.J.; Dallagiovanna, B.; Holetz, F.; et al. Ribosome profiling reveals translation control as a key mechanism generating differential gene expression in *Trypanosoma cruzi*. *BMC Genom.* **2015**, *16*, 443. [CrossRef]

58. Parsons, M.; Myler, P.J. Illuminating parasite protein production by ribosome profiling. *Trends Parasitol.* **2016**, *32*, 446–457. [CrossRef]

59. Altschul, S.F.; Madden, T.L.; Schaffer, A.A.; Zhang, J.; Zhang, Z.; Miller, W.; Lipman, D.J. Gapped BLAST and PSI-BLAST: A new generation of protein database search programs. *Nucleic Acids Res.* **1997**, *25*, 3389–3402. [CrossRef]

60. Savelli, B.; Li, Q.; Webber, M.; Jemmat, A.M.; Robitaille, A.; Zamocky, M.; Mathe, C.; Dunand, C. RedoxiBase: A database for ROS homeostasis regulated proteins. *Redox Biol.* **2019**, *26*, 101247. [CrossRef]

61. Benson, D.A.; Cavanaugh, M.; Clark, K.; Karsch-Mizrachi, I.; Ostell, J.; Pruitt, K.D.; Sayers, E.W. GenBank. *Nucleic Acids Res.* **2018**, *46*, D41–D47. [CrossRef]

62. Jones, P.; Binns, D.; Chang, H.Y.; Fraser, M.; Li, W.; McAnulla, C.; McWilliam, H.; Maslen, J.; Mitchell, A.; Nuka, G.; et al. InterProScan 5: Genome-scale protein function classification. *Bioinformatics* **2014**, *30*, 1236–1240. [CrossRef] [PubMed]

63. Steinegger, M.; Soding, J. MMseqs2 enables sensitive protein sequence searching for the analysis of massive data sets. *Nat. Biotechnol.* **2017**, *35*, 1026–1028. [CrossRef] [PubMed]

64. Katoh, K.; Standley, D.M. MAFFT multiple sequence alignment software version 7: Improvements in performance and usability. *Mol. Biol. Evol.* **2013**, *30*, 772–780. [CrossRef] [PubMed]

65. Capella-Gutiérrez, S.; Silla-Martinez, J.M.; Gabaldon, T. trimAl: A tool for automated alignment trimming in large-scale phylogenetic analyses. *Bioinformatics* **2009**, *25*, 1972–1973. [CrossRef]

66. Wang, H.C.; Minh, B.Q.; Susko, E.; Roger, A.J. Modeling site heterogeneity with posterior mean site frequency profiles accelerates accurate phylogenomic estimation. *Syst. Biol.* **2018**, *67*, 216–235. [CrossRef]

67. Nguyen, L.T.; Schmidt, H.A.; von Haeseler, A.; Minh, B.Q. IQ-TREE: A fast and effective stochastic algorithm for estimating maximum-likelihood phylogenies. *Mol. Biol. Evol.* **2015**, *32*, 268–274. [CrossRef]

68. Hiller, K.; Grote, A.; Scheer, M.; Munch, R.; Jahn, D. PrediSi: Prediction of signal peptides and their cleavage positions. *Nucleic Acids Res.* **2004**, *32*, W375–W379. [CrossRef]

69. Kume, K.; Amagasa, T.; Hashimoto, T.; Kitagawa, H. NommPred: Prediction of mitochondrial and mitochondrion-related organelle proteins of nonmodel organisms. *Evol. Bioinform. Online* **2018**, *14*, 1176934318819835. [CrossRef]

70. Almagro Armenteros, J.J.; Salvatore, M.; Emanuelsson, O.; Winther, O.; von Heijne, G.; Elofsson, A.; Nielsen, H. Detecting sequence signals in targeting peptides using deep learning. *Life Sci. Alliance* **2019**, *2*, e201900429. [CrossRef]

71. Blum, T.; Briesemeister, S.; Kohlbacher, O. MultiLoc2: Integrating phylogeny and Gene Ontology terms improves subcellular protein localization prediction. *BMC Bioinform.* **2009**, *10*, 274. [CrossRef]

72. Krogh, A.; Larsson, B.; von Heijne, G.; Sonnhammer, E.L. Predicting transmembrane protein topology with a hidden Markov model: Application to complete genomes. *J. Mol. Biol.* **2001**, *305*, 567–580. [CrossRef] [PubMed]

73. Käll, L.; Krogh, A.; Sonnhammer, E.L. Advantages of combined transmembrane topology and signal peptide prediction–the Phobius web server. *Nucleic Acids Res.* **2007**, *35*, W429–W432. [CrossRef] [PubMed]

74. Kearse, M.; Moir, R.; Wilson, A.; Stones-Havas, S.; Cheung, M.; Sturrock, S.; Buxton, S.; Cooper, A.; Markowitz, S.; Duran, C.; et al. Geneious Basic: An integrated and extendable desktop software platform for the organization and analysis of sequence data. *Bioinformatics* **2012**, *28*, 1647–1649. [CrossRef] [PubMed]

75. Škodová-Sveráková, I.; Prokopchuk, G.; Peña-Diaz, P.; Záhonová, K.; Moos, M.; Horváth, A.; Šimek, P.; Lukeš, J. Unique dynamics of paramylon storage in the marine euglenozoan *Diplonema papillatum*. *Protist* **2020**, *171*, 125717. [CrossRef]

76. Tashyreva, D.; Prokopchuk, G.; Votýpka, J.; Yabuki, A.; Horák, A.; Lukeš, J. Life cycle, ultrastructure, and phylogeny of new diplonemids and their endosymbiotic bacteria. *mBio* **2018**, *9*, e02447–e17. [CrossRef]

77. Yurchenko, V.; Kostygov, A.; Havlová, J.; Grybchuk-Ieremenko, A.; Ševčíková, T.; Lukeš, J.; Ševčík, J.; Votýpka, J. Diversity of trypanosomatids in cockroaches and the description of *Herpetomonas tarakana* sp. n. *J. Eukaryot. Microbiol.* **2016**, *63*, 198–209. [CrossRef]

78. Changmai, P.; Horáková, E.; Long, S.; Černotíková-Stříbrná, E.; McDonald, L.M.; Bontempi, E.J.; Lukeš, J. Both human ferredoxins equally efficiently rescue ferredoxin deficiency in *Trypanosoma brucei*. *Mol. Microbiol.* **2013**, *89*, 135–151. [CrossRef]

79. Hutner, S.H.; Zahalsky, A.C.; Aronson, S.A.; Baker, H.; Frank, O. Culture media for Euglena gracilis. In *Methods in Cell Physiology*; Prescott, D.M., Ed.; Academic Press: New York, NY, USA; London, UK, 1966; pp. 217–228.

80. Cramer, M.; Myers, J. Growth and photosynthetic characteristics of *Euglena gracilis*. *Arch. Mikrobiol.* **1952**, *17*, 384–402. [CrossRef]

81. Bradford, M.M. A rapid and sensitive method for the quantitation of microgram quantities of protein utilizing the principle of protein-dye binding. *Anal. Biochem.* **1976**, *72*, 248–254. [CrossRef]

82. Monteiro, G.; Horta, B.B.; Pimenta, D.C.; Augusto, O.; Netto, L.E. Reduction of 1-Cys peroxiredoxins by ascorbate changes the thiol-specific antioxidant paradigm, revealing another function of vitamin C. *Proc. Natl. Acad. Sci. USA* **2007**, *104*, 4886–4891. [CrossRef]

 *pathogens*

Article

Inventory and Evolution of Mitochondrion-localized Family A DNA Polymerases in Euglenozoa

Ryo Harada [1], Yoshihisa Hirakawa [2], Akinori Yabuki [3], Yuichiro Kashiyama [4,5], Moe Maruyama [5], Ryo Onuma [6], Petr Soukal [7], Shinya Miyagishima [6], Vladimír Hampl [7], Goro Tanifuji [8] and Yuji Inagaki [1,2,9,*]

[1] Graduate School of Life and Environmental Sciences, University of Tsukuba, Tsukuba 305-8572, Japan; oqo21109797@gmail.com

[2] Faculty of Life and Environmental Sciences, University of Tsukuba, Tsukuba 305-8572, Japan; hirakawa.yoshi.fp@u.tsukuba.ac.jp

[3] Japan Agency for Marine-Earth Science and Technology, Yokosuka 236-0001, Japan; yabukia@jamstec.go.jp

[4] Department of Applied Chemistry and Food Science, Fukui University of Technology, Fukui 910-8505, Japan; yuichirob306@gmail.com

[5] Graduate School of Engineering, Fukui University of Technology, Fukui 910-8505, Japan; harukote.mar@gmail.com

[6] Department of Gene Function and Phenomics, National Institute of Genetics, Mishima 411-8540, Japan; ronuma@nig.ac.jp (R.O.); smiyagis@nig.ac.jp (S.M.)

[7] Department of Parasitology, Charles University, Faculty of Science, BIOCEV, 252 42 Vestec, Czech Republic; petr.soukal@natur.cuni.cz (P.S.); vladimir.hampl@natur.cuni.cz (V.H.)

[8] Department of Zoology, National Museum of Nature and Science, Tsukuba 305-0005, Japan; gorot@kahaku.go.jp

[9] Center for Computational Sciences, University of Tsukuba, Tsukuba 305-8577, Japan

* Correspondence: yuji@ccs.tsukuba.ac.jp

Received: 1 February 2020; Accepted: 30 March 2020; Published: 1 April 2020

Abstract: The order Trypanosomatida has been well studied due to its pathogenicity and the unique biology of the mitochondrion. In *Trypanosoma brucei*, four DNA polymerases, namely PolIA, PolIB, PolIC, and PolID, related to bacterial DNA polymerase I (PolI), were shown to be localized in mitochondria experimentally. These mitochondrion-localized DNA polymerases are phylogenetically distinct from other family A DNA polymerases, such as bacterial PolI, DNA polymerase gamma (Polγ) in human and yeasts, "plant and protist organellar DNA polymerase (POP)" in diverse eukaryotes. However, the diversity of mitochondrion-localized DNA polymerases in Euglenozoa other than Trypanosomatida is poorly understood. In this study, we discovered putative mitochondrion-localized DNA polymerases in broad members of three major classes of Euglenozoa—Kinetoplastea, Diplonemea, and Euglenida—to explore the origin and evolution of trypanosomatid PolIA-D. We unveiled distinct inventories of mitochondrion-localized DNA polymerases in the three classes: (1) PolIA is ubiquitous across the three euglenozoan classes, (2) PolIB, C, and D are restricted in kinetoplastids, (3) new types of mitochondrion-localized DNA polymerases were identified in a prokinetoplastid and diplonemids, and (4) evolutionarily distinct types of POP were found in euglenids. We finally propose scenarios to explain the inventories of mitochondrion-localized DNA polymerases in Kinetoplastea, Diplonemea, and Euglenida.

Keywords: DNA replication; family A DNA polymerase; plant and protist organellar DNA polymerase; Trypanosomatida; Kinetoplastea; Diplonemea; Euglenida; Prokinetoplastina

1. Introduction

Members of the order Trypanosomatida have been extensively studied because of their pathogenicity to humans. *Trypanosoma brucei*, *Trypanosoma cruzi*, and the species belonging to the genus *Leishmania* cause African trypanosomiasis (sleeping sickness), American trypanosomiasis (Chagas disease), and leishmaniasis, respectively [1]. Besides their significance as the causative agents of deadly diseases, trypanosomatids are important for basic biological research due to the complex architecture of their mitochondrial genomes (mtDNAs) and RNA-editing of mitochondrial transcripts [2]. Trypanosomatids possess a unique mtDNA comprising two types of circular DNA molecule—maxicircles and minicircles—interlocked with one another (so-called kinetoplast DNA or kDNA). A single kDNA contains dozens of maxicircles and thousands of minicircles. Maxicircles carry protein-coding genes and ribosomal RNA genes, of which transcripts need to be edited post-transcriptionally by extensive insertions and deletions of uridines with the help of guide RNAs (gRNAs) transcribed from minicircles. The structural complexity of kDNA is seemingly consistent with a unique set of DNA polymerases required for kDNA replication. In the genus *Trypanosoma*, phylogenetically diverse DNA polymerases were experimentally shown to be localized in mitochondria; (i) PolIA, PolIB, PolIC, and PolID [3] belong to family A, the members of which bear the sequence similarity to bacterial DNA polymerase I (PolI) [4], two of DNA polymerase beta in family X [5], and a DNA polymerase kappa in family Y [6]. Besides PolIA-D in trypanosomatids, several DNA polymerases of family A are known to be localized in mitochondria, such as DNA polymerase gamma (Polγ) in animals and yeasts [7] and plant and protist organellar DNA polymerase (POP), which is also targeted to the plastids of plants and algae [8–10].

Trypanosomatida, together with Eubodonida, Parabodonida, Neobodonida, and Prokinetoplastida, are assembled to the class Kinetoplastea [11]. In principal, the characteristics of kDNA (and unique gene expression from kDNA) in trypanosomatids seem to be ubiquitous across the members of Kinetoplastea with modifications [12,13]. In the tree of eukaryotes, Kinetoplastea is further related to the classes Diplonemea and Euglenida, and the family Symbiontida, forming the phylum Euglenozoa [11]. Diplonemid mitochondria contain numerous circular DNA molecules (minicircles) and each chromosome possesses "gene module(s)" that are a piece of the coding regions [14,15]. Both 5′ and 3′ non-coding regions of primary transcripts from gene modules are removed and the resulting transcripts were then assembled into a mature mRNA by *trans*-splicing. After the removal of the 5′ and 3′ non-coding regions described above, transcripts from certain modules undergo substitution RNA editing (cytidine-to-uridine, adenosine-to-inosine and/or guanosine-to-adenosine) and/or appendage RNA editing at the 3′ end (uridine and/or adenosine-appendage) [16]. The architecture of euglenid mtDNA seems to be simpler than those of kinetoplastid/diplonemid mtDNA [17–19]. The mtDNA of *Euglena gracilis*, a representative species of Euglenida, is composed of multiple linear DNA molecules, each of which carries one or two full-length genes [18,19]. Although no sequence data are available, the mitochondrion of the euglenid *Petalomonas cantuscygni* was reported to contain multiple DNA molecules in both linear and circular forms based on electron microscopic observation [17]. Finally, our current knowledge of Symbiontida is restricted to morphological information and small subunit ribosomal RNA gene sequences [20–22]. Importantly, no systematic survey of mitochondrion-localized DNA polymerases has been done for any of the members of Euglenozoa except trypanosomatids.

In this study, we aim to retrace how the current inventory of mitochondrion-localized DNA polymerases in trypanosomatids has been shaped during the evolution of Euglenozoa. We searched for putative mitochondrion-localized DNA polymerases in diverse euglenozoans. Briefly, we detected PolIA in all of the euglenids, diplonemids, and kinetoplastids examined here, except for a single case of putative secondary loss. PolIB, C, and D are seemingly restricted to members of Kinetoplastea. In addition, we detected novel DNA polymerases, named PolI-Perk1 and PolI-Perk2, and PolI-dipl, all of which are apparently related to but distinct from PolIB-D, in the prokinetoplastid *Perkinsela* sp. and diplonemids, respectively. In euglenids, three distinct types of POP were found and at least two of them were most likely to be localized in mitochondria. According to the inventories of family A

DNA polymerases in Euglenida, Diplonemea, and Kinetoplastea, we here discuss the evolution of mitochondrion-localized DNA polymerases in Euglenozoa.

2. Results

Pioneering studies demonstrated that the previously described mitochondrion-localized DNA polymerases, namely Polγ, POP, trypanosomatid PolIA–D belong to family A [10,23]. Thus, we surveyed family A DNA polymerases in both public and in-house transcriptome data of four kinetoplastids, four diplonemids, and six euglenids (14 species in total). We repeated the same search against the genome data of two kinetoplatids, *Bodo saltans* and *Perkinsela* sp. As a result, 37 family A DNA polymerases were identified in 14 euglenozoan species and then subjected to phylogenetic analyses along with their homologs, including Polγ, POP, and trypanosomatid PolIA–D. Based on the phylogenetic affinity, we classified the 37 newly identified family A DNA polymerases into nine POP, 13 PolIA, three PolIB, a single PolIC, five PolID, and six "PolIBCD-related" DNA polymerases, named PolI-Perk1, PolI-Perk2, and PolI-dipl (see below).

2.1. PolIA is Ubiquitous in Euglenida, Diplonemea, and Kinetoplastea

In the trypanosomatid *Trypanosoma brucei*, four distinct types of family A DNA polymerase—PolIA, B, C, and D—are known to be involved in the maintenance of DNA in their mitochondria [3,24–26]. We found that all the species examined in this study (except *Perkinsela* sp.) possess sequences that grouped robustly with trypanosomatid PolIA in the global phylogeny of family A DNA polymerases (Figure 1). We here propose that euglenids, diplonemids, and kinetoplastids (except for *Perkinsela* sp.) possess PolIA, which can be traced back to a single DNA polymerase in the common ancestor of the three classes of Euglenozoa. In the PolIA clade, the homologs of kinetoplastids, diplonemids, and euglenids formed individual subclades, and their monophylies were supported by maximum-likelihood bootstrap values (MLBPs) of 68–93% and Bayesian posterior probabilities (BPPs) of 0.93 to 1.0, while the relationship among the three subclades was not resolved with confidence (Figure 1). The PolIA homologs found in this study were predicted to have the family A DNA polymerase domain (PF00476) at their C-termini (Table S1), as seen in the *Trypanosoma brucei* homolog [3].

We recovered the complete N-termini of PolIA homologs in only four kinetoplastids, a single diplonemid and two euglenids out of the 16 homologs examined in this study. None or only one out of the four in silico programs predicted a mitochondrial targeting signal (MTS) at the N-termini of the four kinetoplastid homologs (Figure 2). Although *Trypanosoma brucei* PolIA was shown to be localized in mitochondria experimentally, its N-terminal MTS was not detected by in silico prediction [3]. This likely stems from the difficulty in predicting the mitochondrion-localized proteins in *Trypanosoma brucei* based on the N-terminal amino acid sequences [27]. In contrast, three out of the four programs predicted an MTS in the homolog of the diplonemid *Flectonema neradi*. For the *Euglena gracilis* homolog, only a single program predicted an MTS in its N-terminus. Nevertheless, the study on the *Euglena gracilis* mitochondrial proteome recognized PolIA as a mitochondrial protein [28]. The N-terminus of the *Peranema* homolog was predicted to have an MTS by all of the four programs. The N-termini of the rest of 14 PolIA homologs were incomplete and thus could not be subjected to the MTS prediction (triangles; Figure 2). Considering the robust affinity between *Trypanosoma brucei* PolIA, of which subcellular localization was experimentally confirmed [3], and the other PolIA homologs, we suspect that all of the PolIA homologs identified in this study are mitochondrion-localized proteins.

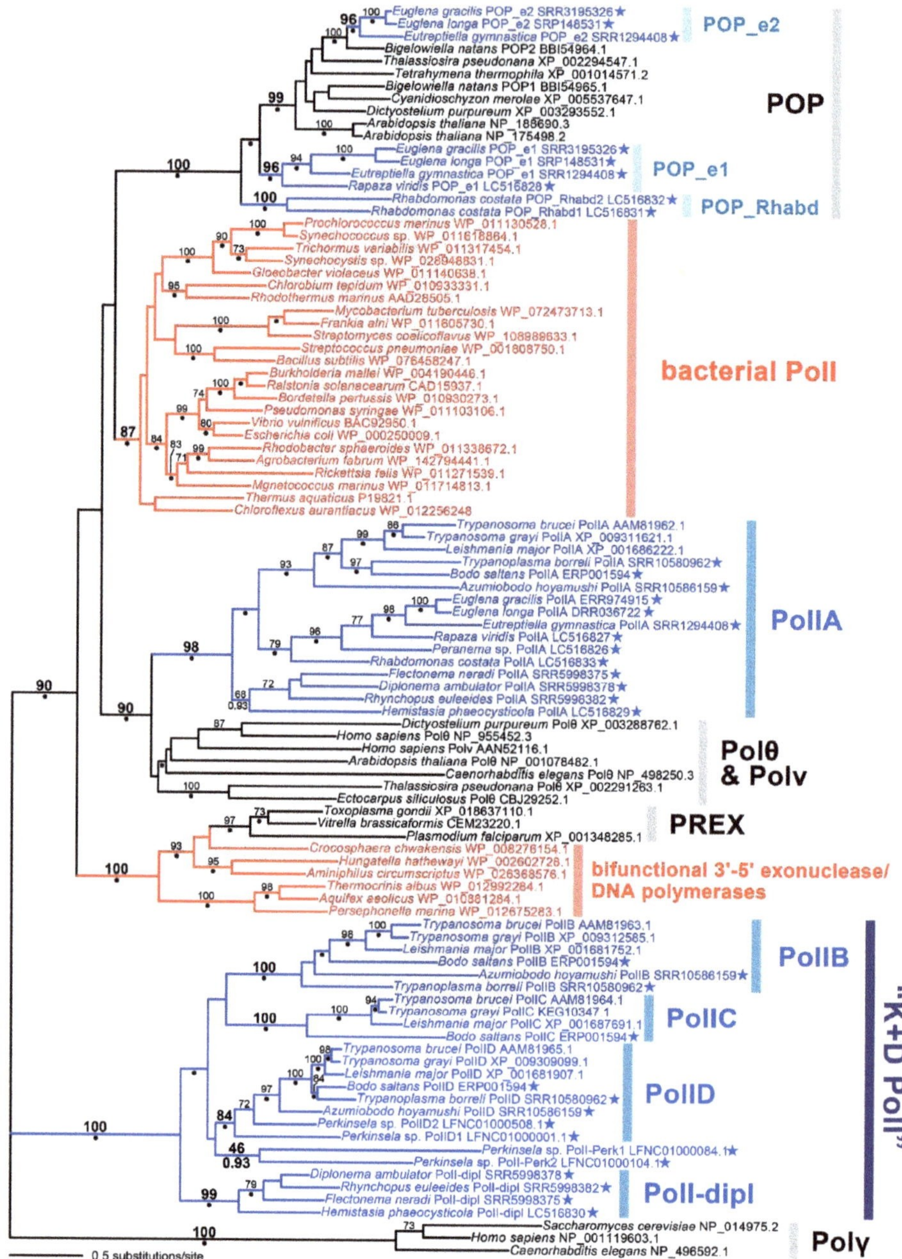

Figure 1. Maximum likelihood (ML) phylogenetic tree of family A DNA polymerases. ML bootstrap values equal to or greater than 70% are shown at the corresponding nodes, except the value for the clade of two *Perkinsela* sequences. Nodes marked by dots were supported by Bayesian posterior probabilities (BPPs) equal to or greater than 0.95, but the BPPs smaller than 0.95 are shown for the nodes of our interest. The bacterial sequences are shown in red. The euglenozoan sequences are in blue. The sequences identified in this study are highlighted by stars.

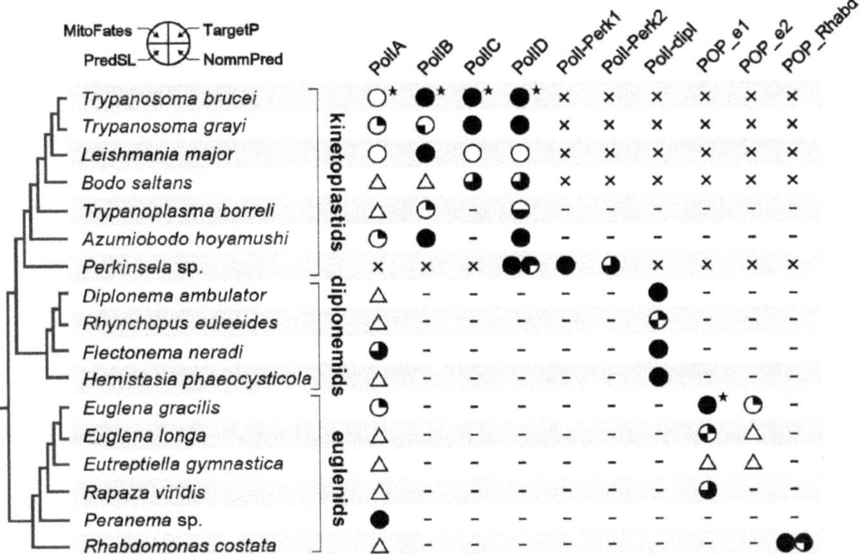

Figure 2. Inventories of family A DNA polymerases in Euglenida, Diplonemea, and Kinetoplastea. For each species examined here, the presence (absence) of each type is displayed by a circle/triangle (a dash/cross). The circles and triangles represent sequences with the complete N-termini and those of which N-termini were absent, respectively. The dashes and crosses represent the absences of homologs in transcriptome and those in both transcriptome and genome, respectively. The sequences with the complete N-termini were subjected to in silico prediction of the mitochondrial targeted signal (MTS) at their N-termini by using TargetP [29], NommPred [30], PredSL [31], and MitoFates [32]. In the case of the MTS being predicted, a subset (or all) of the quarters is filled (upper-right, TargetP; lower-right, NommPred; lower-left, PredSL; upper-left, MitoFates). Stars, which are associated with the four DNA polymerases of *Trypanosoma brucei* and POP_e1 of *Euglena gracilis* indicate the experimentally confirmed mitochondrion-localization. The branching order in the Kinetoplastea clade, that in the Diplonemea clade, and that in the Euglenida clade are based on Yazaki et al. (2017), Tashyreva et al. (2018), and Bicudo and Menezes (2016), respectively [33–35].

The precise function of PolIA of *Trypanosoma brucei* has yet to be clarified experimentally [3]. PolIA showed a clear phylogenetic affinity to Polθ that is involved in DNA repair in the nuclear genome (Figure 1) and was postulated to be involved in mtDNA repair [36]. Based on the proposed function of *Trypanosoma brucei* PolIA, this DNA polymerase may be involved in mtDNA repair in diverse euglenozoans.

2.2. PolIB, C, D, and "PolIBCD-Related" DNA Polymerases in Diplonemea and Kinetoplastea

Trypanosomatid PolIB, C and D were shown to be closely related to each other but remote from PolIA in previous phylogenetic studies [10,23]. In this section, we describe the distribution and evolution of PolIB, C, D and "PolIBCD-related" DNA polymerases in Euglenozoa. In brief, the sequences which grouped directly with trypanosomatid PolIB, C or D were found only in the kinetoplastids but not in the euglenids or diplonemids examined here.

In Figure 1, trypanosomatid PolID are grouped with the homologs of the eubodonid *Bodo saltans*, the parabodonid *Trypanoplasma borreli*, the neobodonid *Azumiobodo hoyamushi*, and the prokinetoplastid *Perkinsela* sp. together with an MLBP of 84% and a BPP of 0.97, indicating that PolID is ubiquitous in Kinetoplastea. PolIB was detected in all of the orders of Kinetoplastea except for Prokinetoplastida, as *B. saltans*, *Trypanoplasma borreli*, and *A. hoyamushi* appeared to possess DNA polymerases that formed a clade with trypanosomatid PolIB with an MLBP of 100% and a BPP of 1.00. The distribution of PolIC is likely restricted to Trypanosomatida and Eubodonida, as only a single *B. saltans* homolog grouped with trypanosomatid PolIC with an MLBP of 100% and a BPP of 1.00. All of the PolIB homologs appeared to possess both 3′-5′ exonuclease domain (PF00929) and polymerase domain (PF00476) (Figure 3; see Table S1 for the details). Only the polymerase domain (PF00476) was found in the four PolIC homologs assessed here (Table S1). Although *Trypanosoma brucei* PolID has been reported to possess both 3′-5′ exonuclease domain (PF01612) and polymerase domain (PF00476) [3], we detected only the latter domain in the rest of the PolID homologs assessed here (including the homologs of *Trypanosoma grayi* and *L. major*; Figure 3 and Table S1).

We identified novel family A DNA polymerases in *Perkinsela* sp. and diplonemids, both of which formed a large clade with PolIB, C, and D (Figure 1; labelled as "K+D PolI"). All of the four diplonemids examined here possess the DNA polymerases that formed a clade with an MLBP of 99% and a BPP of 1.00 (designated as "PolI-dipl"), suggesting that this type of DNA polymerase has been inherited vertically from an ancestral diplonemid. Both 3′-5′ exonuclease domain (PF01612) and polymerase domain (PF00476) were conserved in three out of the four PolI-dipl homologs, while the former domain was absent in the *H. phaeocysticola* homolog (only the domain structure of the *Flectonema neradi* homolog is shown in Figure 3; the domain structures of other homologs are provided in Table S1). We found two DNA polymerases in *Perkinsela* sp. (designated as "PolI-Perk1" and "PolI-Perk2"), which were tied together with an MLBP of 46% and a BPP of 0.93 (Figure 1). Only the polymerase domain (PF00476) was found in PolI-Perk1 and PolI-Perk2 (Figure 3 and Table S1). Overall, our phylogenetic analyses failed to resolve the relationship among PolI-Perk1, PolI-Perk2, and four clades of PolIB, C, D, and -dipl with confidence. If we believe the ML tree topology shown in Figure 1, PolI-Perk1 and -Perk2 belong to a novel type of DNA polymerase that is closely related to but clearly distinct from PolIB, C, D, or -dipl. Alternatively, due to the lack of phylogenetic resolution, we cannot exclude the possibility of PolI-Perk1 and -Perk2 (or PolI-Perk2 and -Perk1) being PolIB and C in *Perkinsela* sp., respectively. Unfortunately, we cannot make any definite conclusions on the origins of PolI-Perk1 and -Perk2 in this study.

We succeeded in recovering the N-termini of all of the PolIC and D homologs examined here, and 10 out of the 12 homologs were predicted in silico to have an MTS by at least two out of the four programs (Figure 2). Among the six PolIB homologs, the complete N-termini were available for all of them except that of *B. saltans*, and MTS was robustly predicted at the N-termini of the *Trypanosoma brucei*, *L. major* and *A. hoyamushi* homologs. Based on their phylogenetic affinity to the homologous sequences in trypanosomatids (Figure 1) and in silico MTS prediction (Figure 2), we propose that the newly identified PolIB, C, and D are localized in their mitochondria. The N-terminal sequences of PolI-Perk1, -Perk2, and -dipl are available and at least two out of the four programs predicted an MTS in their N-termini (Figure 2). Thus, the novel DNA polymerases found in *Perkinsela* sp. and diplonemids are likely to be localized in their mitochondria.

PolIB, C, and D were experimentally shown to be essential for *Trypanosoma brucei* growth and mtDNA replication in both procyclic and bloodstream forms [3,24–26]. Although the difference in function among PolIB, C, and D is poorly understood, their functions in mtDNA replication are unlikely to overlap one another [37]. All we can propose here is the simplest and most conserved scenario—no substantial change in function has occurred to PolIB, C or D through the evolution of Kinetoplastea. Regrettably, the amino acid sequences and domain structures are insufficient to speculate about the precise functions of the novel mitochondrion-localized DNA polymerases in *Perkinsela* sp. and diplonemids, which are absent in trypanosomatids.

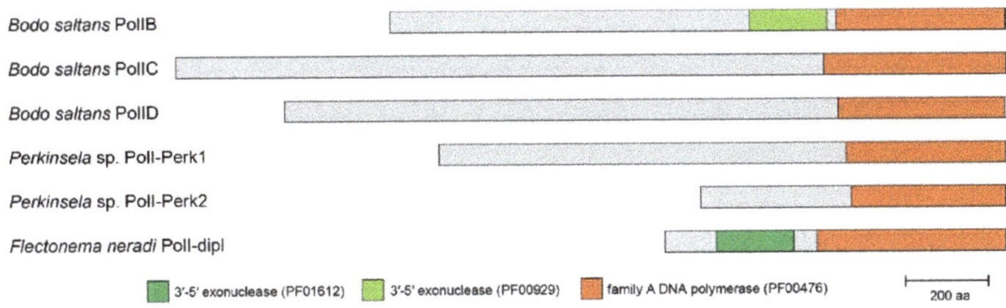

Figure 3. Domain structures of PolIB, C, D, and PolIBCD-related DNA polymerases. The domain structures of PolIB, C, and D are represented by the corresponding homologs of *Bodo saltans* (Note that the N-terminus of *B. saltans* PolIB is incomplete). As PolI-Perk1 and -Perk2 have been undescribed prior to this study, we provide their domain structures. The domain structure of another previously undescribed DNA polymerase identified in diplonemids (PolI-dipl) is represented by the *Flectonema neradi* homolog. Two types of the 3′-5′ exonuclease domains are highlighted in different colors (PF01612 and PF00929 correspond to dark green and light green, respectively). The family A DNA polymerase domains are shown in orange. The detailed domain structures of the kinetoplastid and diplonemid PolIB, C, D, and PolIBCD-related DNA polymerases are described in Table S1.

2.3. POP in Euglenida

We surveyed family A DNA polymerase sequences in six euglenids, and identified POP homologs from all of the species examined here, except *Peranema* sp. In total, nine POP homologs were found in the five euglenids examined in this study. A phylogenetic analysis of POP alignment separated the euglenid homologs into three distinct types, namely, "POP_e1," "POP_e2," and "POP_Rhabd" (Figure 4). *Euglena gracilis*, *Euglena longa*, *Eutreptiella gymnastica*, and *Rapaza viridis* (members of Euglenophyceae) share POP_e1, which grouped together with an MLBP of 79% and a BPP of 0.95. *Rhabdomonas costata* appeared to possess two POP homologs (POP_Rhabd1 and _Rhabd2) that were tied together with an MLBP of 100% and a BPP of 0.99. In the POP phylogeny, the clade of POP_Rhabd1 and _Rhabd2, and that of four POP_e1 homologs were branched subsequently from the root of the entire POP clade (Figure 4). However, because the backbone of the clade is not supported, the sister relationship between POP_e1 and POP_Rhabd cannot be excluded. POP_e2 was identified in *Euglena* spp. and *Eutreptiella gymnastica* and formed a clade with an MLBP of 97% and a BPP of 1.00. The POP_e2 clade was separated from the POP_e1 or POP_Rhabd homologs but grouped with the plastid-localized POP homologs in chlorarachniophytes with an MLBP of 99% and a BPP of 1.00 (Figure 4). As reported for the previously studied POP homologs, POP_e1 and POP_e2 appeared to possess both 3′-5′ exonuclease domain and polymerase domain (Figure 5; see Table S1 for the details). POP_Rhabd1 and _Rhabd2 seemingly lack the 3′-5′ exonuclease domain (Figure 5 and Table S1).

POP homologs are often localized in both mitochondria and plastids in photosynthetic species [38,39]. The N-termini of three out of the four POP_e1 homologs were completed and predicted to function as an MTS by at least two out of the four in silico programs (Figure 2). On the other hand, neither SignalP [40] nor TMHMM [41] predicted the N-terminal amino acid sequence of *Euglena gracilis* POP_e1 as a typical plastid targeting signal (PTS). Importantly, POP_e1 was detected as a part of the mitochondria proteome [28], while not recognized as the plastid-localized protein [42]. These results consistently suggest that the POP_e1 homologs found in this study are mitochondrion-localized. The two POP_Rhabd homologs were predicted to have an MTS by at least three out of the four programs (Figure 2), suggesting that the two DNA polymerases in *Rhabdomonas costata* are localized in the mitochondria. Based solely on the sequence data, we have little insight into the difference in function between the two mitochondrion-localized DNA polymerases in euglenids, PolIA and POP_e1/POP_Rhabd.

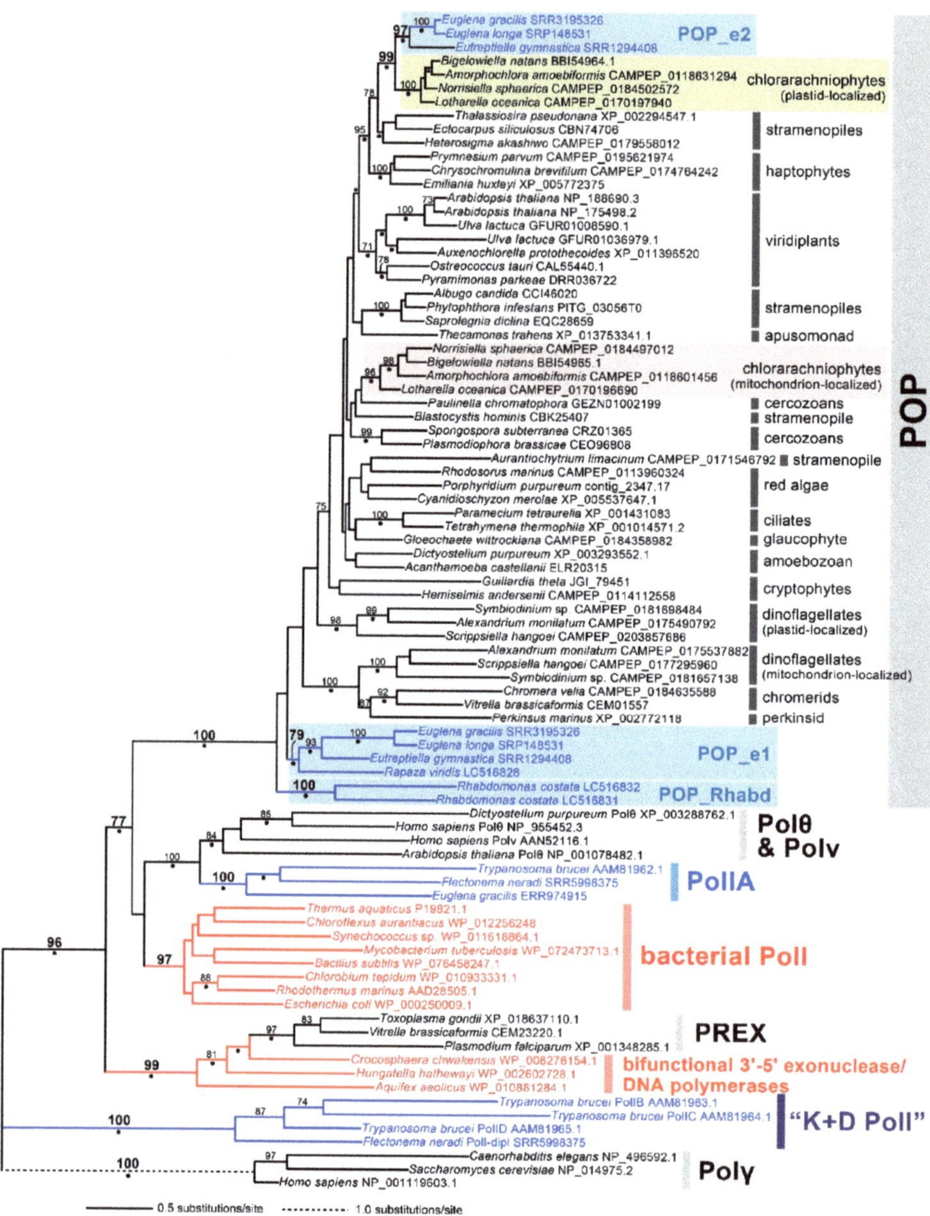

Figure 4. ML phylogenetic tree of plant and protist organellar DNA polymerase (POP). ML bootstrap values equal to or greater than 70% are shown at the corresponding nodes. Nodes marked by dots were supported by Bayesian posterior probabilities equal to or greater than 0.95. The bacterial sequences are shown in red. The euglenozoan sequences are in blue. The three types of POP identified in euglenids (POP_e1, POP_e2, and POP_Rhabd) are shaded in blue. Mitochondrion- and plastid-localized POP in chlorarachniophytes are shaded in orange and light green, respectively.

Among the three POP_e2 homologs, we completed the N-terminus of the *Euglena gracilis* homolog. Only a single program predicted an MTS in the N-terminus of the *Euglena gracilis* POP_e2, and this is insufficient to propose its mitochondrial localization. Likewise, no PTS was predicted at the N-terminus of *Euglena gracilis* POP_e2. Indeed, *Euglena gracilis* POP_e2 was recognized as neither a mitochondrial nor plastid protein in the proteomic studies [42]. Thus, we conclude that the POP_e2 homologs are localized in the cytosol.

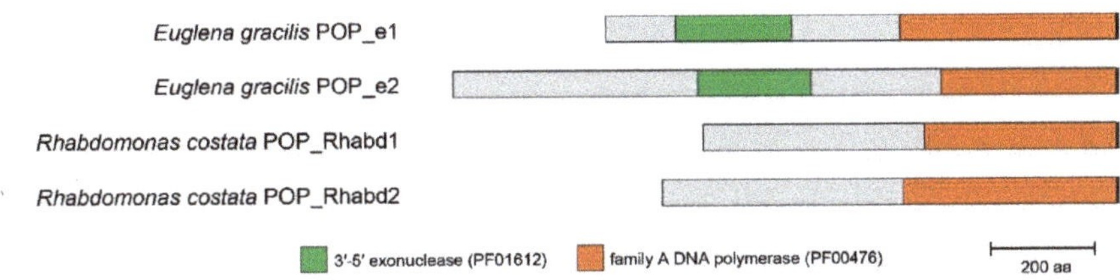

Figure 5. Domain structures of the POP homologs in euglenids. The domain structures of POP_e1 and POP_e2 are represented by the *Euglena gracilis* homologs. The two POP homologs identified in *Rhabdomonas costata* (POP_Rhabd1 and POP_Rhabd2) possess only the family A DNA polymerase domain (shown in orange). The 3′-5′ exonuclease domain (PF01612) is shown in dark green. The detailed domain structures of the POP homologs identified in this study are described in Table S1.

3. Discussion

This study unveiled that PolIA is ubiquitously distributed among Euglenida, Diplonemea, and Kinetoplastea. Thus, we firmly conclude that this type of family A DNA polymerase was obtained in the common ancestor of these three classes and has been inherited vertically to the extant descendants. We also propose that *Perkinsela* sp., which is an obligate intracellular organism of *Paramoeba*, lost PolIA secondarily. There is a room for arguing whether PolIA is absolutely absent in *Perkinsela* sp. However, family A DNA polymerases were surveyed in both the transcriptome and genome data of this species, and there may be little chance to overlook a PolIA gene in the high-quality genome data in particular [43]. Consequently, we propose that a loss of PolIA occurred on the branch leading to *Perkinsela* sp. The conservation of PolIA in Kinetoplastea implies the importance of this DNA polymerase for kDNA maintenance in Euglenozoa. There is a study experimentally demonstrating that PolIA was shown to be dispensable under normal growth conditions in *Trypanosoma brucei* [3]. We suspect that the dispensability of PolIA varies among the life stages of the trypanosome development.

In contrast to the ubiquity of PolIA among Euglenida, Diplonemea, and Kinetoplastea, PolIB, C, and D were identified in the members of Kinetoplastea alone. To our knowledge, no high-quality genome data is publicly available for any of the diplonemids or euglenids. However, it is unlikely that none of the DNA polymerases of interest were overlooked in the transcriptome data of the four diplonemids and six euglenids examined here. Thus, we conclude that PolIB, C, and D are restricted in Kinetoplastea. In addition, we identified PolI-Perk1 and -Perk2 in *Perkinsela* sp., and PolI-dipl in diplonemids, both of which were previously undescribed. As PolIB, C, D, -Perk1, -Perk2, and -dipl formed a "K+D PolI" clade with high statistical support, these DNA polymerases can be traced back to a single ancestral mitochondrion-localized DNA polymerase in the common ancestor of the classes Kinetoplastea and Diplonemea. We here propose that the ancestral DNA polymerase in the two classes was similar to the extant PolI-dipl, and, after the separation of the two classes, the ancestral type of the mitochondrion-localized DNA polymerase has been kept as PolI-dipl in Diplonemea, but has diverged into PolIB, C, D, -Perk1, and -Perk2 during the evolution of Kinetoplastea. So far, it is reasonable to propose that the common ancestor of Kinetoplastea possessed PolID, which was found in all of the members of Kinetoplastea examined in this study. On the other hand, the precise evolutions of PolIB, C, -Perk1, and -Perk2 remain unclear because of two obstacles discussed below.

Firstly, the relationship among PolIB, C, D, -Perk1, and -Perk2 was essentially unresolved in the phylogenetic analyses and makes it difficult to infer how the particular types of DNA polymerase emerged during the divergence of Kinetoplastea. Future phylogenetic analyses with improved sequence sampling may provide a better resolution for the relationship among PolIB, C, D, -Perk1, and -Perk2. Secondly, there is a certain level of uncertainty about the inventories of mitochondrion-localized DNA polymerases in members of Kinetoplastea. The absence of PolIC, -Perk1, and -Perk2 in Neobodonida and Parabodonida needs to be reexamined after the genome data become available from the representative species of the two orders. Likewise, the inventory of mitochondrion-localized DNA polymerases in the class Prokinetoplastida relies entirely on *Perkinsela* sp. in this study. Thus, we need to examine (1) the absence of PolIB and/or C, and (2) the ubiquities of PolI-Perk1 and -Perk2 in members of Prokinetoplastida in the future.

At least one of the three phylogenetically distinct types of POP were detected in all of the euglenids examined in this study, except for *Peranema* sp. It is difficult to conclude that *Peranema* sp. truly lacks any POP homolog, as we only surveyed family A DNA polymerases in its transcriptome data. We predicted that POP_e1 and POP_Rhabd are localized in mitochondria and POP_e2 is a cytosolic protein. Unfortunately, the current data remain uncertain regarding the POP evolution in Euglenida. For instance, we cannot be sure whether the ancestral euglenid possessed a POP homolog for DNA replication in the mitochondrion. If we hypothesize the absence of POP in the ancestral euglenid, a straightforward interpretation of the distribution of the two distinct types of POP over the tree of Euglenida [33] is that POP_Rhabd and POP_e1 emerged separately (1) on the branch leading to *Rhabdomonas costata* and (2) the common ancestor of Euglenophyceae, respectively. In the future, the timing of POP_e1 emergence needs to be revised by incorporating the presence/secondary loss/absence of this type of POP in the early-branching (heterotrophic) species in the tree of Euglenida (e.g., *Peranema* sp.). When the presence/secondary loss of POP_e1 is confirmed in a heterotrophic species, the emergence of POP_e1 should be pushed back to a more ancient branch than that leading to the ancestral euglenophycean species in the tree of Euglenida. In addition, there is a possibility for an alternative scenario assuming that POP_e1 and POP_Rhabd, which were not so distant from each other in the POP phylogeny (Figure 3), evolved from a single POP in the ancestral euglenid. To understand the evolution of mitochondrion-localized POPs in euglenids better, we need to know the inventories of POP in phylogenetically broad euglenids, particularly those of early-branching species.

The proteome data from *Euglena gracilis* suggest that POP_e2 is a cytosolic protein. Although the cytosolic DNA polymerases are not of prime interest in this study, we here discuss the origin of POP_e2 briefly. The POP phylogeny recovered the intimate evolutionary affinity between POP_e2 and the plastid-localized POP in chlorarachniophytes (Figure 3). It is noteworthy that the POP homologs of euglenids and chlorarachniophytes appeared to be distant from those of *Pyramimonas parkeae* and members of Ulvophyceae that are close relatives of the algal endosymbionts which gave rise to the plastids in the two algal groups of interest [44,45]. Thus, no endosymbiotic gene transfer can be invoked in the evolution of POP_e2 in euglenids or plastid-localized POP in chlorarachniophytes. As Euglenida and Chlorarachniophyta are distantly related to each other in the organismal tree of eukaryotes [11], we propose that a POP gene may have been exchanged between the two distant groups, albeit the direction of the gene transfer remains uncertain. Alternatively, an as-yet-unknown eukaryote may have donated a POP gene to Euglenophyceae and Chlorarachniophyta separately. If so, we need to understand the precise diversity of POP in eukaryotes to pinpoint the donor of the ancestral POP_e2 gene.

4. Materials and Methods

4.1. Sequence Data Preparation

We obtained the transcriptome data of the following members of Euglenozoa from NCBI Sequence Read Archive [46]: Three kinetoplastids (*Bodo saltans*, GenBank accession number ERP001594;

Trypanoplasma borreli ATCC 50836, SRR10580962; *Azumiobodo hoyamushi*, SRR10586159), three diplonemids (*Diplonema ambulator*, SRR5998378; *Rhynchopus euleeides*, SRR5998382; *Flectonema neradi*, SRR5998375), four euglenids (*Euglena gracilis*, ERR974915, SRR3195326; *Euglena longa*, SRP148531; *Eutreptiella gymnastica* NIES-381, SRR1294408 and *Rhabdomonas costata* PRJNA550357), and a green alga (*Pyramimonas parkeae*, DRR036722). The raw sequence reads were trimmed by fastp v0.19.7 [47] with the -q 20 -u 80 option and then assembled by Trinity v2.8.4 [48]. The assembled genome data of *Perkinsela* sp. CCAP 1560/4 (LFNC01) and *Bodo saltans* strain LakeKonstanz (CYKH01) were downloaded from the GenBank database [46]. We searched for the nucleotide sequences encoding family A DNA polymerases in the 14 assembled transcriptome/genome data described above by TBLASTN [49] using the DNA polymerase domain of *Escherichia coli* DNA polymerase I (KHH06131; the portion corresponding to the 491th–928th amino acid residues) as a query. We retrieved the sequences matched to the query with E-values equal to or less than 1×10^{-4} as the candidates of family A DNA polymerases.

We repeated the procedures described above on our in-house transcriptome data of the diplonemid *Hemistasia phaeocysticola* and two euglenids (*Rapaza viridis* and *Peranema* sp.). The transcripts encoding the putative family A DNA polymerases were amplified by reverse transcription PCR and the resultant amplicons were sequenced by using the Sanger method. The nucleotide sequences determined in this study were deposited to GenBank/DDBJ/EMBL accession numbers LC516826–LC516833.

4.2. Phylogenetic Analysis of family A DNA Polymerases

We found 37 putative family A DNA polymerase sequences in 14 euglenozoan species (four kinetoplastids, four diplonemids, and 6 euglenids) in this study. These sequences were aligned with other family A DNA polymerases, including PolIA, B, C, and D in *Trypanosoma brucei*, *Trypanosoma grayi*, and *Leishmania major*, DNA polymerase γ, θ, and ν, POP, plastid replication and repair enzyme complex (PREX) in Apicomplexa, bacterial PolI, and bifunctional 3′-5′ exonuclease/DNA polymerase [10,23]. These family A DNA polymerases were sampled to include at least three sequences representing each clade in the previous phylogenetic trees. The amino acid sequences were aligned by MAFFT v7.407 [50] with the L-INS-i model. Ambiguously aligned positions were discarded manually, and gap-containing positions were trimmed by using trimAI v1.4 [51] with the -gt 0.9 option. The final alignment comprised 100 sequences with 426 unambiguously aligned amino acid positions. We subjected this alignment to the maximum-likelihood (ML) phylogenetic analysis by IQ-TREE v1.6.12 [52] using the LG + C20 + F + Γ model. The guide tree was obtained by the LG + I + Γ model that was selected by ModelFinder [53]. The statistical support for each bipartition in the ML tree was calculated by 100 non-parametric bootstrap replicates.

The family A alignment was also analyzed with Bayesian method by PhyloBayes v4.1 [54] using the CAT + GTR model. Four Markov chain Monte Carlo (MCMC) chains were run for 25,000 cycles with burn-in of 2500 (maxdiff = 0.144933). Subsequently, the consensus tree with branch lengths and BPPs was calculated from the remaining trees.

4.3. Phylogenetic Analysis of POP

We found 9 transcripts encoding POP in 5 euglenid species. Their putative amino acid sequences were added to the alignment that was generated and analyzed in Hirakawa and Watanabe (2019) [10]. In total, 58 POP sequences and 28 family A DNA polymerase sequences belonging to non-POP subfamilies were re-aligned by MAFFT v7.407 with L-INS-i model. After the trimming of gap-containing positions by using trimAI with the -gt 0.8 option, the final "POP" alignment comprised 86 sequences with 509 unambiguously aligned amino acid positions. The ML and ML bootstrap analyses were performed as described above.

The POP alignment was also analyzed with Bayesian method by PhyloBayes v4.1 [54] using the CAT + GTR model. Four MCMC chains were run for 100,000 cycles with burn-in of 10,000 (maxdiff = 0.286478). Subsequently, the consensus tree with branch lengths and BPPs was calculated from the remaining trees.

4.4. In silico Prediction of Subcellular Localization and Functional Domains of Family A DNA Polymerases

The mitochondrial localization of the family A DNA polymerases identified in this study were predicted based on their N-terminal sequences by the four different programs, TargetP 1.1 [29], NommPred [30], PredSL [31] and MitoFates [32]. In addition, the *Euglena gracilis* POP_e1 and POP_e2 sequences were subjected to SignalP v3.0 [40] and TMHMM v2.0 [41] to evaluate their plastid localization. Functional domains were searched by HMMER v3.3 [55] with the Pfam database [56].

5. Conclusions

In the current study, we provide the inventory of mitochondrion-localized DNA polymerases in phylogenetically broad members of Euglenozoa. The current study demonstrates that the three major classes of Euglenozoa (i.e., Kinetoplastea, Diplonemea, and Euglenida) possess distinctive sets of mitochondrion-localized family A DNA polymerases (summarized in Figure 2). Unfortunately, the inventory of mitochondrion-localized DNA polymerases lends no direct support to solve a 'big question' in the evolution of Euglenozoa—how have the distinctive mtDNA architectures emerged and been maintained in Euglenozoa? However, we believe that, in the long run, the results presented here can be a foundation for future studies on the evolution of euglenozoan mitochondria.

To further investigate the early evolution of mitochondrion-localized DNA polymerase inventory and mtDNA architecture in Euglenozoa, the sequence data from the family Symbiontida are indispensable. This lineage was seemingly separated prior to the divergence of Kinetoplastea, Diplonemea, and Euglenida [21,22], but neither the mtDNA data nor genome/transcriptome are currently available. In addition, heterotrophic members of Euglenida are necessary to be studied. This study includes only two heterotrophic euglenids (*Peranema* sp. and *Rhabdomonas constata*), but these species may not be sufficient to represent the diversity of the basal branches in the Euglenida tree [33]. Finally, recent culture-independent studies suggested the presence of previously undescribed lineages that branched after the separation of diplonemids but prior to the divergence of the known kinetoplastids [57–61]. One of these undescribed lineages was found to possess a diplonemid-like mtDNA architecture by sequencing its genome amplified from a single cell (isolate D1) [61]. We retrieved a single family A DNA polymerase sequence, which showed a weak phylogenetic affinity to PolID, in the sequence data of isolate D1 (see the Supplementary Materials). The undescribed lineages mentioned above are critical to understanding the transition of the mitochondrion-localized DNA polymerase inventory and mtDNA architecture in Euglenozoa.

Supplementary Materials: The following are available online at http://www.mdpi.com/2076-0817/9/4/257/s1, Figure_S1.pdf: ML phylogenetic tree of family A DNA polymerases including the DNA polymerase of the genome data from a single diplonemid-like cell (isolate D1). Table_S1.xlsx: The results of domain searches on the family A DNA polymerases identified in this study. Harada_FamADNApol_ali.nxs: The amino acid sequence alignment in nexus format used for the global phylogeny of family A DNA polymerases (Figure 1). Harada_FamADNApol_ali2.nxs: The amino acid sequence alignment in nexus format used for the global phylogeny of family A DNA polymerases including the sequence of a diplonemid-like organism (isolate D1; Figure S1). Harada_POP_ali.nxs: The amino acid sequence alignment in nexus format used for the POP phylogeny (Figure 3).

Author Contributions: Data curation, R.H., Y.H., A.Y., Y.K., M.M., R.O., P.S., S.M., V.H., G.T., and Y.I. Formal analysis, R.H., Y.H., P.S., V.H., and Y.I. Funding acquisition, Y.I. Investigation, R.H. and Y.H. Project administration, Y.I. Resources, A.Y., Y.K., M.M., R.O., P.S., S.M., V.H., and G.T. Supervision, Y.H. and Y.I. Visualization, R.H. and Y.I. Writing—original ms, R.H. and Y.I. Writing—review & editing, R.H., Y.H., A.Y., Y.K., M.M., R.O., P.S., S.M., V.H., G.T., and Y.I. All authors have read and agree to the published version of the manuscript.

Funding: This research was funded by the Japan Society for the Promotion of Sciences (18KK0203 and 19H03280, YI; 17H03723 and 26840123, GT; 17K19434, AY), the Czech Science Foundation (16-25280S, VH) and Centre for research of pathogenicity and virulence of parasites) a reg. č.: CZ.02.1.01/0.0/0.0/16_019/0000759.

Acknowledgments: Computational resources were supplied by the Ministry of Education, Youth and Sports of the Czech Republic under the Projects CESNET (Project No. LM2015042) and CERIT-Scientific Cloud (Project No. LM2015085) provided within the program Projects of Large Research, Development and Innovations Infrastructures.

Conflicts of Interest: The authors declare no conflict of interest.

References

1. Barrett, M.P.; Burchmore, R.J.S.; Stich, A.; Lazzari, J.O.; Frasch, A.C.; Cazzulo, J.J.; Krishna, S. The trypanosomiases. *Lancet* **2003**, *362*, 1469–1480. [CrossRef]

2. Verner, Z.; Basu, S.; Benz, C.; Dixit, S.; Dobáková, E.; Faktorová, D.; Hashimi, H.; Horáková, E.; Huang, Z.; Paris, Z.; et al. Malleable mitochondrion of *Trypanosoma brucei*. In *International Review of Cell and Molecular Biology*; Academic Press: Cambridge, MA, USA, 2015; Volume 315, pp. 73–151.

3. Klingbeil, M.M.; Motyka, S.A.; Englund, P.T. Multiple mitochondrial DNA polymerases in *Trypanosoma brucei*. *Mol. Cell* **2002**, *10*, 175–186. [CrossRef]

4. Krasich, R.; Copeland, W.C. DNA polymerases in the mitochondria: A critical review of the evidence. *Physiol. Behav.* **2017**, *22*, 692–709.

5. Saxowsky, T.T.; Choudhary, G.; Klingbeil, M.M.; Englund, P.T. *Trypanosoma brucei* has two distinct mitochondrial DNA polymerase β enzymes. *J. Biol. Chem.* **2003**, *278*, 49095–49101. [CrossRef]

6. Rajão, M.A.; Passos-Silva, D.G.; DaRocha, W.D.; Franco, G.R.; Macedo, A.M.; Pena, S.D.J.; Teixeira, S.M.; Machado, C.R. DNA polymerase kappa from *Trypanosoma cruzi* localizes to the mitochondria, bypasses 8-oxoguanine lesions and performs DNA synthesis in a recombination intermediate. *Mol. Microbiol.* **2009**, *71*, 185–197. [CrossRef] [PubMed]

7. Graziewicz, M.A.; Longley, M.J.; Copeland, W.C. DNA polymerase γ in mitochondrial DNA replication and repair. *Chem. Rev.* **2006**, *106*, 383–405. [CrossRef] [PubMed]

8. Christensen, A.C.; Lyznik, A.; Mohammed, S.; Elowsky, C.G.; Elo, A.; Yule, R.; Mackenzie, S.A. Dual-domain, dual-targeting organellar protein presequences in *Arabidopsis* can use non-AUG start codons. *Plant Cell* **2005**, *17*, 2805–2816. [CrossRef] [PubMed]

9. Moriyama, T.; Sato, N. Enzymes involved in organellar DNA replication in photosynthetic eukaryotes. *Front. Plant Sci.* **2014**, *5*, 480:1–480:12. [CrossRef]

10. Hirakawa, Y.; Watanabe, A. Organellar DNA polymerases in complex plastid-bearing algae. *Biomolecules* **2019**, *9*, 140. [CrossRef]

11. Adl, S.M.; Bass, D.; Lane, C.E.; Lukeš, J.; Schoch, C.L.; Smirnov, A.; Agatha, S.; Berney, C.; Brown, M.W.; Burki, F.; et al. Revisions to the classification, nomenclature, and diversity of eukaryotes. *J. Eukaryot. Microbiol.* **2019**, *66*, 4–119. [CrossRef]

12. Lukeš, J.; Guilbride, D.L.; Votýpka, J.; Zíková, A.; Benne, R.; Englund, P.T. Kinetoplast DNA network: Evolution of an improbable structure. *Eukayot. Cell* **2002**, *1*, 495–502. [CrossRef] [PubMed]

13. David, V.; Flegontov, P.; Gerasimov, E.; Tanifuji, G.; Hashimi, H.; Logacheva, M.D.; Maruyama, S.; Onodera, N.T.; Gray, M.W.; Archibald, J.M.; et al. Gene loss and error-prone RNA editing in the mitochondrion of *Perkinsela*, an endosymbiotic kinetoplastid. *MBio* **2015**, *6*, e01498-15:1–e01498-15:12. [CrossRef] [PubMed]

14. Yabuki, A.; Tanifuji, G.; Kusaka, C.; Takishita, K.; Fujikura, K. Hyper-eccentric structural genes in the mitochondrial genome of the algal parasite *Hemistasia phaeocysticola*. *Genome Biol. Evol.* **2016**, *8*, 2870–2878. [PubMed]

15. Burger, G.; Valach, M. Perfection of eccentricity: Mitochondrial genomes of diplonemids. *IUBMB Life* **2018**, *70*, 1197–1206. [CrossRef]

16. Kaur, B.; Záhonová, K.; Valach, M.; Faktorová, D.; Prokopchuk, G.; Burger, G.; Lukeš, J. Gene fragmentation and RNA editing without borders: Eccentric mitochondrial genomes of diplonemids. *Nucleic Acids Res.* **2020**, *48*, 2694–2708. [CrossRef]

17. Roy, J.; Faktorová, D.; Lukeš, J.; Burger, G. Unusual mitochondrial genome structures throughout the Euglenozoa. *Protist* **2007**, *158*, 385–396. [CrossRef]

18. Spencer, D.F.; Gray, M.W. Ribosomal RNA genes in *Euglena gracilis* mitochondrial DNA: Fragmented genes in a seemingly fragmented genome. *Mol. Genet. Genom.* **2011**, *285*, 19–31. [CrossRef]

19. Dobáková, E.; Flegontov, P.; Skalický, T.; Lukeš, J. Unexpectedly streamlined mitochondrial genome of the euglenozoan *Euglena gracilis*. *Genome Biol. Evol.* **2015**, *7*, 3358–3367. [CrossRef]

20. Simpson, A.G.B.; Hoff, J.V.D.; Bernard, C.; Burton, H.R.; Patterson, D.J. The ultrastructure and systematic position of the euglenozoon *Postgaardi mariagerensis*. *Arch. Protistenkd.* **1997**, *147*, 213–225. [CrossRef]

21. Yubuki, N.; Edgcomb, V.P.; Bernhard, J.M.; Leander, B.S. Ultrastructure and molecular phylogeny of *Calkinsia aureus*: Cellular identity of a novel clade of deep-sea euglenozoans with epibiotic bacteria. *BMC Microbiol.* **2009**, *9*, 16:1–16:22. [CrossRef]

22. Breglia, S.A.; Yubuki, N.; Hoppenrath, M.; Leander, B.S. Ultrastructure and molecular phylogenetic position of a novel euglenozoan with extrusive episymbiotic bacteria: *Bihospites bacati* n. gen. et sp. (Symbiontida). *BMC Microbiol.* **2010**, *10*, 145:1–145:21. [CrossRef]

23. Moriyama, T.; Terasawa, K.; Fujiwara, M.; Sato, N. Purification and characterization of organellar DNA polymerases in the red alga *Cyanidioschyzon merolae*. *FEBS J.* **2008**, *275*, 2899–2918. [CrossRef] [PubMed]

24. Chandler, J.; Vandoros, A.V.; Mozeleski, B.; Klingbeil, M.M. Stem-loop silencing reveals that a third mitochondrial DNA polymerase, POLID, is required for kinetoplast DNA replication in trypanosomes. *Eukaryot. Cell* **2008**, *7*, 2141–2146. [CrossRef] [PubMed]

25. Bruhn, D.F.; Mozeleski, B.; Falkin, L.; Klingbeil, M.M. Mitochondrial DNA polymerase POLIB is essential for minicircle DNA replication in African trypanosomes. *Mol. Microbiol.* **2010**, *75*, 1414–1425. [CrossRef] [PubMed]

26. Bruhn, D.F.; Sammartino, M.P.; Klingbeil, M.M. Three mitochondrial DNA polymerases are essential for kinetoplast DNA replication and survival of bloodstream form *Trypanosoma brucei*. *Eukaryot. Cell* **2011**, *10*, 734–743. [CrossRef] [PubMed]

27. Panigrahi, A.K.; Ogata, Y.; Zíková, A.; Anupama, A.; Dalley, R.A.; Acestor, N.; Myler, P.J.; Stuart, K.D. A comprehensive analysis of *Trypanosoma brucei* mitochondrial proteome. *Proteomics* **2009**, *9*, 434–450. [CrossRef]

28. Hammond, M.J.; Nenarokova, A.; Butenko, A.; Zoltner, M.; Dobáková, E.L.; Field, M.C.; Lukeš, J. A uniquely complex mitochondrial proteome from *Euglena gracilis*. *Mol. Biol. Evol.* **2020**. Epub ahead of print. [CrossRef]

29. Emanuelsson, O.; Nielsen, H.; Brunak, S.; Heijne, G.V. Predicting subcellular localization of proteins based on their N-terminal amino acid sequence. *J. Mol. Biol.* **2000**, *300*, 1005–1016. [CrossRef]

30. Kume, K.; Amagasa, T.; Hashimoto, T.; Kitagawa, H. NommPred: Prediction of mitochondrial and mitochondrion-related organelle proteins of nonmodel organisms. *Evol. Bioinform.* **2018**, *14*, 1–12. [CrossRef]

31. Petsalaki, E.I.; Bagos, P.G.; Litou, Z.I.; Hamodrakas, S.J. PredSL: A tool for the N-terminal sequence-based prediction of protein subcellular localization. *Genom. Proteom. Bioinf.* **2006**, *4*, 48–55. [CrossRef]

32. Fukasawa, Y.; Tsuji, J.; Fu, S.C.; Tomii, K.; Horton, P.; Imai, K. MitoFates: Improved prediction of mitochondrial targeting sequences and their cleavage sites. *Mol. Cell. Proteomics* **2015**, *14*, 1113–1126. [CrossRef] [PubMed]

33. Bicudo, C.E.d.M.; Menezes, M. Phylogeny and classification of Euglenophyceae: A brief review. *Front. Ecol. Evol.* **2016**, *4*, 17:1–17:15. [CrossRef]

34. Tashyreva, D.; Prokopchuk, G.; Yabuki, A.; Kaur, B.; Faktorová, D.; Votýpka, J.; Kusaka, C.; Fujikura, K.; Shiratori, T.; Ishida, K.I.; et al. Phylogeny and morphology of new diplonemids from Japan. *Protist* **2018**, *169*, 158–179. [CrossRef] [PubMed]

35. Yazaki, E.; Ishikawa, S.A.; Kume, K.; Kumagai, A.; Kamaishi, T.; Tanifuji, G.; Hashimoto, T.; Inagaki, Y. Global kinetoplastea phylogeny inferred from a large-scale multigene alignment including parasitic species for better understanding transitions from a free-living to a parasitic lifestyle. *Genes Genet. Syst.* **2017**, *92*, 35–42. [CrossRef] [PubMed]

36. Tosal, L.; Comendador, M.A.; Sierra, L.M. The mus308 locus of *Drosophila melanogaster* is implicated in the bypass of ENU-induced *O*-alkylpyrimidine adducts. *Mol. Gen. Genet.* **2000**, *263*, 144–151. [CrossRef] [PubMed]

37. Concepción-Acevedo, J.; Miller, J.C.; Boucher, M.J.; Klingbeil, M.M. Cell cycle localization dynamics of mitochondrial DNA polymerase IC in African trypanosomes. *Mol. Biol. Cell* **2018**, *29*, 2540–2552. [CrossRef] [PubMed]

38. Ono, Y.; Sakai, A.; Takechi, K.; Takio, S.; Takusagawa, M.; Takano, H. NtPolI-like1 and NtPolI-like2, bacterial DNA polymerase I homologs isolated from BY-2 cultured tobacco cells, encode DNA polymerases engaged in DNA replication in both plastids and mitochondria. *Plant Cell Physiol.* **2007**, *48*, 1679–1692. [CrossRef]

39. Moriyama, T.; Tajima, N.; Sekine, K.; Sato, N. Localization and phylogenetic analysis of enzymes related to organellar genome replication in the unicellular rhodophyte *Cyanidioschyzon merolae*. *Genome Biol. Evol.* **2014**, *6*, 228–237. [CrossRef]

40. Bendtsen, J.D.; Nielsen, H.; Heijne, G.V.; Brunak, S. Improved prediction of signal peptides: SignalP 3.0. *J. Mol. Biol.* **2004**, *340*, 783–795. [CrossRef]

41. Krogh, A.; Larsson, B.; Heijne, G.V.; Sonnhammer, E.L.L. Predicting transmembrane protein topology with a hidden Markov model: Application to complete genomes. *J. Mol. Biol.* **2001**, *305*, 567–580. [CrossRef]

42. Novák Vanclová, A.M.G.; Zoltner, M.; Kelly, S.; Soukal, P.; Záhonová, K.; Füssy, Z.; Ebenezer, T.E.; Lacová Dobáková, E.; Eliáš, M.; Lukeš, J.; et al. Metabolic quirks and the colourful history of the *Euglena gracilis* secondary plastid. *New Phytol.* **2020**, *225*, 1578–1592. [CrossRef] [PubMed]

43. Tanifuji, G.; Cenci, U.; Moog, D.; Dean, S.; Nakayama, T.; David, V.; Fiala, I.; Curtis, B.A.; Sibbald, S.J.; Onodera, N.T.; et al. Genome sequencing reveals metabolic and cellular interdependence in an amoeba-kinetoplastid symbiosis. *Sci. Rep.* **2017**, *7*, 11688:1–11688:13. [CrossRef] [PubMed]

44. Turmel, M.; Gagnon, M.C.; O'Kelly, C.J.; Otis, C.; Lemieux, C. The chloroplast genomes of the green algae *Pyramimonas*, *Monomastix*, and *Pycnococcus* shed new light on the evolutionary history of prasinophytes and the origin of the secondary chloroplasts of euglenids. *Mol. Biol. Evol.* **2009**, *26*, 631–648. [CrossRef] [PubMed]

45. Suzuki, S.; Hirakawa, Y.; Kofuji, R.; Sugita, M.; Ishida, K. Plastid genome sequences of *Gymnochlora stellata*, *Lotharella vacuolata*, and *Partenskyella glossopodia* reveal remarkable structural conservation among chlorarachniophyte species. *J. Plant Res.* **2016**, *129*, 581–590. [CrossRef]

46. Sayers, E.W.; Beck, J.; Brister, J.R.; Bolton, E.E.; Canese, K.; Comeau, D.C.; Funk, K.; Ketter, A.; Kim, S.; Kimchi, A.; et al. Database resources of the National Center for Biotechnology Information. *Nucleic Acids Res.* **2020**, *48*, D9–D16. [CrossRef]

47. Chen, S.; Zhou, Y.; Chen, Y.; Gu, J. fastp: An ultra-fast all-in-one FASTQ preprocessor. *Bioinformatics* **2018**, *34*, i884–i890. [CrossRef]

48. Grabherr, M.G.; Haas, B.J.; Yassour, M.; Levin, J.Z.; Thompson, D.A.; Amit, I.; Adiconis, X.; Fan, L.; Raychowdhury, R.; Zeng, Q.; et al. Trinity: Reconstructing a full-length transcriptome without a genome from RNA-Seq data. *Nat. Biotechnol.* **2011**, *29*, 644–652. [CrossRef]

49. Camacho, C.; Coulouris, G.; Avagyan, V.; Ma, N.; Papadopoulos, J.; Bealer, K.; Madden, T.L. BLAST+: Architecture and applications. *BMC Bioinform.* **2009**, *10*, 421:1–421:9. [CrossRef]

50. Katoh, K.; Standley, D.M. MAFFT multiple sequence alignment software version 7: Improvements in performance and usability. *Mol. Biol. Evol.* **2013**, *30*, 772–780. [CrossRef]

51. Capella-Gutiérrez, S.; Silla-Martínez, J.M.; Gabaldón, T. trimAl: A tool for automated alignment trimming in large-scale phylogenetic analyses. *Bioinformatics* **2009**, *25*, 1972–1973. [CrossRef]

52. Nguyen, L.T.; Schmidt, H.A.; Von Haeseler, A.; Minh, B.Q. IQ-TREE: A fast and effective stochastic algorithm for estimating maximum-likelihood phylogenies. *Mol. Biol. Evol.* **2015**, *32*, 268–274. [CrossRef] [PubMed]

53. Kalyaanamoorthy, S.; Minh, B.Q.; Wong, T.K.F.; Von Haeseler, A.; Jermiin, L.S. ModelFinder: Fast model selection for accurate phylogenetic estimates. *Nat. Methods* **2017**, *14*, 587–589. [CrossRef] [PubMed]

54. Lartillot, N.; Lepage, T.; Blanquart, S. PhyloBayes 3: Bayesian software package for phylogenetic reconstruction and molecular dating. *Bioinformatics* **2009**, *25*, 2286–2288. [CrossRef] [PubMed]

55. Eddy, S.R. HMMER: Biosequence Analysis Using Profile Hidden Markov Models. Available online: http://hmmer.org/ (accessed on 10 December 2019).

56. El-Gebali, S.; Mistry, J.; Bateman, A.; Eddy, S.R.; Luciani, A.; Potter, S.C.; Qureshi, M.; Richardson, L.J.; Salazar, G.A.; Smart, A.; et al. The Pfam protein families database in 2019. *Nucleic Acids Res.* **2019**, *47*, D427–D432. [CrossRef]

57. López-García, P.; Duperron, S.; Philippot, P.; Foriel, J.; Susini, J.; Moreira, D. Bacterial diversity in hydrothermal sediment and epsilonproteobacterial dominance in experimental microcolonizers at the Mid-Atlantic Ridge. *Environ. Microbiol.* **2003**, *5*, 961–976. [CrossRef]

58. Scheckenbach, F.; Hausmann, K.; Wylezich, C.; Weitere, M.; Arndt, H. Large-scale patterns in biodiversity of microbial eukaryotes from the abyssal sea floor. *Proc. Natl. Acad. Sci. USA* **2010**, *107*, 115–120. [CrossRef]

59. Marquardt, M.; Vader, A.; Stübner, E.I.; Reigstad, M.; Gabrielsen, T.M. Strong seasonality of marine microbial eukaryotes in a high-Arctic fjord (Isfjorden, in West Spitsbergen, Norway). *Appl. Environ. Microbiol.* **2016**, *82*, 1868–1880. [CrossRef]

60. Gawryluk, R.M.R.; del Campo, J.; Okamoto, N.; Strassert, J.F.H.; Lukeš, J.; Richards, T.A.; Worden, A.Z.; Santoro, A.E.; Keeling, P.J. Morphological identification and single-cell genomics of marine diplonemids. *Curr. Biol.* **2016**, *26*, 3053–3059. [CrossRef]

61. Wideman, J.G.; Lax, G.; Leonard, G.; Milner, D.S.; Rodríguez-Martínez, R.; Simpson, A.G.B.; Richards, T.A. A single-cell genome reveals diplonemid-like ancestry of kinetoplastid mitochondrial gene structure. *Philos. Trans. R. Soc. B Biol. Sci.* **2019**, *374*, 20190100. [CrossRef]

Article

Common Structural Patterns in the Maxicircle Divergent Region of Trypanosomatidae

Evgeny S. Gerasimov [1,2,3,*], Ksenia A. Zamyatnina [1], Nadezda S. Matveeva [1,2], Yulia A. Rudenskaya [1], Natalya Kraeva [4], Alexander A. Kolesnikov [1] and Vyacheslav Yurchenko [2,4,*]

1   Faculty of Biology, M. V. Lomonosov Moscow State University, Moscow 119991, Russia; zamksju@rambler.ru (K.A.Z.); ucheb.mn@gmail.com (N.S.M.); illuvi@yandex.ru (Y.A.R.); aak330@yandex.ru (A.A.K.)
2   Martsinovsky Institute of Medical Parasitology, Tropical and Vector Borne Diseases, Sechenov University, Moscow 119435, Russia
3   Institute for Information Transmission Problems, Russian Academy of Sciences, Moscow 127051, Russia
4   Life Science Research Centre, Faculty of Science, University of Ostrava, 710 00 Ostrava, Czech Republic; luzikhina@gmail.com
*   Correspondence: jalgard@gmail.com (E.S.G.); vyacheslav.yurchenko@osu.cz (V.Y.)

Received: 9 January 2020; Accepted: 3 February 2020; Published: 5 February 2020

Abstract: Maxicircles of all kinetoplastid flagellates are functional analogs of mitochondrial genome of other eukaryotes. They consist of two distinct parts, called the coding region and the divergent region (DR). The DR is composed of highly repetitive sequences and, as such, remains the least explored segment of a trypanosomatid genome. It is extremely difficult to sequence and assemble, that is why very few full length maxicircle sequences were available until now. Using PacBio data, we assembled 17 complete maxicircles from different species of trypanosomatids. Here we present their large-scale comparative analysis and describe common patterns of DR organization in trypanosomatids.

Keywords: divergent region; maxicircle; kinetoplast; trypanosomatids; repeats; genomic rearrangements; mitochondrion

1. Introduction

Trypanosomatids are unicellular parasitic organisms with a single large mitochondrion per cell [1,2]. Mitochondrial DNA of trypanosomatids is a network of concatenated circular molecules of two types: maxicircles and minicircles [3]. Minicircles are exclusive to trypanosomatids, these relatively small, but numerous, molecules are directly involved in U-insertion/deletion editing system as they encode gRNAs [4,5]. There are thousands of minicircles present in a single mitochondrion [6–8]. In contrast, maxicircles are much larger (25–50 kbps) and present in up to 100 copies [9]. It is highly likely that both molecule populations are heterogeneous in the cell [10–17]. Maxicircles are functional equivalents of mitochondrial genome of other eukaryotes. Mitochondrial genes are compactly located in the so-called coding region (CR) of a maxicircle, which is typically about 16 kbps, and demonstrate high level of synteny between species [18–21]. The rest of maxicircle's molecule is termed the divergent region (DR), sometimes also called the variable region [22–25].

The DR was shown to be composed of repetitive sequence elements and, therefore, extremely hard to sequence and assemble [26–28]. Most of the previous comparative studies were based not on the direct nucleotide sequence analysis, but rather on hybridization techniques, which allowed to estimate sequence homology between fragments of maxicircles only roughly [19,25,29]. These experiments showed that the DR sequences of maxicircles vary significantly between species/strains, while, in contrast, the CRs preserve a high level of synteny. The nature of this variation remained unclear. Only few fragments from DR were directly sequenced following PCR amplification, molecular cloning and Sanger sequencing. For example, the analysis of 2.76 kb fragment from DR of *Leishmania tarentolae* revealed that various repetitive elements are grouped into larger clusters, which, in turn, are repeats themselves. Most repetitive units within clusters are not perfect repeats, as they share 65–100% sequence identity [26]. The DR was proposed to regulate gene expression, however, a putative origin of replication, topoisomerase II binding sites, and heterogeneously sized transcripts were documented in this region, arguing that its role may be more complex [22].

With the development of sequencing technologies some maxicircles were sequenced completely [12,30–32]. However, even paired-end Illumina sequencing was often not sufficient to assemble the complete maxicircle, and, because of that, most studies were focused only on coding region with short flanks [11,13,18], and even the most recent assembly from Illumina data had recovered only 3.5 kb of DR (as in [32]). Analysis of 3.5-kb and 5-kb regions of DR of *Trypanosoma lewisi*, flanking 12S and ND5 genes, respectively, revealed that these loci have different patterns of repeats [31]. Keeping up with authors' terminology, we will call them sections I and II, respectively. Section I is composed of short highly repetitive units, while section II is a series of tandem duplications of a longer sequence. The repeat units in DR tend to be organized in a head-to-tail orientation, frequently forming nearly perfect clusters of tandem repeats.

Arguably, an even more interesting observation is the ability of DR to evolve quickly. Massive structural rearrangements were observed in cultures of *Leishmania* spp. under various conditions [33–35]. Of note, molecular studies of these rearrangements have never been done with full length maxicircle sequences.

Repeated sequences, especially tandem repeats and large duplications, complicate the assembly process and result in errors even in the coding sequences. Nevertheless, sufficiently long reads can help to overcome this hurdle [36]. In the current work, we used PacBio reads to assemble maxicircles, thus, contributing to the collection of known DR structures. We present here sequences of 17 new full-length maxicircles, more than double of what has been available before, and brief comparative analysis of their DR structures, revealing common patterns and specific features of this region.

2. Results

2.1. Complete Maxicircle Assembly Overview

Using our custom pipeline to assemble PacBio reads containing 12S and/or ND5 mitochondrial genes we recovered 17 full-length mitochondrial maxicircle sequences of different trypanosomatid species.

Maxicircles have different sizes. The shortest sequence in this study of 23,201 bp belongs to *Trypanosoma brucei* (Lister 427), while the longest one of 47,384 bp is from *T. cruzi* (Dm28c). Table 1 demonstrates that size variation of maxicircles primarily reflects the differences in length of the DR, which is consistent with previous observations [18,19,21]. The CRs of all assembled maxicircles are similar in length and exhibit a high level of synteny. Some minor length variations between the CRs can be explained by the editing patterns (variable length of the edited domains) of some cryptogenes. As such, the shortest CR sequences were documented in *Trypanosoma* spp., where cryptogenes undergo the most extensive RNA editing [4,5], while the longest CR sequences were in monoxenous trypanosomatids, whose cryptogenes have reduced editing domains [37–39].

Table 1. Overview of assembled maxicircles. Brief statistics shows the basic assembly parameters: total maxicircle length, divergent region length, coding region length, average coverage per nucleotide, and total number of reads, included in the assembly and GenBank accession number.

Species (Strain)	Length, bp	DR, bp	CR, bp	Coverage	Reads	GenBank Accession
Leishmania (Viannia) guyanensis (204-365)	27,631	11,485	16,146	8.2	39	MN904521
Leishmania (Leishmania) mexicana (215-49)	27,138	10,860	16,278	10.6	36	MN904523
Leishmania (L.) aethiopica (209-622)	29,037	12,978	16,059	5.1	20 *	MN904514
Leishmania (L.) infantum (193-S1775)	29,512	13,313	16,199	21.7	94	MN904522
Leishmania (L.) tropica (216-162)	29,557	13406	16,151	20.1	64	MN904525
Leishmania (L.) amazonensis (210-660)	33,779	17,521	16,258	10.0	37	MN904515
Leishmania (L.) donovani (193-S616)	34,088	17,883	16,205	8.9	31	MN904519
Leishmania (L.) donovani (FDAARGOS_361)	33,278	17,238	16,040	16.0	61	MN904520
Leishmania (L.) donovani (Pasteur)	36,676	20,521	16,155	17.2	62	MN904518
Leishmania (V.) braziliensis (208-905)	26,728	10,568	16,160	13.9	89	MN904516
Leishmania (V.) braziliensis (208-954)	29,618	13,442	16,176	20.1	87	MN904517
Trypanosoma brucei (Lister 427)	23,201	8391	14,810	9.1	33 *	MN904526
Trypanosoma cruzi (TCC)	39,883	24,498	15,385	25.7	93	MN904528
Trypanosoma cruzi (Dm28c)	47,384	32,012	15,372	27.9	98	MN904527
Leptomonas pyrrhocoris (H10)	31,564	15,450	16,114	19.9	47	MN904524
Herpetomonas megaseliae	29,680	14,463	15,217	17.7	57	MN904513
Crithidia expoeki (BJ08_175)	25,722	8594	17,128	26.78	57	MN904512

* Assembled maxicircle contig was not marked as 'circular' by Canu assembler.

The DRs of maxicircles are much more variable in size, reflecting the dynamics of this structure in evolution. This difference is over 20 kbps for maxicircles of different trypanosomatid species. Even within the phylogenetically compact group of *Leishmania* spp., the DR length between the two strains of *L. braziliensis* and three strains of *L. donovani* vary by ≈3 kbps.

2.2. Nucleotide-Level Analysis of the DR Sequences

PCA (Principal Component Analysis) of triplet frequencies demonstrated that even on the level of very small sequence elements (triplets) the DR has species-specific composition and the clustering is generally consistent with phylogenetic relationships of the species (Figure 1). Importantly, despite the documented length variation between strains and species, their triplet spectrum remains very stable. For example, while the length variation between the DRs of *Leishmania donovani* strains and *L. infantum* is about 4500 bp, they cluster together on the PCA plot. The same is true for *T. cruzi* strains, which possess even more dramatic DR length difference. The most parsimonious explanation for this observation is that the level of repeat arrays is very unstable, despite the fact that the structure of small repetitive units of the DR is species-specific. Rearrangements at this level may account for the DR length variation and can happen frequently, due to the recombination between stable repetitive units.

Next, we analyzed the pattern of k-mer frequencies in maxicircles for small values of k (Figure 2). This analysis revealed the regularity of DR structure at a very basic level of short perfect repeats. For example, *Leishmania amazonensis* has 160 exact matches of 5'-TTAAATTAAATTAAATTAAATTAAA-3' sequence in the 33,779 bp-long maxicircle, and this sequence is a pentamer of the 'TTAAA' block itself. The DR of *L. pyrrhocoris* H10 has only 45 perfect repeats of 5'-ATATTGAAAATAAAGTGCTAGATA-3' sequence in its maxicircle of approximately the same length, and this sequence is not composed of smaller units. In the DRs of *T. cruzi* TCC maxicircles, the most frequent 24-mer is repeated only 25 times, implying more diffused organization of repeated units.

Such a small-scale regularity appears to be changing quickly and does not reflect phylogenetic relationships between species. Three *L. donovani* strains have maxicircles, which are located close to each other on the PCA axis (Figure 1), exhibit different patterns of k-mer occurrences (Figure 2).

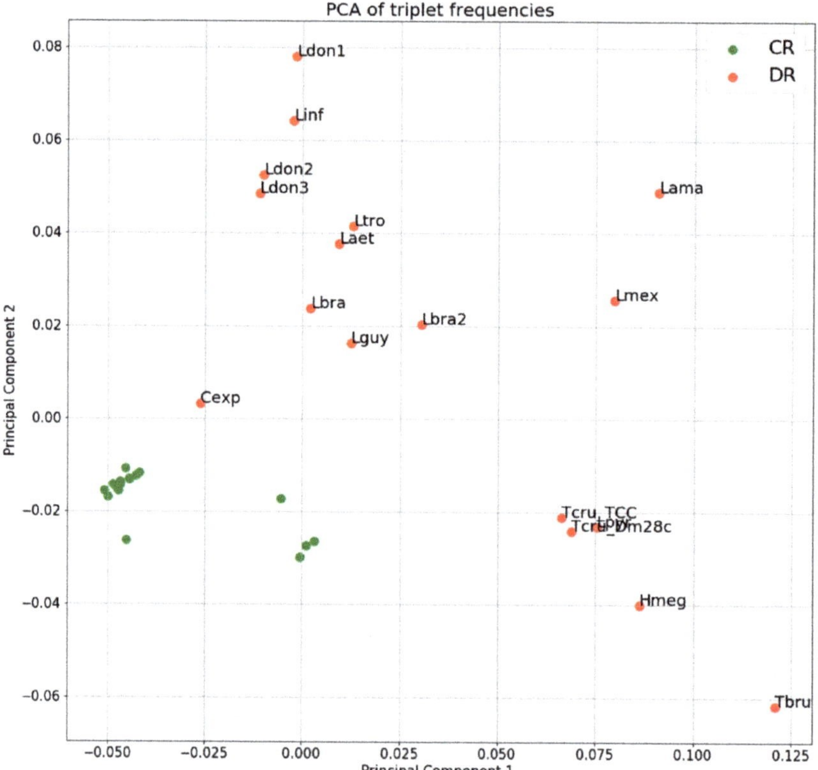

Figure 1. Principal component analysis of triplet spectra of maxicircles. Triplet frequency vectors were calculated for the coding region (green) and divergent region (red) of each maxicircle. One and two principal components were chosen to show on the scatter plot. Lama = *Leishmania amazonensis*; Hmeg = *Herpetomonas megaseliae*; Laet = *Leishmania aethiopica*; Cexp = *Crithidia expoeki*; Lbra2 = *Leishmania braziliensis* (208-905); Lbra = *Leishmania braziliensis* (208-954); Ldon1 = *Leishmania donovani* (Pasteur); Ldon2 = *Leishmania donovani* (193-S616); Ldon3 = *Leishmania donovani* (FDAARGOS_361); Lguy = *Leishmania guyanensis*; Linf = *Leishmania infantum*; Lmex = *Leishmania mexicana*; Lpyr = *Leptomonas pyrrhocoris*; Ltro = *Leishmania tropica*; Tbru = *Trypanosoma brucei*; Dm28c = *Trypanosoma cruzi* (Dm28c); TCC = *Trypanosoma cruzi* (TCC).

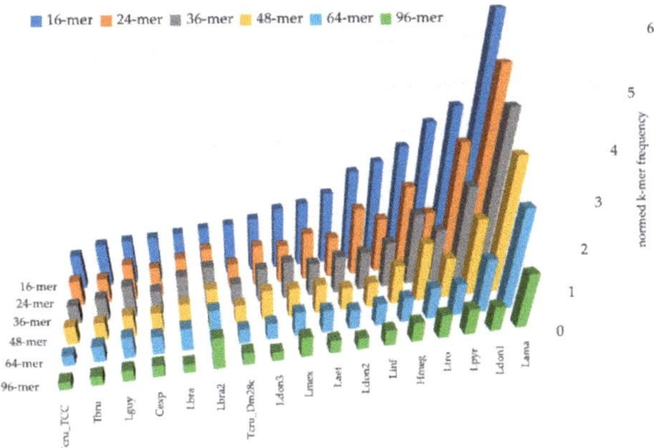

Figure 2. Number of occurrences of most frequent k-mer in maxicircle for values of k = 16, 24, 36, 48, 64, and 96. Species abbreviations are the same as for Figure 1.

2.3. Large-Scale Divergent Region Architecture

Previous studies of sequences, derived from the DRs of *Leishmania* spp., revealed that they are composed almost exclusively of repeats. These repeats are grouped in arrays or clusters [26,40]. In addition to *Leishmania*, these were also studies in *Strigomonas oncopelti* (called then *Crithidia oncopelti*) [41]. Here, we built full maps of divergent region of 17 different species using the MUMmer package to identify homologous regions and to show global patterns of DR organization.

Dotplots of six maxicircles, depicting variants of DR organization in studied species, are shown in Figure 3. Dotplots of other assembled maxicircles are shown in Figure A1. There are some common traits of the DR architecture. Firstly, repetitive elements are always arranged in a head-to-tail way. Secondly, there are two elements, which can be usually distinguished in the DR sequences. Hereafter, we will denote them as P5 and P12, according to their proximity (P) to the *ND5* and *12S* genes, respectively. Both elements are composed of repeated sequences, but these sequences have a different length of the repeat (periodicity), structure, and pattern of repetitions. The P5 is a tandem repeat with a large period, but small number of repetitions. In contrast, the P12 element is composed of highly repetitive units with a small period, which are organized in repeat arrays of varying length. Arrays are frequently interspersed by repeat units of a radically different type, the I-elements, which are dissimilar to repeats forming the arrays (Figure 3b, red arrow). The I-elements are common in *Leishmania* maxicircles, where they clearly demarcate arrays' borders. According to the number of arrays in the P12 element, the DRs can be pentameric (*Leishmania infantum* or *Leptomonas pyrrhocoris*; see Figure 3b,d), or tetrameric (*Leishmania guyanensis*, see Figure 3a). In some species the I-element is absent. In these cases, arrays of repeats are not visually separated on dotplots and appear as single-tandem clusters (for example, the P12 element of *L. amazonensis* and *T.cruzi*, Figure 3c,e, respectively). Such clusters, however, are not perfect tandem repeats, as the repeated units diverge with varying percent of sequence identity.

The general architecture of the DR, described above, however, undergoes major changes in some species. In *T. brucei*, which seems to have rather small DR, the P5 element is reduced to a short non-repetitive sequence. Very similar reduction of the P5 element was also documented in *Herpetomonas megaseliae* (Figure A2c).

The complete maps of the DRs (Circos plots) for the six species are shown in Figure 4; maps for the rest of the assembled genomes are presented in Figure A2. The repeats have various degree of sequence similarity. This can be exemplified by the P12 element of, for example, *Leishmania guyanensis* (Figure 4a) or *L. aethiopica* (Figure A2a). The most striking case in this regard is *T. brucei* (Figure 4f), in which the sequence similarity between most repeats is below 70%, with only few repeat units in the middle of the DR sharing about 90% of sequence identity. The P5 element of most *Leishmania* spp. ends with an almost perfect tandem repeat, which is very GC-rich and is dimeric in most genomes. A single copy of array repeat unit at the beginning of the P12 element is present in all *Leishmania* genome analyzed. This unit is marked by the black arrows in Figures 3b and 4b.

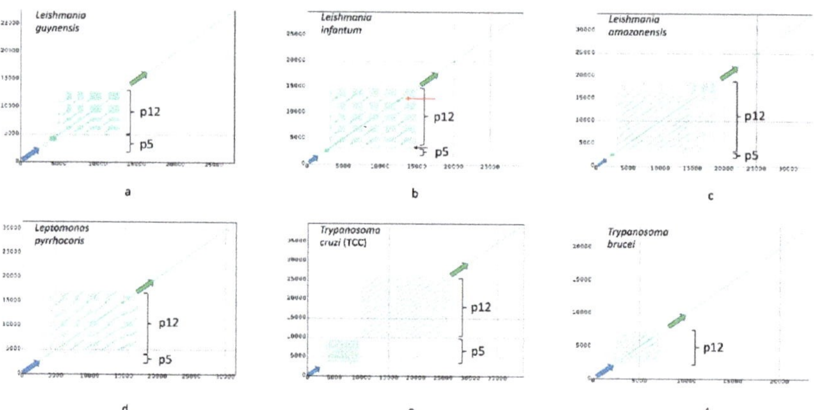

Figure 3. Dotplots of maxicircles of *Leishmania guyanensis* (**a**); *L. infantum* (**b**); *L. amazonensis* (**c**); *Leptomonas pyrrhocoris* (**d**); *Trypanosoma cruzi* strain TCC (**e**); *T. brucei* (**f**). Green and blue arrows denote *12S rRNA* and *ND5* genes, respectively. All plots show full maxicircle starting with the ND5 gene at (0,0). Red and black arrows in (**b**) indicate the I-element and a single-copy array repeat unit, respectively. Dot plots of other assembled maxicircles can be found in Figure A1.

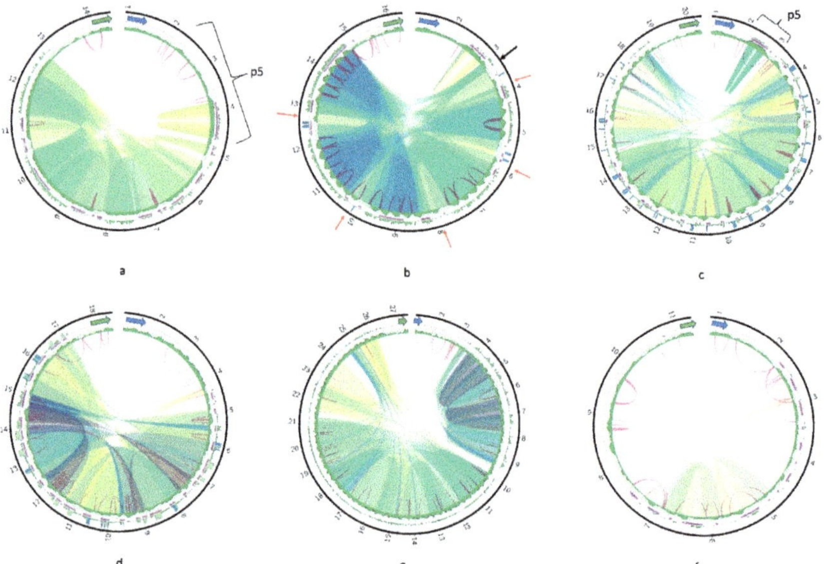

Figure 4. Circos plots of the divergent region (DR) of *Leishmania guyanensis* (**a**); *L. infantum* (**b**); *L. amazonensis* (**c**); *Leptomonas pyrrhocoris* (**d**); *Trypanosoma cruzi* strain TCC (**e**); *T. brucei* (**f**). Green and blue arrows denote the *12S rRNA* and *ND5* genes, respectively. Outer tracks are histograms of 24-mer frequency; middle tracks indicate tandem repeats; and the inner tracks are GC-content profiles. Ribbons inside the circle connect homologous regions; color represents percent of sequence identity in range [80%; 100%] in the order red, yellow, green, and blue. Violet arcs denote inverted repeats.

In contrast, there are some species with almost identical regions in their DRs. For example, five arrays of repeats from the pentameric P12 element of *Leishmania infantum* share over 95% identity in their sequences (Figure 4b). In *L. amazonensis*, highly similar blocks of much smaller size are also detected in the P12 element (Figure 4c).

Of note, the GC profiles of most *Leishmania* spp. DRs are very regular (e.g., the repeat arrays within the P12 element have a distinct pattern), while three *Trypanosoma* spp. and three monoxenous trypanosomatids analyzed here have less regular GC profiles of their P12 element.

Repeat units within arrays in the P12 element can form either perfect or separated by short indels tandems. Interestingly, no tandems can be detected in both *T. cruzi* maxicircles (Figures 4e and A2d) even with rather relaxed mreps settings. Inverted repeats could be detected in all DR sequences. There was no discernable common pattern of their localization. Almost invariably, these repeats are very short with identity between complementary parts reaching 85–91%. Very frequently, the inverted repeats were detected only in some copies of repeat units within arrays, and were missing from neighboring units because of the sequence divergence.

Special kinds of sequences in the DR of some species can be revealed with the k-mer frequency plots. Blue bars on outer track histograms in Figure 4 indicate that a 24-mer starting at the current sequence position is repeated over 40 times in *Leishmania infantum*, *L. amazonensis*, and *Leptomonas pyrrhocoris*. These are always extremely AT-rich sequences composed of a single repeated motif. Interestingly, unique non-repetitive sequences can also be present in the DRs, mostly on the border between the P5 and P12 elements.

2.4. Comparative Analysis and Dynamics of DR Structure

Structural rearrangements in the DRs have been described earlier [33–35]. Here, we compared the full DRs of maxicircles in order to reveal structural variations between strains and species. The DR structures for three strains of *L. donovani* are presented in Figure 5. The P12 element is octameric, pentameric, and trimeric in the Pasteur, 193-S616, and FDAARGOS_361 strains, respectively. Both the number of arrays in the P12 element and the size of each array vary significantly. Most proximal to the *12S* rRNA gene array in the FDAARGOS_361 strain is big, while in the Pasteur strain all eight arrays are rather small in size. Thus, the P12 element appears mainly responsible for the overall maxicircle size variation.

Sequence similarity between units of arrays also differ between three *L. donovani* strains. For example, in the Pasteur strain, all eight repeat arrays are almost identical (Figure 5d), while in the 193-S616 strain most repeats share less than 90% identity, with only three blocks of tandem repeats having higher similarity (Figure 5e, blue ribbons).

Next, we compared the DR structures in different strains and species in order to reveal most conservative sequence elements. In all the cases, the highest similarity was between the *12S* rRNA and *ND5* genes, flanking the DR sequences. In addition, there were also sequence elements inside the DR, which shared over 80% identity between the strains (Figure 6a,b) or even the species (Figure 6c). These elements were usually localized in the P12 element of the DR, and only in *L. braziliensis* low similarity was detected between tandem repeats of the P5 element. The locations of homologous elements in the P12 coincide with the layout of repeat arrays, indicating that repeat arrays are built from repeat units with different degrees of conservation. More conserved units tend to be in the beginning of each array. Different strains of *Leishmania donovani* demonstrate the highest degree of sequence similarity in their DRs (Figure 6d,e).

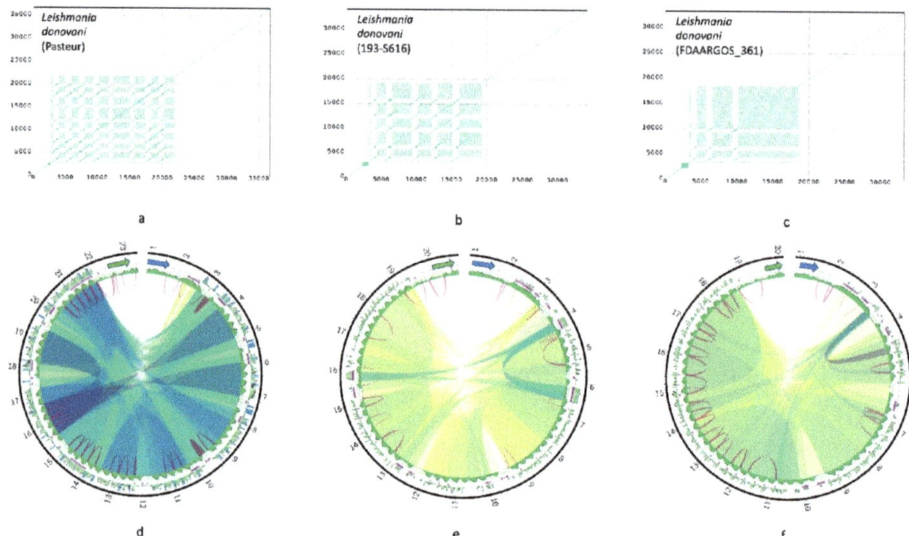

Figure 5. Dotplots and corresponding Circos plots of the DR of *Leishmania donovani* strains: Pasteur (**a,d**); 193-S616 (**b,e**); FDAARGOS_361 (**c,f**) Circos plots are presented in the same way as in Figure 4.

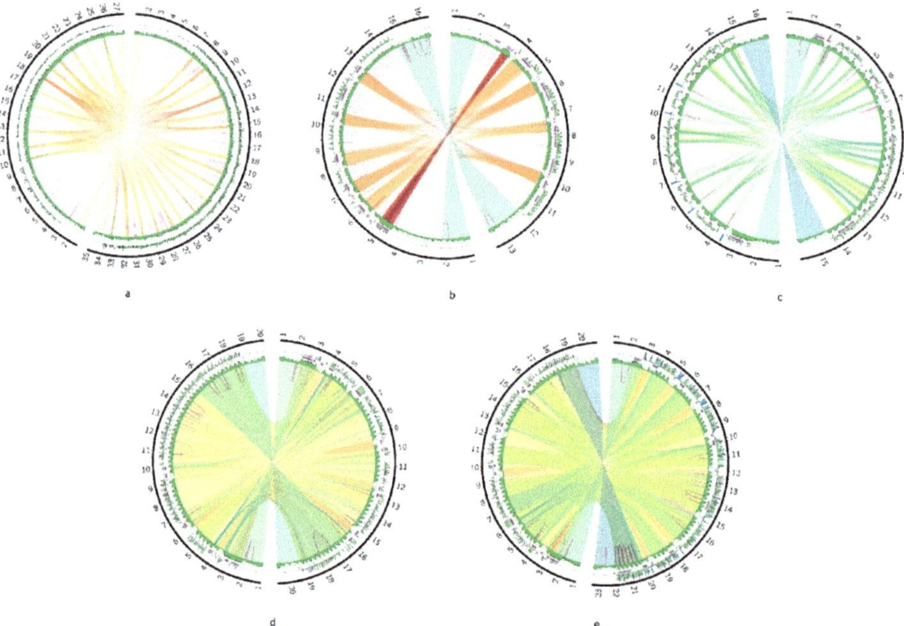

Figure 6. Circos plots comparing the DRs of two species or strains: *T. cruzi* strains TCC (left contig) and Dm28c (right contig) (**a**); *L. braziliensis* strains (**b**); *Leishmania tropica* (left contig) and *L. aethiopica* (right contig) (**c**); *L. donovani* strains 193-S616 (left contig) and FDAARGON_361 (right contig) (**d**); *L. donovani* strains 193-S616 (left contig) and Pasteur (right contig) (**e**).

2.5. Quality Control of Maxicircle Assemblies

Since PacBio sequencing datasets, used in this study, targeted primarily nuclear genomes and, therefore, were not enriched for mitochondrial DNA, we occasionally observed low coverage of the final maxicircle assembly (below 10 reads). Inspecting coverage profiles in Integrated Genome Browser (IGB), we noticed that generally, but not always, less covered regions of the maxicircle were from the DR. We carefully examined all the assemblies and discarded one for *L. panamensis*, which did not meet our quality criteria. Two maxicircle contigs (marked in Table 1) were not circularized by the Canu assembler, but had a detectable sequence overlap, which included a copy of the *ND5* gene and this allowed us to close the circle correctly.

For two species (*L. donovani* FDAARGOS_361 and *L. pyrrhocoris* H10) we used available Illumina paired-end sequencing reads to assess the quality of assemblies. We found that Illumina reads align well on both of our assemblies, supporting all nucleotides of the maxicircle sequences. For *L. donovani* strain FDAARGOS_361, mapping the paired-end reads (150 + 150 with average insert size of 180 bp) onto our assembly produced a profile with uneven coverage of the maxicircle sequence. For *Leptomonas pyrrhocoris* H10 we used the 150 + 150 paired-end reads with average insert size of 940 bp. In this case the coverage profile did not differ between the coding and divergent regions [42] (Supplementary Figure S1). We explain this by the cumulative effect of the sequencing insert size (longer reads improve repeat resolution) and the complexity of the DR (the *L. donovani* DR has more homogeneous repeat units and, as such, is a more difficult reference for the read mapping).

We also compared the maxicircle of *Leishmania donovani* strain Pasteur with that from the GenBank (CP022652), which was produced by the HGAP3 assembly algorithm from the same PacBio reads. The GenBank sequence contains duplicated nucleotides on the 5′ end, which overlap with ≈4000 bp on the 3′ end of the assembly. After trimming, both sequences had nearly equal length and high level of sequence identity, though identity percent dropped slightly in the DR, indicating that the choice of the PacBio assembly method may influence the resultant sequence, especially in the low-covered regions. Supplementary Figure S2 demonstrates synteny of both assemblies.

3. Discussion

Here we present the sequences of 17 new full trypanosomatid maxicircles. The big problem of maxicircle sequencing is the divergent region, which remains difficult to assemble even using the modern NGS techniques [31,32]. In the current work we used PacBio datasets to assemble maxicircles, but even with this technique the assemblies often suffered from the low coverage in their DR. Nevertheless, in most cases we had a sufficient number of reads to reconstruct the sequence of a full maxicircle.

Our analysis revealed the common architectural pattern of the DR organization. We described two regions with different structure, the P5 and P12 elements, according to their proximity to the *ND5* and *12S* genes, respectively. Previously, similar structures have been described for *T. lewisi* [19] as sections I and II [31]. The section I (which is P12 in our terms) was composed of short, highly repetitive units; while section II (P5 in our terms), consisted of several large duplications. This appears to be a common architecture of the DR of maxicircles.

The difference in DR length is connected to the massive rearrangements in the P12 element, which is built from the repetitive elements, further organized in arrays. Numbers and lengths of such arrays are species-specific and can change quickly. Sequence rearrangements in the *12S*-proximal clusters of the *Leishmania* spp. DRs were investigated previously [34,40] and the current work confirms previous findings, stipulating that the P12 compartment is very dynamic. Of note, differences in the P12 region also determine DR length variation in different isolates of *Trypanosoma brucei* [22]. The driving forces of such quick rearrangements remain to be investigated further. While the mechanism of these changes remains unclear, the DR will stay unreliable as a putative genetic marker for trypanosomatid identification, because its stability in isolated cultures over time has never been investigated.

It has been also proposed that there can be different structural classes of maxicircles within the same pool of kDNA [10,18,34], but our findings can neither confirm, nor disprove this. In most cases we might have insufficient coverage to detect alternative maxicircle variants that can be present, taking into account that the proportion of alternative variant can be as low as 2%. On the other hand, our results confirm that the *12S*-proximal region is very dynamic, varying between phylogenetically close strains of *Leishmania donovani* and *L. braziliensis*. Inverted repeats are detected frequently in this region, possibly indicating its high recombination potential. Moreover, we detected a few PacBio reads bearing *ND5* and/or *12S* rRNA sequences, which were not included in the final assembly by the software used. These reads can originate from rare maxicircle variants, but current datasets have insufficient sequencing depth to confirm this.

Repeat units in the P12 locus are both species-specific and conservative, strongly supporting the conclusion that these arrays are not junk DNA. It is plausible to suggest that these repeat arrays contain binding sites for some transcription factors or DNA maintenance proteins, which may stabilize such DNA-protein complexes. In this case, structural rearrangements of the P12 element can play a role in nucleoid remodeling, controlling access of proteins to the coding region loci. The P5 element appears to be more conservative in terms of organization, and in *Leishmania* spp. it is present in a form of a GC-rich tandem repeat. In some species it can be reduced or absent, indicating that this structural pattern is not crucial for maxicircle replication or transcription.

It has been suggested that a putative replication origin and promotor sequences are located in the DR [12,22]. For example, imperfect palindrome sequences and motifs with high sequence similarity to the CSB3 blocks of minicircles were documented in some repeat copies of the *T. brucei* P12 element [22]. It has been proposed that they can serve as maxicircle replication origins. In our work, we did not detect any sequence homology with minicircular CBS3 blocks in the DRs of assembled maxicircles. We did detect various inverted repeats and imperfect palindromes, scattered in the DR, but their exact biological role remains to be investigated further.

4. Materials and Methods

4.1. Complete Maxicircle Assembly

In order to assemble full-length maxicircles, we used the available PacBio datasets deposited in the SRA NCBI database. Information on accession numbers of sequences, used in this work, is summarized in Table 2.

Table 2. Sources of used datasets. Accession numbers for raw data used to assemble maxicircle sequences in this study. All accessions numbers are for the PacBio reads, if not specified otherwise.

Species (Strain)	BioProject	Run Accession Number
Leishmania (L.) *mexicana* (215-49)		SRR7867272, SRR7867273, SRR7867284,SRR7867285
Leishmania guyanensis (204-365)		SRR7867261, SRR7867262, SRR7867269-SRR7867271
Leishmania (L.) *aethiopica* (209-622)	PRJNA484340	SRR8377733, SRR7867274, SRR7867278-SRR7867281
Leishmania (L.) *infantum* (193-S1775)		SRR7867264-SRR7867268
Leishmania (L.) *tropica* (216-162)		SRR7867286-SRR7867292
Leishmania (L.) *amazonensis* (210-660)		SRR7867275-SRR7867277, SRR8377732
Leishmania (V.) *braziliensis* (208-905)		SRR7880312; SRR7880319; SRR7880320

Table 2. *Cont.*

Species (Strain)	BioProject	Run Accession Number
Leishmania (V.) *braziliensis* (208-954)		SRR7880309-SRR7880311
Leishmania (L.) *donovani* (193-S616)		SRR7880313, SRR7880314, SRR7880316
Leishmania donovani (1S line, FDAARGOS_361)	PRJNA231221	SRR5932752-SRR5932754, SRR5932751 (Illumina, paired-end)
Leishmania donovani (1S2D line, Pasteur) [2]	PRJNA396645	SRR5902665-SRR5902672
Trypanosoma brucei (Lister 427)	PRJEB18945	ERR1794935-ERR1794915
Trypanosoma cruzi (TCC)	PRJNA432753	SRR6822075
Trypanosoma cruzi (Dm28c)	PRJNA433042	SRR6809376
Leptomonas pyrrhocoris (H10)	PRJNA598933 [1]	
Herpetomonas megaseliae	PRJEB7883	ERR1036240-ERR1036242, ERR1046607-ERR1046611
Crithidia expoeki (BJ08_175)	Assembled sequence [3]	

[1] Sequence was assembled from PacBio data generated in our lab. [2] GenBank CP022652 (Ramasamy, G.; McDonald, J.; Sur, A.; and Myler, P., BioProject PRJNA396645, unpublished). [3] Sequence was assembled from PacBio data generated in [43].

Reads were downloaded and converted to the fasta format. For quality control, all reads below 10 kbps were discarded. A local database for Blastn (from local installation of NCBI-Blast suite 2.3.0+) was built from these reads and searched for the *12S* and *ND5* maxicircle gene sequences, using sequences of *Leptomonas pyrrhocoris*, *Trypanosoma cruzi*, *T. brucei*, *Leishmania amazonensis*, *L. tarentolae*, and *Crithidia expoeki* as queries. Parameters for Blastn search were "-evalue 1e-10 -word_size 5 -outfmt 6". Reads producing alignments with alignment length over 1000 were selected for further analysis. Assembly was performed with the Canu v1.8 assembler [44] with the "genomeSize = 50 k". For *Leishmania donovani* and *Leptomonas pyrrhocoris* H10, we also tested the assembly protocol with different values of the "genomeSize" parameter (in a range of 20–100 k with a step of 10 k) and it did not influence the final contig sequence. To eliminate overlaps, the resultant sequences were trimmed using the nucmer tool from MUMmer v3.23 package [45] with settings "nucmer -maxmatch -nosimplify contigs.fasta contigs.fasta; show-coords -lrcTH out.delta".

4.2. Maxicircle Annotation

Assembled maxicircles were annotated in a semi-automatic mode. We checked the synteny of maxicircle genes using Blastn searches for *ND1*, *ND4*, *COI*, *Cyb*, *9S*, *MURF2*, and *COII* genes. These genes are not (or almost not) edited and, therefore, are good markers for automated annotation. We concluded that all assembled maxicircles preserved gene order documented for trypanosomatid maxicircles [21].

The best scoring match was used to determine the initial annotation coordinates for the *12S* and *ND5* genes, followed by the manual curation. For the convenience of further analysis, we rotated all maxicircle sequences, so that each assembly started from the *ND5* gene, followed be the DR, *12S*, and *rps12* genes.

4.3. Quality Control of the Assemblies

Assemblies were checked by two criteria: (i) read coverage is sufficient throughout the assembly, and (ii) final maxicircle has perfectly assembled CR. The PacBio reads were mapped on the assembly with bwa mem algorithm of bwa v0.7.12 [46], with the settings "-x pacbio -B 3 -O 3,3 -L 7,7". Alignment files were processed with SAMtools v1.9 [47]. For two isolates, *Leishmania donovani* FDAARGOS_361 and *Leptomonas pyrrhocoris* H10, the Illumina paired-end sequencing data were used to verify PacBio assemblies. Illumina reads were mapped with Bowtie2 v2.3.4.1 [48] with the following settings "-X 700

–very-sensitive –end-to-end –no-unal". Coverage profiles were analyzed with SAMtools and custom python script, sorted bam files were inspected in IGB [49].

4.4. Analysis of Repeats

Frequency of each k-mer in a sequence was calculated using a custom python script. Repeated sequence regions were determined with the "repeat-match" from the MUMmer package. Regions with high level of sequence identity were found by the nucmer tool of the same package. Tandem repeats were identified using mreps v2.6 [50] with "-res 3 -minsize 40 -maxsize 6000 -minperiod 10 -maxperiod 4000" options. Inverted repeats were found with einverted from the EMBOSS package v6.6.0.0 [51] with "-match 3 -mismatch -4 -gap 12 -threshold 33".

4.5. Plotting DR Maps

Resulting outputs of repeat and homology search algorithms were plotted using Circos [52]. Format conversion was performed with the python/awk scripts. The outer track of each Circos plot is a histogram showing the number of times a 24-mer, staring at the current position, appears in a maxicircle (24-mer frequency plot). Track min and max values were adjusted to show all 24-mers that appear 4–40 times (drawn with green colors), 24-mers with frequency value above 40 were also shown, but the height of the bars was fixed (drawn with blue). Tandem repeats were drawn as purple segments on the middle track of each Circos plot. The GC profiles (in window of 100 bp with a step of 1), were plotted in green on the inner track. Regions with sufficient sequence identity were drawn as colored ribbons, denoting the percent of sequence identity. Only regions over 100 bp and with minimal sequence identity of 80% were shown. The color of each ribbon was calculated by mapping sequence identity value in range [80, 100] to 11-class 'Spectral' color palette of ColorBrewer2 (http://colorbrewer2.org/#type=diverging&scheme=Spectral&n=9), which sets red, green, and blue colors for values around 80%, 90%, and 100%, respectively. Dotplot diagrams of maxicircles were generated with mummerplot in MUMmer v3.23.

5. Conclusions

We assembled and analyzed 17 complete maxicircles from different trypanosomatid species. Our analysis was focused on most arcane part of a maxicircle—the divergent region. This region is composed of repeated sequences, difficult to sequence and assemble, even with the modern NGS methods. Comparative analysis allowed us to infer the high-level architecture of the DR and describe some species-specific structural features. Our findings are consistent with previous works, which were focused on low-level sequence features, and shed light on how these low-level elements are organized and how they evolve.

Supplementary Materials: The following are available online at http://www.mdpi.com/2076-0817/9/2/100/s1, **Figure S1**: Assembly quality control by read mapping on assembled contig. Screenshots from IGB with opened coverage tracks for bam files are shown. Red segment shows the borders of divergent region. (**a**) Mapping Illumina HiSeq paired-end reads (SRA: SRR5932751) with average insert size of 180 bp on *Leishmania donovani* strain FDAARGOS_361 maxicircle. (**b**) Mapping Illumina HiSeq paired-end reads with average insert size of 940 bp on *Leptomonas pyrrhocoris* H10 maxicircle. **Figure S2**: Comparison of *L. donovani* strain Pasteur assemblies. Assembly produced by our pipeline (with Canu) is on the left, maxicircle assembled with HGAP3 (GeneBank: CP022652) is on the right. Both assembled sequences are rotated to begin with the *ND5* gene (0 coordinate), so the upper part of the diagram corresponds to the CR and the lower one is the DR. Ribbon colors reflects the percent of sequence identity between regions in a range between 98% and 100%. Dark green corresponds to nearly 100% sequence identity, light green is the middle of interval (99%), red—identity near 98%. Only regions longer than 500 bp are shown.

Author Contributions: Conceptualization, E.S.G., A.A.K. and V.Y.; methodology, E.S.G.; formal analysis, E.S.G.; investigation, E.S.G., K.A.Z., N.S.M., Y.A.R. and N.K.; resources, E.S.G.; data curation, E.S.G.; writing—original draft preparation, E.S.G.; writing—review and editing, E.S.G. and V.Y.; visualization, E.S.G.; supervision, E.S.G. and V.Y.; project administration, E.S.G.; funding acquisition, E.S.G. All authors have read and agreed to the published version of the manuscript.

Funding: This research was funded by Russian Science Foundation (RSF) grant number 19-74-10008 to E.S.G., A.A.K., K.A.Z., N.S.M. and Y.A.R.; and European Regional Development Fund (ERDF) grant number CZ.02.1.01/0.0/ 0.0/16_019/0000759 to V.Y.

Conflicts of Interest: V.Y. is an Academic Editor of Pathogens. Other authors declare no conflict of interest.

Appendix A Appendix

Dotplots and Circos plots for species not included into the main text.

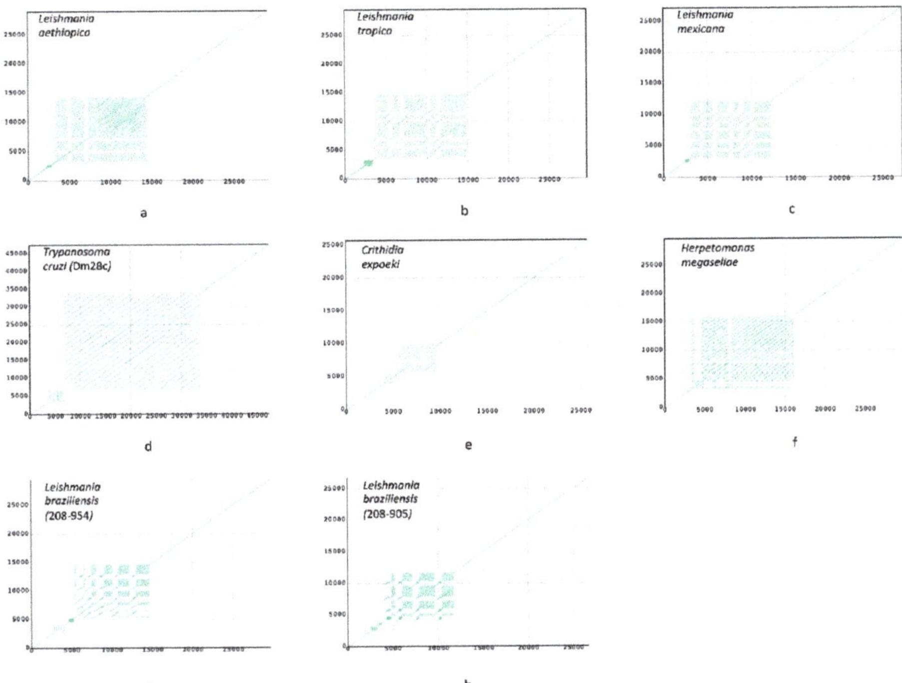

Figure A1. Dotplots of maxicircles of Leishmania aethiopica (**a**); L. tropica (**b**); L. mexicana (**c**); Trypanosoma cruzi strain Dm28c (**d**); Crithidia expoeki (**e**); Herpetomonas megaseliae (**f**); L. braziliensis strain 208-954 (**g**); L. braziliensis strain 208-905 (**h**). All plots show full maxicircle starting with the ND5 gene at (0, 0).

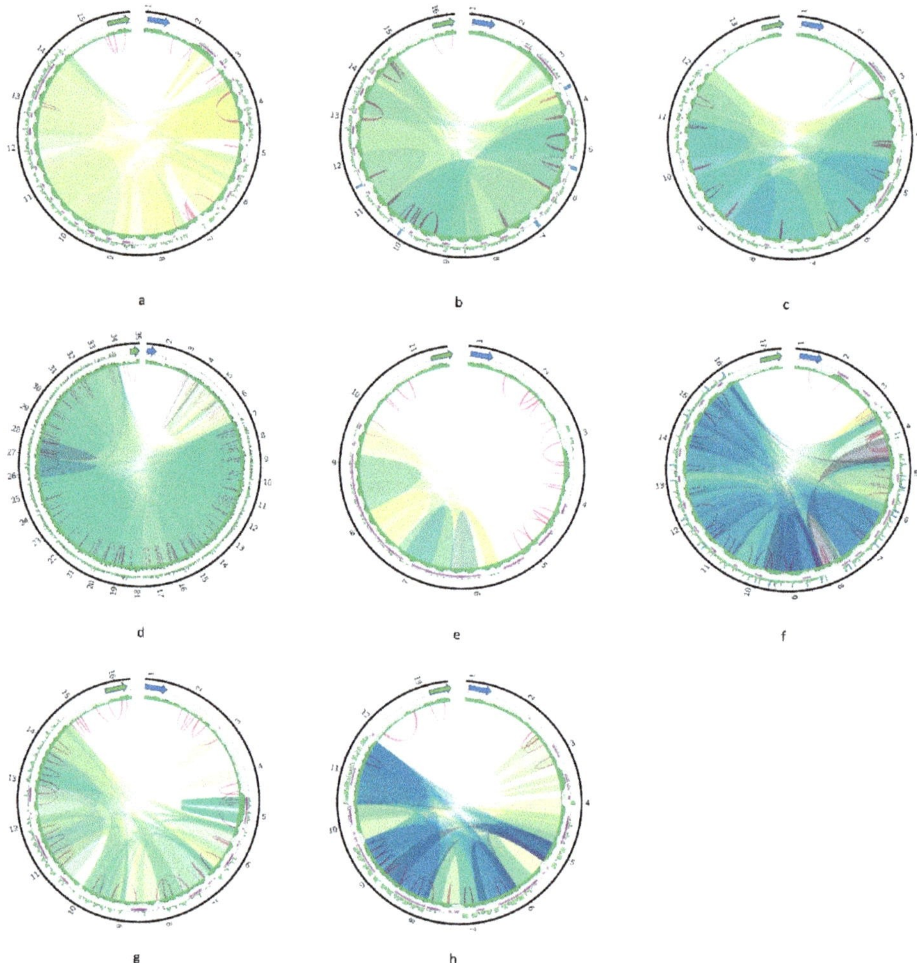

Figure A2. Circos plots of the DR of Leishmania aethiopica (**a**); Leishmania tropica (**b**); Leishmania mexicana (**c**); Trypanosoma cruzi strain Dm28c (**d**); Crithidia expoeki (**e**); Herpetomonas megaseliae (**f**); Leishmania braziliensis strain 208-954 (**g**); Leishmania braziliensis strain 208-905 (**h**). Green and blue arrows indicate 12S and ND5 genes, respectively. The outer track is a histogram of 24-mer frequency; the middle track shows tandem repeats; the inner track is a GC content profile. Ribbons inside the circle connect homologous regions; the color represents percent of sequence identity in the range [80%; 100%]. Violet arcs denote inverted repeats.

References

1. Maslov, D.A.; Opperdoes, F.R.; Kostygov, A.Y.; Hashimi, H.; Lukeš, J.; Yurchenko, V. Recent advances in trypanosomatid research: Genome organization, expression, metabolism, taxonomy and evolution. *Parasitology* **2019**, *146*, 1–27. [CrossRef] [PubMed]
2. Lukeš, J.; Butenko, A.; Hashimi, H.; Maslov, D.A.; Votýpka, J.; Yurchenko, V. Trypanosomatids are much more than just trypanosomes: Clues from the expanded family tree. *Trends Parasitol.* **2018**, *34*, 466–480. [CrossRef] [PubMed]
3. Lukeš, J.; Guilbride, D.L.; Votýpka, J.; Zíková, A.; Benne, R.; Englund, P.T. Kinetoplast DNA network: Evolution of an improbable structure. *Eukaryot Cell* **2002**, *1*, 495–502. [CrossRef] [PubMed]

4. Aphasizhev, R.; Aphasizheva, I. Mitochondrial RNA editing in trypanosomes: Small RNAs in control. *Biochimie* **2014**, *100*, 125–131. [CrossRef] [PubMed]

5. Simpson, L.; Sbicego, S.; Aphasizhev, R. Uridine insertion/deletion RNA editing in trypanosome mitochondria: A complex business. *RNA* **2003**, *9*, 265–276. [CrossRef] [PubMed]

6. Yurchenko, V.; Kolesnikov, A.A. Minicircular kinetoplast DNA of Trypanosomatidae. *Mol. Biol. (Mosk)* **2001**, *35*, 3–13. (In Russian) [CrossRef]

7. Thomas, S.; Martinez, L.L.; Westenberger, S.J.; Sturm, N.R. A population study of the minicircles in Trypanosoma cruzi: Predicting guide RNAs in the absence of empirical RNA editing. *BMC Genom.* **2007**, *8*, 133. [CrossRef]

8. Hong, M.; Simpson, L. Genomic organization of Trypanosoma brucei kinetoplast DNA minicircles. *Protist* **2003**, *154*, 265–279. [CrossRef]

9. Shapiro, T.A. Kinetoplast DNA maxicircles: Networks within networks. *Proc. Natl. Acad. Sci. USA* **1993**, *90*, 7809–7813. [CrossRef]

10. Flegontov, P.N.; Kolesnikov, A.A. Radically different maxicircle classes within the same kinetoplast: An artefact or a novel feature of the kinetoplast genome? *Kinetoplastid Biol. Dis.* **2006**, *5*, 5. [CrossRef]

11. Messenger, L.A.; Llewellyn, M.S.; Bhattacharyya, T.; Franzen, O.; Lewis, M.D.; Ramirez, J.D.; Carrasco, H.J.; Andersson, B.; Miles, M.A. Multiple mitochondrial introgression events and heteroplasmy in Trypanosoma cruzi revealed by maxicircle MLST and Next-Generation Sequencing. *PLoS Negl. Trop. Dis.* **2012**, *6*, e1584. [CrossRef] [PubMed]

12. Westenberger, S.J.; Cerqueira, G.C.; El-Sayed, N.M.; Zingales, B.; Campbell, D.A.; Sturm, N.R. Trypanosoma cruzi mitochondrial maxicircles display species- and strain-specific variation and a conserved element in the non-coding region. *BMC Genom.* **2006**, *7*, 60. [CrossRef] [PubMed]

13. Carranza, J.C.; Valadares, H.M.; D'Avila, D.A.; Baptista, R.P.; Moreno, M.; Galvao, L.M.; Chiari, E.; Sturm, N.R.; Gontijo, E.D.; Macedo, A.M.; et al. Trypanosoma cruzi maxicircle heterogeneity in Chagas disease patients from Brazil. *Int. J. Parasitol.* **2009**, *39*, 963–973. [CrossRef] [PubMed]

14. Cooper, S.; Wadsworth, E.S.; Ochsenreiter, T.; Ivens, A.; Savill, N.J.; Schnaufer, A. Assembly and annotation of the mitochondrial minicircle genome of a differentiation-competent strain of Trypanosoma brucei. *Nucleic Acids Res.* **2019**, *47*, 11304–11325. [CrossRef] [PubMed]

15. Kleisen, C.M.; Borst, P. Sequence heterogeneity of the mini-circles of kinetoplast DNA of Crithidia luciliae and evidence for the presence of a component more complex than mini-circle DNA in the kinetoplast network. *Biochim. Biophys. Acta* **1975**, *407*, 473–478. [CrossRef]

16. Pestov, D.G.; Gladkaya, L.A.; Maslov, D.A.; Kolesnikov, A.A. Characterization of kinetoplast minicircle DNA in the lower trypanosomatid Crithidia oncopelti. *Mol. Biochem. Parasitol.* **1990**, *41*, 135–145. [CrossRef]

17. Challberg, S.S.; Englund, P.T. Heterogeneity of minicircles in kinetoplast DNA of Leishmania tarentolae. *J. Mol. Biol.* **1980**, *138*, 447–472. [CrossRef]

18. Camacho, E.; Rastrojo, A.; Sanchiz, A.; Gonzalez-de la Fuente, S.; Aguado, B.; Requena, J.M. Leishmania mitochondrial genomes: Maxicircle structure and heterogeneity of minicircles. *Genes (Basel)* **2019**, *10*, 758. [CrossRef]

19. Maslov, D.A.; Kolesnikov, A.A.; Zaitseva, G.N. Conservative and divergent base sequence regions in the maxicircle kinetoplast DNA of several trypanosomatid flagellates. *Mol. Biochem. Parasitol.* **1984**, *12*, 351–364. [CrossRef]

20. Simpson, L.; Neckelmann, N.; de la Cruz, V.F.; Simpson, A.M.; Feagin, J.E.; Jasmer, D.P.; Stuart, J.E. Comparison of the maxicircle (mitochondrial) genomes of Leishmania tarentolae and Trypanosoma brucei at the level of nucleotide sequence. *J. Biol. Chem.* **1987**, *262*, 6182–6196.

21. Kaufer, A.; Barratt, J.; Stark, D.; Ellis, J. The complete coding region of the maxicircle as a superior phylogenetic marker for exploring evolutionary relationships between members of the Leishmaniinae. *Infect. Genet. Evol.* **2019**, *70*, 90–100. [CrossRef]

22. Myler, P.J.; Glick, D.; Feagin, J.E.; Morales, T.H.; Stuart, K.D. Structural organization of the maxicircle variable region of Trypanosoma brucei: Identification of potential replication origins and topoisomerase II binding sites. *Nucleic Acids Res.* **1993**, *21*, 687–694. [CrossRef] [PubMed]

23. Borst, P.; Weijers, P.J.; Brakenhoff, G.J. Analysis by electron microscopy of the variable segment in the maxicircle of kinetoplast DNA from Trypanosoma brucei. *Biochim. Biophys. Acta* **1982**, *699*, 272–280. [CrossRef]

24. Borst, P.; Fase-Fowler, F.; Hoeijmakers, J.H.; Frasch, A.C. Variations in maxicircle and minicircle sequences in kinetoplast DNAs from different Trypanosoma brucei strains. *Biochim. Biophys. Acta* **1980**, *610*, 197–210. [CrossRef]
25. Muhich, M.L.; Simpson, L.; Simpson, A.M. Comparison of maxicircle DNAs of Leishmania tarentolae and Trypanosoma brucei. *Proc. Natl. Acad. Sci. USA* **1983**, *80*, 4060–4064. [CrossRef] [PubMed]
26. Muhich, M.L.; Neckelmann, N.; Simpson, L. The divergent region of the Leishmania tarentolae kinetoplast maxicircle DNA contains a diverse set of repetitive sequences. *Nucleic Acids Res.* **1985**, *13*, 3241–3260. [CrossRef]
27. Sloof, P.; de Haan, A.; Eier, W.; van Iersel, M.; Boel, E.; van Steeg, H.; Benne, R. The nucleotide sequence of the variable region in Trypanosoma brucei completes the sequence analysis of the maxicircle component of mitochondrial kinetoplast DNA. *Mol. Biochem. Parasitol.* **1992**, *56*, 289–299. [CrossRef]
28. Lee, S.T.; Liu, H.Y.; Chu, T.; Lin, S.Y. Specific A+T-rich repetitive DNA sequences in maxicircles from wildtype Leishmania mexicana amazonensis and variants with DNA amplification. *Exp. Parasitol.* **1994**, *79*, 29–40. [CrossRef]
29. Borst, P.; Hoeijmakers, J.H.; Hajduk, S. Structure, function and evolution of kinetoplast DNA. *Parasitology* **1981**, *82*, 81–93.
30. Ruvalcaba-Trejo, L.I.; Sturm, N.R. The Trypanosoma cruzi Sylvio X10 strain maxicircle sequence: The third musketeer. *BMC Genom.* **2011**, *12*, 58. [CrossRef]
31. Lin, R.H.; Lai, D.H.; Zheng, L.L.; Wu, J.; Lukes, J.; Hide, G.; Lun, Z.R. Analysis of the mitochondrial maxicircle of Trypanosoma lewisi, a neglected human pathogen. *Parasit Vectors* **2015**, *8*, 665. [CrossRef] [PubMed]
32. Urrea, D.A.; Triana-Chavez, O.; Alzate, J.F. Mitochondrial genomics of human pathogenic parasite Leishmania (Viannia) panamensis. *PeerJ* **2019**, *7*, e7235. [CrossRef] [PubMed]
33. Lee, S.T.; Tarn, C.; Wang, C.Y. Characterization of sequence changes in kinetoplast DNA maxicircles of drug-resistant Leishmania. *Mol. Biochem. Parasitol.* **1992**, *56*, 197–207. [CrossRef]
34. Flegontov, P.N.; Strelkova, M.V.; Kolesnikov, A.A. The Leishmania major maxicircle divergent region is variable in different isolates and cell types. *Mol. Biochem. Parasitol.* **2006**, *146*, 173–179. [CrossRef] [PubMed]
35. Lee, S.Y.; Lee, S.T.; Chang, K.P. Transkinetoplastidy—A novel phenomenon involving bulk alterations of mitochondrion-kinetoplast DNA of a trypanosomatid protozoan. *J. Protozool.* **1992**, *39*, 190–196. [CrossRef]
36. Tørresen, O.K.; Star, B.; Mier, P.; Andrade-Navarro, M.A.; Bateman, A.; Jarnot, P.; Gruca, A.; Grynberg, M.; Kajava, A.V.; Promponas, V.J.; et al. Tandem repeats lead to sequence assembly errors and impose multi-level challenges for genome and protein databases. *Nucleic Acids Res.* **2019**, *47*, 10994–11006. [CrossRef]
37. Gerasimov, E.S.; Gasparyan, A.A.; Kaurov, I.; Tichy, B.; Logacheva, M.D.; Kolesnikov, A.A.; Lukes, J.; Yurchenko, V.; Zimmer, S.L.; Flegontov, P. Trypanosomatid mitochondrial RNA editing: Dramatically complex transcript repertoires revealed with a dedicated mapping tool. *Nucleic Acids Res.* **2018**, *46*, 765–781. [CrossRef]
38. Aravin, A.A.; Yurchenko, V.; Merzlyak, E.; Kolesnikov, A.A. The mitochondrial ND8 gene from Crithidia oncopelti is not pan-edited. *FEBS Lett.* **1998**, *431*, 457–460. [CrossRef]
39. Gerasimov, E.S.; Kostygov, A.Y.; Yan, S.; Kolesnikov, A.A. From cryptogene to gene? ND8 editing domain reduction in insect trypanosomatids. *Eur. J. Protistol.* **2012**, *48*, 185–193. [CrossRef]
40. Flegontov, P.; Guo, Q.; Ren, L.; Strelkova, M.V.; Kolesnikov, A.A. Conserved repeats in the kinetoplast maxicircle divergent region of Leishmania sp. and Leptomonas seymouri. *Mol. Genet. Genom.* **2006**, *276*, 322–333. [CrossRef]
41. Horváth, A.; Maslov, D.A.; Peters, L.S.; Haviernik, P.; Wuestenhagen, T.; Kolesnikov, A.A. Analysis of the sequence repeats in the divergent region of maxicircle DNA from kinetoplasts of Crithidia oncopelti. *Mol. Biol. (Mosk)* **1990**, *24*, 1539–1548. (In Russian)
42. Flegontov, P.; Butenko, A.; Firsov, S.; Kraeva, N.; Eliáš, M.; Field, M.C.; Filatov, D.; Flegontova, O.; Gerasimov, E.S.; Hlaváčová, J.; et al. Genome of Leptomonas pyrrhocoris: A high-quality reference for monoxenous trypanosomatids and new insights into evolution of Leishmania. *Sci. Rep.* **2016**, *6*, 23704. [CrossRef]
43. Gerasimov, E.; Zemp, N.; Schmid-Hempel, R.; Schmid-Hempel, P.; Yurchenko, V. Genomic variation among strains of *Crithidia bombi* and *C. expoeki. mSphere* **2019**, *4*, e00482-19. [CrossRef] [PubMed]

44. Koren, S.; Walenz, B.P.; Berlin, K.; Miller, J.R.; Bergman, N.H.; Phillippy, A.M. Canu: Scalable and accurate long-read assembly via adaptive k-mer weighting and repeat separation. *Genome Res.* **2017**, *27*, 722–736. [CrossRef] [PubMed]

45. Kurtz, S.; Phillippy, A.; Delcher, A.L.; Smoot, M.; Shumway, M.; Antonescu, C.; Salzberg, S.L. Versatile and open software for comparing large genomes. *Genome Biol.* **2004**, *5*, R12. [CrossRef]

46. Li, H.; Durbin, R. Fast and accurate long-read alignment with Burrows-Wheeler transform. *Bioinformatics* **2010**, *26*, 589–595. [CrossRef]

47. Ramirez-Gonzalez, R.H.; Bonnal, R.; Caccamo, M.; Maclean, D. Bio-SAMtools: Ruby bindings for SAMtools, a library for accessing BAM files containing high-throughput sequence alignments. *Source Code Biol. Med.* **2012**, *7*, 6. [CrossRef]

48. Langmead, B.; Salzberg, S.L. Fast gapped-read alignment with Bowtie 2. *Nat. Methods* **2012**, *9*, 357–359. [CrossRef]

49. Freese, N.H.; Norris, D.C.; Loraine, A.E. Integrated genome browser: Visual analytics platform for genomics. *Bioinformatics* **2016**, *32*, 2089–2095. [CrossRef]

50. Kolpakov, R.; Bana, G.; Kucherov, G. Mreps: Efficient and flexible detection of tandem repeats in DNA. *Nucleic Acids Res.* **2003**, *31*, 3672–3678. [CrossRef]

51. Rice, P.; Longden, I.; Bleasby, A. EMBOSS: The European Molecular Biology Open Software Suite. *Trends Genet.* **2000**, *16*, 276–277. [CrossRef]

52. Krzywinski, M.; Schein, J.; Birol, I.; Connors, J.; Gascoyne, R.; Horsman, D.; Jones, S.J.; Marra, M.A. Circos: An information aesthetic for comparative genomics. *Genome Res.* **2009**, *19*, 1639–1645. [CrossRef] [PubMed]

 *pathogens*

Article

Frequent Recombination Events in *Leishmania donovani*: Mining Population Data

Igor B. Rogozin [1], **Arzuv Charyyeva** [2], **Ivan A. Sidorenko** [3], **Vladimir N. Babenko** [3] and **Vyacheslav Yurchenko** [2,4,*]

1   National Center for Biotechnology Information, National Library of Medicine, National Institutes of Health, Bethesda, MD 20894, USA; rogozin@ncbi.nlm.nih.gov
2   Life Science Research Centre, Faculty of Science, University of Ostrava, 710 00 Ostrava, Czech Republic; arzuvc@gmail.com
3   Institute of Cytology and Genetics, 630090 Novosibirsk, Russia; vanyasidorenko22@gmail.com (I.A.S.); bob@bionet.nsc.ru (V.N.B.)
4   Martsinovsky Institute of Medical Parasitology, Tropical and Vector Borne Diseases, Sechenov University, 119435 Moscow, Russia
*   Correspondence: vyacheslav.yurchenko@osu.cz

Received: 10 April 2020; Accepted: 13 July 2020; Published: 15 July 2020

Abstract: The *Leishmania donovani* species complex consists of all *L. donovani* and *L. infantum* strains mainly responsible for visceral leishmaniasis (VL). It was suggested that genome rearrangements in *Leishmania* spp. occur very often, thus enabling parasites to adapt to the different environmental conditions. Some of these rearrangements may be directly linked to the virulence or explain the reduced efficacy of antimonial drugs in some isolates. In the current study, we focused on a large-scale analysis of putative gene conversion events using publicly available datasets. Previous population study of *L. donovani* suggested that population variability of *L. donovani* is relatively low, however the authors used masking procedures and strict read selection criteria. We decided to re-analyze DNA-seq data without masking sequences, because we were interested in the most dynamic fraction of the genome. The majority of samples have an excess of putative gene conversion/recombination events in the noncoding regions, however we found an overall excess of putative intrachromosomal gene conversion/recombination in the protein coding genes, compared to putative interchromosomal gene conversion/recombination events.

Keywords: gene conversion; *Leishmania donovani* species complex; whole-genome sequencing; concerted evolution

1. Introduction

The *Leishmania donovani* species complex consists of all *L. donovani* and *L. infantum* strains responsible for visceral leishmaniasis (VL) [1,2]. Besides VL, atypical cutaneous manifestations caused by both species of the complex have been also reported [3,4]. While *L. donovani* is considered to be an anthroponotic agent, *L. infantum* is zoonotic, with dogs and numerous wild mammals being involved in the disease transmission [5]. Both species are widespread with known major foci for *L. donovani* and *L. infantum* being the India/East Africa and the Middle East, respectively [6]. Although *L. chagasi* in the New World was historically referred to as a separate species, recent studies have demonstrated that it is a mere subpopulation of *L. infantum* [7], which had been probably spread in the Americas via migration in the 15th–16th century [8]. The clinical manifestations of the leishmaniasis caused by *L. donovani* spp. complex differ depending on the immune status of the infected individuals, parasite strains and transmitting sand fly vector's species [4,9].

It was suggested that genome rearrangements in *Leishmania* spp. occur very often, thus enabling parasites to adapt to the different environmental conditions [10,11]. Some of these rearrangements may be directly linked to the virulence [12,13] or explain the reduced efficacy of antimonial drugs in some isolates [14]. Below, we will discuss some prominent examples attributed to genome rearrangements in trypanosomatids, with a particular focus on *Leishmania* spp.

Lipophosphoglycan (LPG) is one of the most abundant surface glycoconjugates, which is mainly involved in parasite colonization of the vector's midgut [15]. The LPG molecules are differentially modified during the development, facilitating proper parasites' migration, evasion of the host immune system, and promoting the host specificity [16,17]. Tandem arrangement of the genes, encoding LPG modifying enzymes, provides a strong evidence of gene conversion in *Leishmania* spp. [18,19]. Of note, this model of organization is also conserved in monoxenous (=one host [20,21]) relatives of *Leishmania* of the subfamily Leishmaniinae [22].

Variant surface glycoproteins (VSGs) facilitate immune evasion, while VSG-encoding genes define antigenic variation in trypanosomes [23]. These genes have likely evolved as a result of several gene conversion events [24,25].

One of the essential virulence factors, the glycoprotein 63 (GP63), is encoded in a variable number of copies in different species of *Leishmania* [26]. All *Leishmania* spp., sequenced thus far, harbor at least two long, as well as variable numbers of short GP63-encoding sequences. This suggests that mosaic gene conversion has a high impact on the evolutionary history of these species [26,27]. Other prominent examples of this process in *Leishmania* are genes encoding cysteine proteases [28], *hsp70* gene cluster [29], amastins and A2-A2rel gene clusters [30,31].

In the current study we focused on a large-scale analysis of putative gene conversion events using publicly available datasets. Previous population study of *L. donovani* suggested that population variability of *L. donovani* is fairly low, however the authors used masking procedures and strict read selection criteria [14,32]. We decided to reanalyze DNA-seq data without masking sequences, because we were interested in the most dynamic fraction of the genome. A substantial variability of some regions of the *L. donovani* genome was documented. The majority of samples have an excess of putative gene conversion events in the noncoding regions, however we found an overall excess of putative intrachromosomal gene conversion/recombination in the protein coding genes, compared to putative interchromosomal gene conversion/recombination events.

2. Results

2.1. Analysis of Leishmania donovani DNA-seq Data

We studied five types of read configurations (Figure 1) in 28 *Leishmania donovani* genomes. The number of B-type reads (putative intra-chromosomal gene conversion events, see Materials and Methods for definitions) is approximately equal to that of the S-type reads (sole mapped reads) and substantially greater than that for the C- and D-type reads (inversion and putative inter-chromosomal gene conversion events, respectively) (Tables 1 and 2). The fraction of B-type and S-type reads that overlap with protein-coding genes is also similar and significantly larger than the corresponding values for the C- and D-type reads.

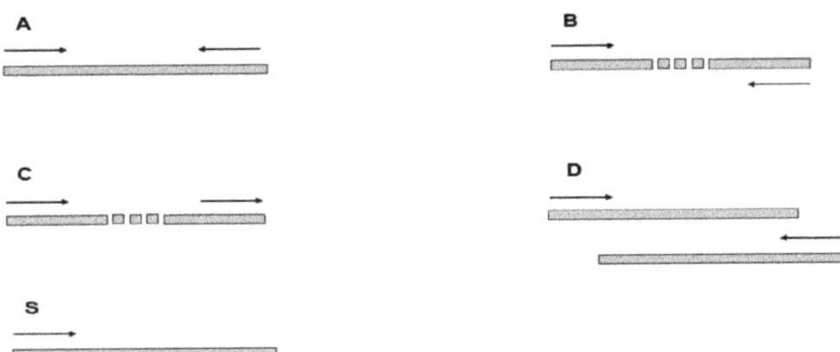

Figure 1. Schematic representation of **A**—(paired end reads), **B**—(putative intrachromosomal gene conversion events), **C**—(inversions), **D**—(putative interchromosomal gene conversion events), and **S**—(sole mapped) type reads.

Table 1. The numbers and fractions of filtered reads in *Leishmania donovani* (28 samples). The total number of unfiltered A-type reads is 459,792,248.

Read Types	Overlap with Protein-Coding Region (Fraction)	No Overlaps with Protein-Coding Region (Fraction)
B	489 (35%)	927 (65%)
C	117 (27%)	308 (73%)
D	125 (21%)	473 (79%)
S	513 (36%)	901 (64%)

Table 2. Pairwise comparisons of different types of filtered reads in *Leishmania donovani*. The 2-tail Fisher exact test was used.

Read Types	C	D	S
B	0.0068	6.9×10^{-10}	0.3460
C		0.0168	0.0009
D			5.3×10^{-12}

To test the possibility that the reason for the observed B-type reads prevalence are because of errors of read mapping with numerous mismatches (for example, low-quality sequencing), we have analyzed reads with the similarity level 90–95%. We found only a few reads that were largely located in noncoding DNA (Table 3). Thus, it is unlikely that low-quality reads make a substantial contribution to the B-type reads located in protein-coding genes.

Table 3. The number of filtered reads with numerous mismatches (the similarity level 90–95%).

Read Types	Overlap with Protein-Coding Region	No Overlaps with Protein-Coding Region
B	1	21
C	1	12
D	2	17
S	14	45

2.2. Gradient of Putative Recombination Events Across Protein-Coding Genes

Polarity is one of the properties of gene conversion: in almost all cases the frequency of conversion exhibits a gradient "across" the gene [33–35]. Here, we studied a distribution of reads across protein-coding genes and 500 bases of flanking regions (Figure 2). The frequency of reads (B, C, D, and S) in the 3′ and 5′ UTRs was substantially lower, compared to the coding regions. We observed a polarity of B- and D-type reads with an increased number of reads toward 3′ end of the protein-coding sequences. C- and S-type reads are distributed more uniformly (Figure 2). We compared the distribution of reads in protein coding regions (bins 2, 3 and 4), and the difference between the distributions of B- and S-type reads is statistically significant ($P_{\chi2}$ = 0.029, a modified χ^2 test, see Materials and Methods). No other significant pairwise differences were detected. We also compared the distribution of reads in protein coding regions (bins 2, 3 and 4) to the uniform distribution using the Pearson's χ^2 test, the B-type read distribution was significantly different from the uniform distribution ($P_{\chi2}$ = 0.006) whereas C-, D- and S-type read distributions were not significantly different from the uniform distribution.

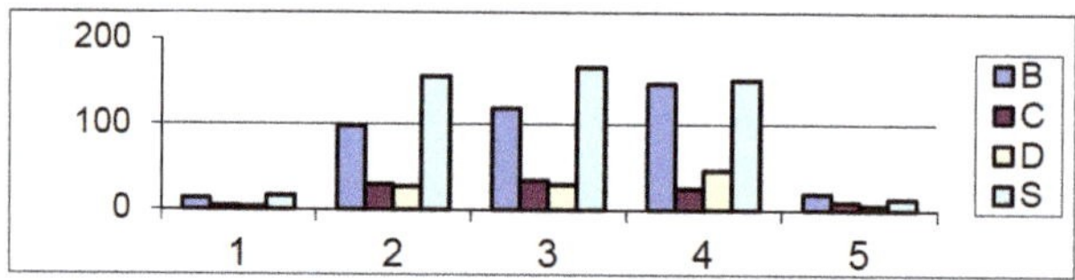

Figure 2. Distribution of B-, C-, D- and S-type reads across protein-coding genes. Number of reads is shown on the Y axis. Bins 1 and 5 correspond to 500 bases of 5′ and 3′ noncoding regions. Protein-coding regions were split into three 33% terciles. Bins 2, 3 and 4 correspond to the first, second and third terciles. We allowed overlaps with protein coding regions for bins 1 and 5 (5′ and 3′ flanking regions), these bins were not used for statistical analyses. All reads that overlap with more than one bin were removed from bins 2, 3 and 4.

2.3. Analyis of B-Type Reads and Putative Adenylate Cyclase Proteins

We extracted B-type reads that overlap with protein-coding genes and do not overlap with S/C/D-type reads. A total of 137 protein-coding genes that overlap with such B-type reads were identified (Supplementary file Table S2). A total of 67 protein-coding genes overlap with two or more B-type reads (Supplementary file Table S2). The gene encoding the XP_003861613.1 protein (conserved in many Kinetoplastida genomes) has the highest number of overlaps with B-type reads (16 reads, Supplementary file Table S2). This gene does not have paralogous sequences in the *Leishmania donovani* genome and it varies across different *L. donovani* strains (~1–2% divergence at the protein level and 18 amino acid deletion/insertion in the first half, Supplementary file Figure S1). The detected B-type reads are likely to represent recombination events across analyzed strains. These B-type reads are located in the 3′ flanking region and at the very end of the coding region (Supplementary file Table S2). The presence of structural variations in the 3′ region of the protein-coding gene is the most likely explanation of the observed pattern of numerous overlapping B-type reads [14,32]. Indeed, systematic analysis of the raw reads suggested that a long insertion (approximately 740 bases) in the 3′ flanking region near the end of the protein-coding region explains the observed B-type reads pattern (Supplementary file Figure S2).

Next we performed BLASTP searches of 137 protein-coding genes that overlap with such B-type reads (Supplementary file Table S2) against all proteins from the *Leishmania donovani* genome. Each BLASTP output was analyzed manually. Some of these protein-coding genes represent multigene families. An example of such a family is a putative adenylate cyclase protein (XP_003858707.1, XP_003858708.1, XP_003858709.1; the protein annotation was taken from a GenBank description of the homolog KPI87396; the BLASTP output is shown in the Supplementary Figure S3). One B-type read was found in the gene encoding the XP_003858707.1 sequence and another B-type read was found in gene encoding the XP_003858708.1 sequence (Figure 3 and Supplementary file Table S2). We extracted members of this protein family in other representatives of the subfamily Leishmaniinae [36]. A reconstructed phylogenetic tree suggested a complex evolutionary history of this multigene family, exemplified by the presence of multiple paralogs in the analyzed genomes (Figure 4).

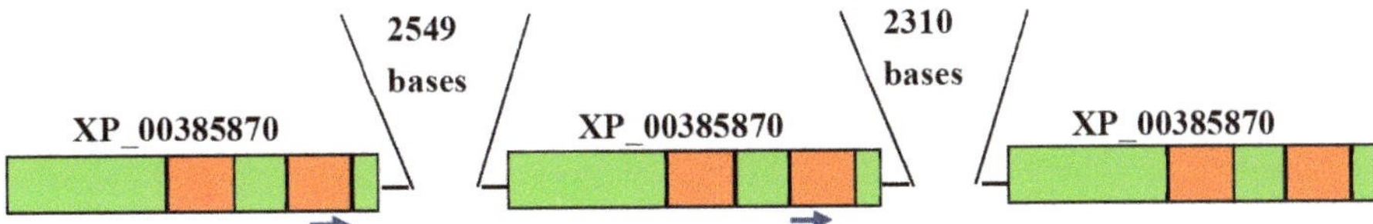

Figure 3. Schematic representation of regions of high and low similarity between XP_003858707.1, XP_003858708.1 and XP_003858709.1 sequences. Details are shown in the Supplementary Figure S3. Low similarity regions (less than 25% identity) are shown in green and high similarity regions (over 75% identity) are shown in red. Blue arrows represent two detected B-type reads.

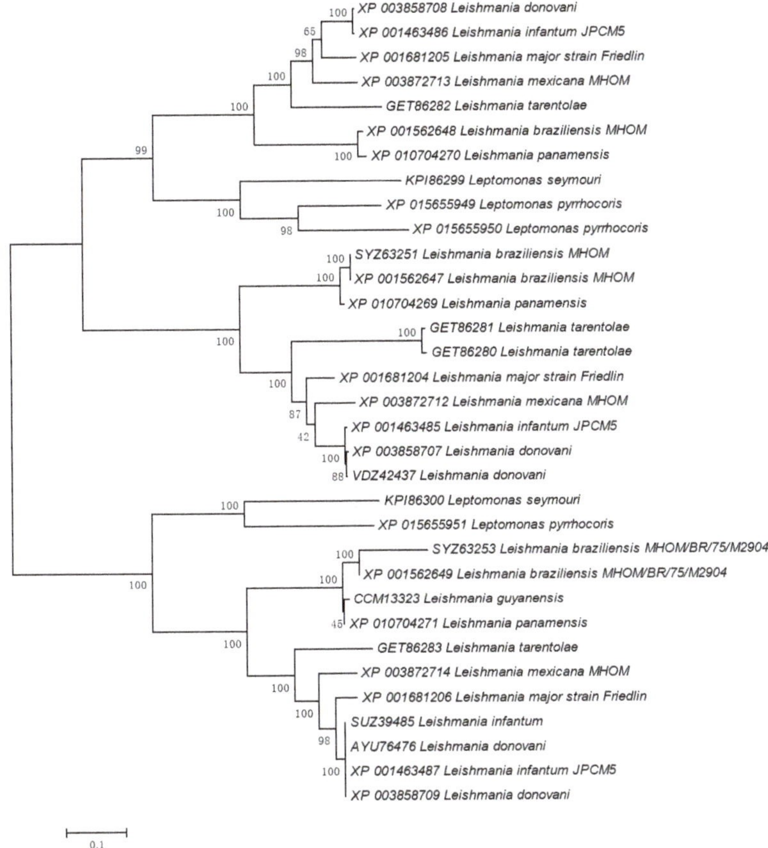

Figure 4. Maximum-likelihood phylogeny for the complete sequence alignment of the adenylate cyclase protein family. GenBank accession numbers precede species/isolate designations. The tree is drawn to scale, the scale bar denotes the number of substitutions per site. The bootstrap values (1000 replicates) are shown at the nodes.

We also performed a phylogenetic analysis of two regions of unusually high conservation (XP_003858708.1 positions 336–510 and 721–806, Figure 3 and Supplementary file Figure S3). Analysis of the reconstructed phylogenetic tree (Figure 5) detected a group of nearly identical *Leishmania donovani* paralogous genes, this group was expected due to the choice of two analyzed regions (Figure 3 and

Supplementary file Figure S3). In addition, three species-specific groups of nearly identical sequences from *L. tarentolae*, *L. major* and *L. panamensis* have been documented (Figure 5). Such tight species-specific clusters are likely to be the result of recurrent gene conversion events [37–39]. Of note, the vast majority of these adenylate cyclase genes form tandemly arranged paralogous gene clusters which may promote gene conversion events (as evident from the TriTryp.DB IDs: *L. donovani* BPK282A1: LdBPK_090280.1, LdBPK_090290.1, LdBPK_090300.1; *L. tarentolae*: LtaP09.0320, LtaP09.0330, LtaP19.0800, LtaP30.0010; *L. major* Friedlin: LmjF.09.0320, LmjF.09.0330, LmjF.09.0340; *L. panamensis* MHOM/COL/81/L13: LPAL13_090008100, LPAL13_090008200, LPAL13_090008300; all TriTryp.DB names are listed in the Supplementary file Table S3). The detected four species-specific groups of nearly identical sequences (Figure 5) do not exist in the phylogenetic tree for the complete sequence alignment (Figure 4), suggesting that putative recurrent gene events are highly specific for the two regions of this multigene family (Figure 3).

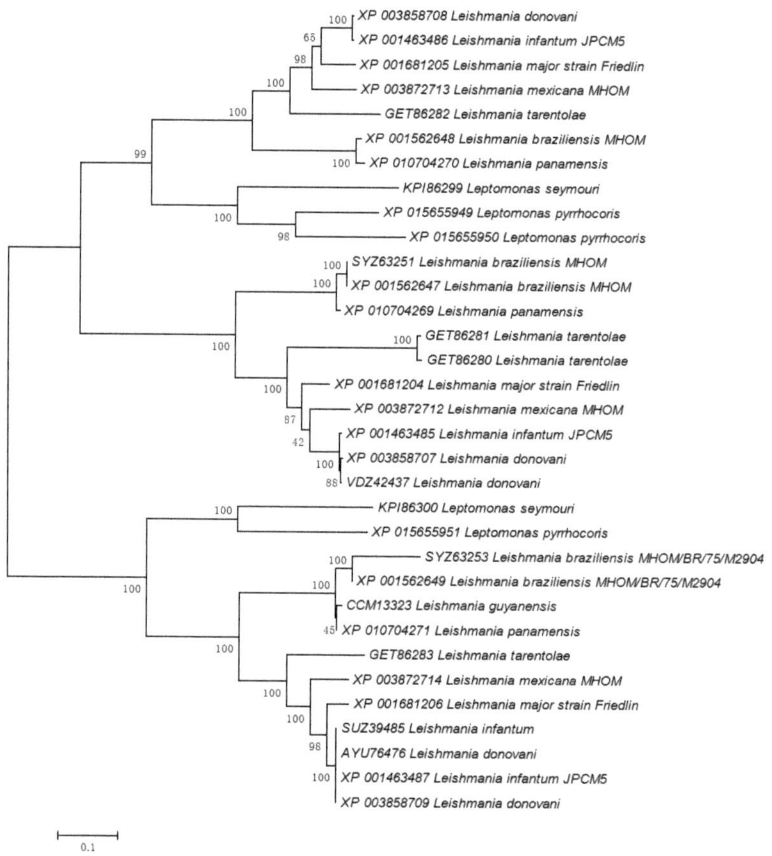

Figure 5. Maximum-likelihood phylogeny for two highly similar regions (positions 457–625 and 1114–1212) of the adenylate cyclase proteins. GenBank accession numbers precede species/isolate designations. The tree is drawn to scale, the scale bar denotes the number of substitutions per site. The bootstrap values (1000 replicates) are shown at the nodes. Major deviations (putative gene conversion events) from the tree inferred from the complete sequence alignment (Figure 4) are marked by black lines.

2.4. Analysis of Sacharomyces cerevisiae

We found only a few filtered reads in the yeast dataset (Table 4). This result suggests that the employed filtering procedure produces results that cannot be explained by whole-genome sequencing artifacts, because sequencing procedures and platforms were highly similar for both studied species.

Table 4. The number of filtered reads in *Saccharomyces cerevisiae* (10 samples). The total number of unfiltered A-type reads is 158,344,992.

Read Types	Overlap with Protein-Coding Region	No Overlaps with Protein-Coding Region
B	0	0
C	0	3
D	1	0
S	0	1

3. Discussion

Comparison of the *Saccharomyces cerevisiae* and *Leishmania donovani* datasets suggests that the observed excess of B-type reads in *L. donovani* is not the result of sequencing/mapping artifacts. We propose that this excess is due to the recombination events. At least some of these events are likely to be gene conversion between members of multigene families, as illustrated by the example of the adenylate cyclase proteins (Figures 4 and 5). This frequent recombination events are consistent with the recent study [32], which revealed greater genetic diversity, including extensive structural variation, than was previously suggested by geographically-focused studies [40]. It should be noted that there is an important methodological difference between our and previously published studies—the filtering procedures used in the latter were rather strict. Removal of these filters uncovered a substantial genomic variability of *Leishmania* isolates (Table 1). However, the usage of the strict filters is a justified and correct approach in order to estimate genetic distances between various samples.

Genetic recombination involves classical crossing-over and gene conversion. Polarity is one of the properties of gene conversion: in many cases the frequency of gene conversion exhibits a gradient across the gene monitored [33–35,41–43]. The frequency of conversion is usually dependent on its location. An interpretation of conversion polarity is that it is caused by the existence of specific initiation sites for recombination, located at the high end of the polarity gradient [34,43,44]. Here we show that the polarity gradient for the studied *L. donovani* is high at the 3′ end of the gene, implying that promoters of protein-coding genes less frequently contain initiation sites compared to the 3′ ends. An example of a putative gene conversion event is shown in Figure 3, where almost identical regions are located in the second half of the protein alignment. Gene conversion was observed in several Leishmaniinae species (Figure 5), implying a high frequency of these events. A substantially higher frequency of B-type reads compared to D-type reads (Table 1) is likely to be explained by the expected higher frequency of intrachromosomal gene conversion/recombination in the protein coding genes compared to interchromosomal gene conversion/recombination events. This is likely to be due to the presence of tandemly arranged multigene families that is a well-known property of the *L. donovani* genome [40].

The functional importance of so-called "concerted evolution" (frequent gene conversion events and unequal crossing-over) remains obscure [45]. For example, it appears that the rate of unequal crossing-over is much higher for rodent polyubiquitin genes than for their human kin, although there is no doubt that the function and conservation of these genes remain exactly the same during the evolution of mammals [46]. Proponents of the concerted evolution hypothesis suggest that the concerted pattern of fixation permits the establishment of biological novelty and species discontinuities in a manner not predicted by the classical genetics that is largely based on concepts of natural selection and genetic drift [47]. However the functional importance of frequent gene conversion events is still an important evolutionary question and systematic analysis of these events in *Leishmania* spp. may help to answer it.

This is further exemplified by the case of trypanosomatid receptor adenylate cyclases [21], which are predicted to govern parasite–host interactions [48]. These proteins are extremely well studied in *Trypanosoma brucei*, where they have been implicated in *quorum sensing* regulation of differentiation in this species [49]. Some members of this protein family negatively regulate social motility in the procyclic stage of the trypanosome life cycle [50]. Nevertheless, the role of these proteins in *Leishmania* biology is under researched. Their expression was documented to be restricted to the sandfly-dwelling promastigotes in the case of *L. donovani* [51]. Our finding that gene conversion may have shaped the repertoire of these receptor proteins in different *Leishmania* spp. testifies to their importance and warrants future investigations into their functional role(s).

4. Materials and Methods

4.1. Datasets

We studied DNA-seq data for *Leishmania donovani* and *Saccharomyces cerevisiae*. For *L. donovani*, we used complete genomic data for the isolate BPK282A1 [52]. Reads were downloaded from the European Nucleotide Archive (ENA) (www.ebi.ac.uk/ena/data/view/PRJEB2086, 30 genomes). In the original study [40] the reference genome was masked at regions that were repetitive, duplicated, close to contig edges, structurally variable, or potentially mis-assembled. Five criteria masked a total of 6,358,203 bases out of the 32,444,998 bases reference genome sequence for *L. donovani* BPK282, resulting in SNPs (single-nucleotide polymorphisms) being called at 26,086,795 or 80.4% of the total nuclear genome [53]. We did not implement any of those filters because we were interested in the most dynamic fraction of the genome. For *S. cerevisiae*, we used DNA-seq data for isolate S288C (assembly R64, www.ncbi.nlm.nih.gov/genome/15). Read data are available under study ERP000140 at the ENA (www.ebi.ac.uk/ena/data/view/SRX155705, 10 genomes). In both cases, sequence reads were generated using the same Illumina HiSeq 2000 platform and standard protocols.

4.2. Data Binning and Filtering

We used the SMALT program (www.sanger.ac.uk/science/tools/smalt-0) for the mapping of paired reads. Firstly, we indexed the reference genomic sequences (ref_genome): smalt_x86_64 index -k 20 -s 13 ref_genome ref_genome.fas and then used paired reads (sampl1 and sample2) with the following set of parameters: smalt_x86_64 map -n 8 -f cigar -o output_file ref_genome sample1.fastq sample2.fastq

We analyzed five SMALT mapping configurations (A/B/S/C/D-types): (A-type) mates are in proper orientation within the limits specified by the -i and -j options (the control set of properly aligned reads, i = 500 and j = 0); (B-type) mates in proper orientation outside the limits specified by the -i and -j options, but on the same chromosome (putative intra-chromosomal gene conversion events and sequencing errors); (C-type) mates are not in proper orientations, but on the same chromosome (mostly sequencing errors and inversion events that are known to be rare); (D-type) mates are mapped to different chromosomes (putative inter-chromosomal gene conversion events and sequencing errors); (S-type) a read was mapped as a single read (sole mapped read of a pair, sequencing errors although this type of errors may be different from other configurations). Manual inspection of the B-type reads suggested that two samples were obvious outliers (Supplementary Table S1). They were removed from further analysis.

We applied only one filter. We removed all reads with 95% identity, min 60 bases ungapped region; the percent identity for a B-type (or C-, or D-, or S-type) read should be equal or more 95% and more than the best overlapping A type read (the minimal mapped ungapped region ≥60 bases for all reads). This filter effectively removed errors of whole-genome sequencing procedures. All types of reads were analyzed as single reads (for example, if one B-type read was filtered out and its mate was not filtered, the second read was included in further analyses). After filtering, all overlapping reads were merged. If a merged set of reads (a merged read) overlapped with a known protein-coding sequence, we assigned this merged read to the set of reads that overlap with protein coding genes. We used the 2-tail Fisher exact test to evaluate homogeneity of 2×2 contingency tables [54]. A modified χ^2 test was used for analyses of 2×3 contingency tables [54].

4.3. Phylogenetic Analysis

Evolutionary analyses were conducted in the package MEGA X [55] as described previously [56]. The phylogenies were inferred using the Maximum Likelihood method. The "Find Best Model (ML)" function was used to determine the appropriate substitution model. The model with the lowest Bayesian Information Criterion score was considered to best describe the substitution pattern for that dataset and was subsequently chosen for phylogenetic analysis. Initial trees for the heuristic search were obtained automatically by applying the Neighbor-Join and BioNJ algorithms to a matrix of pairwise distances estimated using a JTT model, and then selecting the topology with superior log likelihood value. A discrete Gamma distribution was used to model evolutionary rate differences among sites (2 categories (+G, parameter = 1.1990)). All positions with less than 90% site coverage were eliminated. That is, fewer than 10% alignment gaps, missing data, and ambiguous bases were allowed at any position. The multiple sequence alignment is presented in Supplementary file Figure S4.

Supplementary Materials: The following are available online at http://www.mdpi.com/2076-0817/9/7/572/s1, Table S1: *L. donovani* DNA-seq samples, Table S2: 137 *L. donovani* protein-coding genes that overlap with merged B-type reads, Table S3: TriTryp.DB names of the putative adenylate cyclase proteins (Figures 4 and 5), Figure S1: A BLASTP output for the XP_003861613.1 protein sequence (the multiple alignment format), Figure S2: Schematic representation of an insertion in the 3′ flanking region near the end of the gene encoding XP_003861613.1. Figure S3: A BLASTP output showing two conserved regions in XP_003858707.1, XP_003858708.1 and XP_003858709.1 protein sequences, Figure S4: Multiple sequence alignment used in this study.

Author Contributions: Conceptualization, I.B.R. and V.Y.; methodology, I.B.R.; software, I.B.R., I.A.S. and V.N.B.; validation, I.B.R and V.Y.; formal analysis, I.B.R., I.A.S. and V.N.B.; investigation, I.B.R., A.C., I.A.S., V.N.B. and V.Y.; resources, I.B.R., I.A.S., V.N.B. and V.Y.; data curation, I.B.R., A.C., I.A.S. and V.N.B.; writing—original draft preparation, I.B.R., A.C. and V.Y.; writing—review and editing, I.B.R., A.C., I.A.S., V.N.B. and V.Y. All authors have read and agreed to the published version of the manuscript.

Funding: This work was supported by the Intramural Research Program of the National Library of Medicine at the National Institutes of Health (to I.B.R.); the European Regional Funds (CZ.02.1.01/16_019/0000759) and the Russian Science Foundation (grant 19-15-00054 /analysis of adenylate cyclases) to V.Y.

Acknowledgments: We thank members of our laboratories for stimulating discussions.

Conflicts of Interest: Other authors declare no conflict of interest. The funders had no role in the design of the study; in the collection, analyses, or interpretation of data; in the writing of the manuscript, or in the decision to publish the results.

References

1. Bruschi, F.; Gradoni, L. *The Leishmaniases: Old Neglected Tropical Diseases*; Springer: Cham, Switzerland, 2018; p. 245.
2. Burza, S.; Croft, S.L.; Boelaert, M. Leishmaniasis. *Lancet* **2018**, *392*, 951–970. [CrossRef]
3. Guerbouj, S.; Guizani, I.; Speybroeck, N.; Le Ray, D.; Dujardin, J.C. Genomic polymorphism of *Leishmania infantum*: A relationship with clinical pleomorphism? *Infect. Genet. Evol.* **2001**, *1*, 49–59. [CrossRef]
4. Thakur, L.; Singh, K.K.; Shanker, V.; Negi, A.; Jain, A.; Matlashewski, G.; Jain, M. Atypical leishmaniasis: A global perspective with emphasis on the Indian subcontinent. *PLoS Negl. Trop. Dis.* **2018**, *12*, e0006659. [CrossRef] [PubMed]
5. Quinnell, R.J.; Courtenay, O. Transmission, reservoir hosts and control of zoonotic visceral leishmaniasis. *Parasitology* **2009**, *136*, 1915–1934. [CrossRef] [PubMed]
6. Ready, P.D. Epidemiology of visceral leishmaniasis. *Clin. Epidemiol.* **2014**, *6*, 147–154. [CrossRef]

7. Lukeš, J.; Mauricio, I.L.; Schonian, G.; Dujardin, J.C.; Soteriadou, K.; Dedet, J.P.; Kuhls, K.; Tintaya, K.W.; Jirků, M.; Chocholova, E.; et al. Evolutionary and geographical history of the *Leishmania donovani* complex with a revision of current taxonomy. *Proc. Natl. Acad. Sci. USA* **2007**, *104*, 9375–9380. [CrossRef]

8. Leblois, R.; Kuhls, K.; Francois, O.; Schonian, G.; Wirth, T. Guns, germs and dogs: On the origin of *Leishmania chagasi*. *Infect. Genet. Evol.* **2011**, *11*, 1091–1095. [CrossRef]

9. Zhang, W.W.; Ramasamy, G.; McCall, L.I.; Haydock, A.; Ranasinghe, S.; Abeygunasekara, P.; Sirimanna, G.; Wickremasinghe, R.; Myler, P.; Matlashewski, G. Genetic analysis of *Leishmania donovani* tropism using a naturally attenuated cutaneous strain. *PLoS Pathog.* **2014**, *10*, e1004244. [CrossRef]

10. Laffitte, M.N.; Leprohon, P.; Papadopoulou, B.; Ouellette, M. Plasticity of the *Leishmania* genome leading to gene copy number variations and drug resistance. *F1000Research* **2016**, *5*, 2350. [CrossRef]

11. Sádlová, J.; Svobodová, M.; Volf, P. *Leishmania major*: Effect of repeated passages through sandfly vectors or murine hosts. *Ann. Trop. Med. Parasitol.* **1999**, *93*, 599–611. [CrossRef]

12. Lypaczewski, P.; Hoshizaki, J.; Zhang, W.W.; McCall, L.I.; Torcivia-Rodriguez, J.; Simonyan, V.; Kaur, A.; Dewar, K.; Matlashewski, G. A complete *Leishmania donovani* reference genome identifies novel genetic variations associated with virulence. *Sci. Rep.* **2018**, *8*, 1–14. [CrossRef] [PubMed]

13. Fiebig, M.; Kelly, S.; Gluenz, E. Comparative life cycle transcriptomics revises *Leishmania mexicana* genome annotation and links a chromosome duplication with parasitism of vertebrates. *PLoS Pathog.* **2015**, *11*, e1005186. [CrossRef] [PubMed]

14. Rastrojo, A.; Garcia-Hernandez, R.; Vargas, P.; Camacho, E.; Corvo, L.; Imamura, H.; Dujardin, J.C.; Castanys, S.; Aguado, B.; Gamarro, F.; et al. Genomic and transcriptomic alterations in *Leishmania donovani* lines experimentally resistant to antileishmanial drugs. *Int. J. Parasitol. Drugs Drug. Resist.* **2018**, *8*, 246–264. [CrossRef]

15. Dostálová, A.; Volf, P. *Leishmania* development in sand flies: Parasite-vector interactions overview. *Parasit. Vectors* **2012**, *5*, 1–12. [CrossRef]

16. Forestier, C.L.; Gao, Q.; Boons, G.J. *Leishmania* lipophosphoglycan: How to establish structure-activity relationships for this highly complex and multifunctional glycoconjugate? *Front. Cell. Infect. Microbiol.* **2014**, *4*, 193. [CrossRef] [PubMed]

17. Turco, S.J.; Spath, G.F.; Beverley, S.M. Is lipophosphoglycan a virulence factor? A surprising diversity between Leishmania species. *Trends Parasitol.* **2001**, *17*, 223–226. [CrossRef]

18. Dobson, D.E.; Scholtes, L.D.; Valdez, K.E.; Sullivan, D.R.; Mengeling, B.J.; Cilmi, S.; Turco, S.J.; Beverley, S.M. Functional identification of galactosyltransferases (SCGs) required for species-specific modifications of the lipophosphoglycan adhesin controlling *Leishmania major*-sand fly interactions. *J. Biol. Chem.* **2003**, *278*, 15523–15531. [CrossRef]

19. Dobson, D.E.; Mengeling, B.J.; Cilmi, S.; Hickerson, S.; Turco, S.J.; Beverley, S.M. Identification of genes encoding arabinosyltransferases (SCA) mediating developmental modifications of lipophosphoglycan required for sand fly transmission of *Leishmania major*. *J. Biol. Chem.* **2003**, *278*, 28840–28848. [CrossRef]

20. Maslov, D.A.; Opperdoes, F.R.; Kostygov, A.Y.; Hashimi, H.; Lukeš, J.; Yurchenko, V. Recent advances in trypanosomatid research: Genome organization, expression, metabolism, taxonomy and evolution. *Parasitology* **2019**, *146*, 1–27. [CrossRef]

21. Lukeš, J.; Butenko, A.; Hashimi, H.; Maslov, D.A.; Votýpka, J.; Yurchenko, V. Trypanosomatids are much more than just trypanosomes: Clues from the expanded family tree. *Trends Parasitol.* **2018**, *34*, 466–480. [CrossRef]

22. Butenko, A.; Vieira, T.D.S.; Frolov, A.O.; Opperdoes, F.R.; Soares, R.P.; Kostygov, A.Y.; Lukeš, J.; Yurchenko, V. *Leptomonas pyrrhocoris*: Genomic insight into parasite's physiology. *Curr. Genom.* **2018**, *19*, 150–156. [CrossRef] [PubMed]

23. Manna, P.T.; Boehm, C.; Leung, K.F.; Natesan, S.K.; Field, M.C. Life and times: Synthesis, trafficking, and evolution of VSG. *Trends Parasitol.* **2014**, *30*, 251–258. [CrossRef] [PubMed]

24. McCulloch, R.; Rudenko, G.; Borst, P. Gene conversions mediating antigenic variation in *Trypanosoma brucei* can occur in variant surface glycoprotein expression sites lacking 70-base-pair repeat sequences. *Mol. Cell. Biol.* **1997**, *17*, 833–843. [CrossRef] [PubMed]

25. Robinson, N.P.; Burman, N.; Melville, S.E.; Barry, J.D. Predominance of duplicative VSG gene conversion in antigenic variation in African trypanosomes. *Mol. Cell. Biol.* **1999**, *19*, 5839–5846. [CrossRef]

26. Castro Neto, A.L.; Brito, A.; Rezende, A.M.; Magalhaes, F.B.; de Melo Neto, O.P. In silico characterization of multiple genes encoding the GP63 virulence protein from *Leishmania braziliensis*: Identification of sources of variation and putative roles in immune evasion. *BMC Genom.* **2019**, *20*, 1–17. [CrossRef]

27. Mauricio, I.L.; Gaunt, M.W.; Stothard, J.R.; Miles, M.A. Glycoprotein 63 (*gp63*) genes show gene conversion and reveal the evolution of Old World *Leishmania*. *Int. J. Parasitol.* **2007**, *37*, 565–576. [CrossRef]

28. Mottram, J.C.; Frame, M.J.; Brooks, D.R.; Tetley, L.; Hutchison, J.E.; Souza, A.E.; Coombs, G.H. The multiple *cpb* cysteine proteinase genes of *Leishmania mexicana* encode isoenzymes that differ in their stage regulation and substrate preferences. *J. Biol. Chem.* **1997**, *272*, 14285–14293. [CrossRef]

29. Folgueira, C.; Cañavate, C.; Chicharro, C.; Requena, J.M. Genomic organization and expression of the *hsp70* locus in New and Old World *Leishmania* species. *Parasitology* **2007**, *134*, 369–377. [CrossRef]

30. Jackson, A.P. The evolution of amastin surface glycoproteins in trypanosomatid parasites. *Mol. Biol. Evol.* **2010**, *27*, 33–45. [CrossRef]

31. Zhang, W.W.; Matlashewski, G. Characterization of the A2-A2rel gene cluster in *Leishmania donovani*: Involvement of A2 in visceralization during infection. *Mol. MicroBiol.* **2001**, *39*, 935–948. [CrossRef]

32. Franssen, S.U.; Durrant, C.; Stark, O.; Moser, B.; Downing, T.; Imamura, H.; Dujardin, J.C.; Sanders, M.J.; Mauricio, I.; Miles, M.A.; et al. Global genome diversity of the *Leishmania donovani* complex. *eLife* **2020**, *9*, e51243. [CrossRef] [PubMed]

33. Malone, R.E.; Bullard, S.; Lundquist, S.; Kim, S.; Tarkowski, T. A meiotic gene conversion gradient opposite to the direction of transcription. *Nature* **1992**, *359*, 154–155. [CrossRef] [PubMed]

34. Detloff, P.; White, M.A.; Petes, T.D. Analysis of a gene conversion gradient at the *his4* locus in *Saccharomyces cerevisiae*. *Genetics* **1992**, *132*, 113–123.

35. Nicolas, A.; Petes, T.D. Polarity of meiotic gene conversion in fungi: Contrasting views. *Experientia* **1994**, *50*, 242–252. [CrossRef] [PubMed]

36. Kostygov, A.Y.; Yurchenko, V. Revised classification of the subfamily Leishmaniinae (Trypanosomatidae). *Folia Parasitol.* **2017**, *64*, 20. [CrossRef] [PubMed]

37. Ohta, T. On the evolution of multigene families. *Theor. Popul. Biol.* **1983**, *23*, 216–240. [CrossRef]

38. Koop, B.F.; Miyamoto, M.M.; Embury, J.E.; Goodman, M.; Czelusniak, J.; Slightom, J.L. Nucleotide sequence and evolution of the orangutan epsilon globin gene region and surrounding Alu repeats. *J. Mol. Evol.* **1986**, *24*, 94–102. [CrossRef]

39. Nei, M.; Rogozin, I.B.; Piontkivska, H. Purifying selection and birth-and-death evolution in the ubiquitin gene family. *Proc. Natl. Acad. Sci. USA* **2000**, *97*, 10866–10871. [CrossRef]

40. Imamura, H.; Downing, T.; Van den Broeck, F.; Sanders, M.J.; Rijal, S.; Sundar, S.; Mannaert, A.; Vanaerschot, M.; Berg, M.; De Muylder, G.; et al. Evolutionary genomics of epidemic visceral leishmaniasis in the Indian subcontinent. *eLife* **2016**, *5*, e12613. [CrossRef] [PubMed]

41. Eickbush, T.H.; Burke, W.D. The silkmoth *late chorion* locus. II. Gradients of gene conversion in two paired multigene families. *J. Mol. Biol.* **1986**, *190*, 357–366. [CrossRef]

42. Alani, E.; Reenan, R.A.; Kolodner, R.D. Interaction between mismatch repair and genetic recombination in *Saccharomyces cerevisiae*. *Genetics* **1994**, *137*, 19–39. [PubMed]

43. Dooner, H.K.; He, L. Polarized gene conversion at the *bz* locus of maize. *Proc. Natl. Acad. Sci. USA* **2014**, *111*, 13918–13923. [CrossRef]

44. Palmer, S.; Schildkraut, E.; Lazarin, R.; Nguyen, J.; Nickoloff, J.A. Gene conversion tracts in *Saccharomyces cerevisiae* can be extremely short and highly directional. *Nucleic. Acids. Res.* **2003**, *31*, 1164–1173. [CrossRef] [PubMed]

45. Wang, S.; Chen, Y. Phylogenomic analysis demonstrates a pattern of rare and long-lasting concerted evolution in prokaryotes. *Commun. Biol.* **2018**, *1*, 1–11. [CrossRef] [PubMed]

46. Perelygin, A.A.; Kondrashov, F.A.; Rogozin, I.B.; Brinton, M.A. Evolution of the mouse polyubiquitin-C gene. *J. Mol. Evol.* **2002**, *55*, 202–210. [CrossRef]

47. Dover, G. Molecular drive: A cohesive mode of species evolution. *Nature* **1982**, *299*, 111–117. [CrossRef]

48. Makin, L.; Gluenz, E. cAMP signalling in trypanosomatids: Role in pathogenesis and as a drug target. *Trends Parasitol.* **2015**, *31*, 373–379. [CrossRef] [PubMed]

49. Mony, B.M.; MacGregor, P.; Ivens, A.; Rojas, F.; Cowton, A.; Young, J.; Horn, D.; Matthews, K. Genome-wide dissection of the quorum sensing signalling pathway in *Trypanosoma brucei*. *Nature* **2014**, *505*, 681–685. [CrossRef]

50. Imhof, S.; Knusel, S.; Gunasekera, K.; Vu, X.L.; Roditi, I. Social motility of African trypanosomes is a property of a distinct life-cycle stage that occurs early in tsetse fly transmission. *PLoS Pathog.* **2014**, *10*, e1004493. [CrossRef]
51. Sanchez, M.A.; Zeoli, D.; Klamo, E.M.; Kavanaugh, M.P.; Landfear, S.M. A family of putative receptor-adenylate cyclases from *Leishmania donovani*. *J. Biol. Chem.* **1995**, *270*, 17551–17558. [CrossRef]
52. Downing, T.; Imamura, H.; Decuypere, S.; Clark, T.G.; Coombs, G.H.; Cotton, J.A.; Hilley, J.D.; de Doncker, S.; Maes, I.; Mottram, J.C.; et al. Whole genome sequencing of multiple *Leishmania donovani* clinical isolates provides insights into population structure and mechanisms of drug resistance. *Genome Res.* **2011**, *21*, 2143–2156. [CrossRef] [PubMed]
53. Downing, T.; Stark, O.; Vanaerschot, M.; Imamura, H.; Sanders, M.; Decuypere, S.; de Doncker, S.; Maes, I.; Rijal, S.; Sundar, S.; et al. Genome-wide SNP and microsatellite variation illuminate population-level epidemiology in the *Leishmania donovani* species complex. *Infect. Genet. Evol.* **2012**, *12*, 149–159. [CrossRef] [PubMed]
54. Khromov-Borisov, N.N.; Rogozin, I.B.; Pegas Henriques, J.A.; de Serres, F.J. Similarity pattern analysis in mutational distributions. *Mutat. Res.* **1999**, *430*, 55–74. [CrossRef]
55. Kumar, S.; Stecher, G.; Li, M.; Knyaz, C.; Tamura, K. MEGA X: Molecular Evolutionary Genetics Analysis across computing platforms. *Mol. Biol. Evol.* **2018**, *35*, 1547–1549. [CrossRef]
56. Kostygov, A.Y.; Grybchuk-Ieremenko, A.; Malysheva, M.N.; Frolov, A.O.; Yurchenko, V. Molecular revision of the genus *Wallaceina*. *Protist* **2014**, *165*, 594–604. [CrossRef]

MDPI
St. Alban-Anlage 66
4052 Basel
Switzerland
Tel. +41 61 683 77 34
Fax +41 61 302 89 18
www.mdpi.com

Pathogens Editorial Office
E-mail: pathogens@mdpi.com
www.mdpi.com/journal/pathogens